AF500423

Reliability and Optimization of Structural Systems

IFIP – The International Federation for Information Processing

IFIP was founded in 1960 under the auspices of UNESCO, following the First World Computer Congress held in Paris the previous year. An umbrella organization for societies working in information processing, IFIP's aim is two-fold: to support information processing within its member countries and to encourage technology transfer to developing nations. As its mission statement clearly states,

> IFIP's mission is to be the leading, truly international, apolitical organization which encourages and assists in the development, exploitation and application of information technology for the benefit of all people.

IFIP is a non-profitmaking organization, run almost solely by 2500 volunteers. It operates through a number of technical committees, which organize events and publications. IFIP's events range from an international congress to local seminars, but the most important are:

- the IFIP World Computer Congress, held every second year;
- open conferences;
- working conferences.

The flagship event is the IFIP World Computer Congress, at which both invited and contributed papers are presented. Contributed papers are rigorously refereed and the rejection rate is high.

As with the Congress, participation in the open conferences is open to all and papers may be invited or submitted. Again, submitted papers are stringently refereed.

The working conferences are structured differently. They are usually run by a working group and attendance is small and by invitation only. Their purpose is to create an atmosphere conducive to innovation and development. Refereeing is less rigorous and papers are subjected to extensive group discussion.

Publications arising from IFIP events vary. The papers presented at the IFIP World Computer Congress and at open conferences are published as conference proceedings, while the results of the working conferences are often published as collections of selected and edited papers.

Any national society whose primary activity is in information may apply to become a full member of IFIP, although full membership is restricted to one society per country. Full members are entitled to vote at the annual General Assembly, National societies preferring a less committed involvement may apply for associate or corresponding membership. Associate members enjoy the same benefits as full members, but without voting rights. Corresponding members are not represented in IFIP bodies. Affiliated membership is open to non-national societies, and individual and honorary membership schemes are also offered.

Reliability and Optimization of Structural Systems

Proceedings of the sixth IFIP WG7.5 working conference on reliability and optimization of structural systems 1994

Edited by

Rüdiger Rackwitz
Technical University of Munich
Munich
Germany

Guiliano Augusti
Università degli Studi di Roma 'La Sapienza'
Rome
Italy

and

Antonio Borri
Università di Perugia
Perugia
Italy

SPRINGER-SCIENCE+BUSINESS MEDIA, B.V.

First edition 1995

Originally published by Chapman & Hall in 1995
MyCopy version of the original edition 1995

DOI 10.1007/978-0-387-34866-7

A catalogue record for this book is available from the British Library

∞ Printed on permanent acid-free text paper, manufactured in accordance with ANSI/NISO Z39.48-1992 and ANSI/NISO Z39.48-1984 (Permanence of Paper).
www.springer.com/mycopy

CONTENTS

PREFACE

This proceedings volume contains 33 papers presented at the 6th Working Conference on "Reliability and Optimization of Structural Systems" held at the Monastery of the "Basilica di San Francesco", Assisi (Perugia), Italy, on September 7-9, 1994. The Working Conference was organized by the IFIP (International Federation for Information Processing) Working Group 7.5. of Technical Committee 7 and was the 6th in a series, following similar conferences held at the University of Aalborg, Denmark, May 1987, at the Imperial College, London, UK, September 1988, at the University of California, Berkeley, California, USA, March 1990, at the Technical University of Munich, Germany, September 1991, and at the General Education & Training Center Shikoku Electric Power Co., Takamatsu-Shi, Kagawa, Japan, March 1993. The Working Conference was attended by 47 participants from 12 countries.

The objectives of Working Group 7.5 are:

- to promote modern structural systems optimization and reliability theory
- to advance international co-operation in the field of structural system optimization and reliability theory,
- to stimulate research, development and application of structural system optimization and reliability theory,
- to further the dissemination and exchange for information on reliability and optimization of structural systems,
- to encourage education in structural system optimization and reliability theory

At present the members of the Working Group are:

A. H.-S. Ang., USA
G. Augusti, Italy
M.J. Baker, United Kingdom
P. Bjerager, Norway
C.A. Cornell, USA
R.B. Corotis, USA
A. Der Kiureghian, USA
O. Ditlevsen, Denmark
L. Esteva, Mexiko
D.M. Frangopol, USA
H. Furuta, Japan
M. Grigoriu, USA
M. Grimmelt, Germany
C. Guedes Soares, Portugal
H. Ishikawa, Japan
S. Jendo, Poland
N.C. Lind, Canada
H.O. Madsen, Japan
K. Marti, Germany
R.E. Melchers, Australia
F. Moses, USA
Y. Murotsu, Japan
A.S. Nowak (vice-chairman), USA
R. Rackwitz (chairman), Germany
P. Sniady, Poland
J.D. Sorensen, Denmark
P. Thoft-Christensen, Denmark
Y.-K. Wen, USA

Members of the Organizing Committee were:
G. Augusti, Italy (co-chairman)
A. Borri, Italy (co-chairman)
H. Ishikawa, Japan
R. Melcher, Australia
A. Nowak, USA (co-chairman)
R. Rackwitz, Germany (chairman)
P. Thoft-Christensen, Denmark

The Working Conference received financial support from the University of Perugia, several Italian organizations and the Technical University Munich.

On behalf of WG 7.5 and TC-7 the co-chairmen of the Conference would like to express their sincere thanks to the sponsors, to the members of the Oganizing Committee for their valuable assistance, and to the authors for their contributions to these proceedings. Special thanks are due to Mrs. Gisela Kick, Technical University Munich, for her efficient work as conference secretary.

January 1995

R. Rackwitz G. Augusti A. Borri

PART ONE

Keynote Lectures

1

Optimal resource allocation for seismic reliability upgrading of existing structures and lifeline networks

G. Augusti[a] A. Borri[b] and M. Ciampoli[a]

[a] Università di Roma 'La Sapienza', Dipartimento di Ingegneria Strutturale e Geotecnica, Via Eudossiana 18; I-00184 Roma , Italy

[b] Università di Perugia, Facoltà di Ingegneria, Via Santa Lucia Canetola; I-06126 Perugia, Italy

SUMMARY

A campaign of preventive measures aimed at upgrading an ensemble of buildings or other constructed facilities, whose seismic reliability is considered unsatisfactory, faces two difficulties: the limitation of the available resources, that must be therefore used in an optimized way, and the multiplicity of the aspects that should be taken into account in their allocation (direct and indirect economic losses, number of endangered persons, severance of a lifeline network, etc.). This lecture will first discuss in general terms how to approach these problems (in particular how to describe the seismic vulnerability in a way that can be used in such multi-objective optimization), then will present examples of optimal allocation of the available resources to seismic upgrading of existing structures and lifelines. The examples refer to masonry buildings, whose damages can be summed with each other to obtain the total damage of the ensemble, and to highway *networks*, i.e., *systems* whose critical elements are reinforced concrete bridges. Two objective functions are considered for each example and realistic values are attributed to the relevant parameters, so that the examples confirm the feasibility and applicability of the optimal allocation.

INTRODUCTION

Two significant earthquakes hit Italy in the last decades, procuring widespread damage and many victims: the Friuli earthquake of 1976 and the Irpinia earthquake of 1980. The epicentral area of both quakes was a rural but densely populated area, and most of the damages and victims were caused by the collapse of old masonry buildings; the disruption of the transportation networks increased much the difficulties, in particular in the latter case, also because of the collapse of the largest local hospital.

These events spurred much research and led to widespread surveys on the *vulnerability* of existing buildings. Apart from the still open problems concerning the elaboration of significant statistics from the survey data, two evident difficulties now face the exploitation of the collected information to formulate a rational strategy for reduction of earthquake losses, namely: the limited amount of resources that may be available for any preventive upgrading programme, and the multiplicity of the quantities whose reduction should be pursued in any such programme, like direct and indirect economic losses, casualties and deaths, damage to artistic and cultural heritage, environmental damages, deterioration of the *quality of life*.

This lecture will summarize a series of researches, already presented elsewhere (cf. [1-3] and the previous papers therein quoted), on the techniques that can be used to formulate and solve the problems of optimal resource allocation in a campaign for seismic risk reduction, taking account of several objective functions. This type of optimization, which can be a decisive help in the formulation of a rational strategy for seismic risk reduction, had previously received comparatively little attention: the aim of this lecture is not only to illustrate the problems again, but also to stimulate a discussion among specialists on structural reliability on the development and the actual exploitation of the results already obtained.

1. SEISMIC VULNERABILITY AND UPGRADING

The prerequisite for the optimal allocation of the available resources is of course the availability of sufficient statistical data on seismic vulnerability and hazard: much work has indeed be made on both these aspects in recent years. In particular, significant statistics have been and are being collected on the *seismic vulnerability* of buildings and constructed facilities, which is defined as *the probability of damage under an earthquake of given intensity.*

Several alternative ways of describing such *vulnerability* exist, which can be divided into three categories. In fact, the seismic vulnerability of a structure is fully described by a set of *fragility curves*, that relate the probability of reaching a certain *degree* (or *level*) *of damage* (or a well defined *limit state*) with the *intensity* (i.e., the dangerousness) of the earthquake.

However, due to the lack of sufficient data and the difficulties of using directly the fragility curves, seismic vulnerability is often measured in an approximate way by a number (the *vulnerability index*) or - even more simply - by including the structure in a *vulnerability class*. Each description has its appropriate field of application and can be associated to a different way of describing quantitatively the degree of *upgrading*, which is necessary for evaluating the effectiveness of preventive retrofitting measures: the three examples of optimal allocation procedures, presented in the following Sections 3 and 4, will make respectively use of each description of vulnerability and upgrading.

1.1 Fragility curves

Fragility curves require first the definition of the relevant *limit state(s)* or of quantitative measures of the damage, and an analogous definition of the intensity of the action. A set of *fragility curves* refers to a specific construction, and can be obtained by statistics on similar constructions or by numerical calculations. They are therefore used for important structures: for instance, in the following they will be applied to examples of reinforced concrete girder bridges: the *damage* shall be measured by an indicator of the required ductility and of the energy dissipated in the critical zones of the substructure, and the *earthquake intensity* by the peak ground acceleration. To evaluate the effectiveness of an *upgrading* intervention in this approach, a new set of fragility curves must be evaluated for the retrofitted structure, and compared with the initial one.

1.2 Vulnerability index

A set of fragility curves can be replaced, in an approximate way, by a number: the *vulnerability index,* which characterizes a building without explicit reference to earthquake intensity and level of damage. The vulnerability index can be obtained in several ways. For instance, G.N.D.T. (the Italian National Group for Earthquake Loss Reduction) elaborated a form, described in [4] and elsewhere, for surveying existing masonry buildings, which has experienced many variants over the years: the latest version of the form is schematically reproduced in Table 1. The survey team evaluates the quality condition of each item on a four-level scale *(a)*; the *vulnerability index* is then obtained by summing up the values associated to the condition of each item, multiplied by the weights indicated in column *(b)*: with this edition of the form, the index is comprised in the range 0-382.5, the higher values corresponding to the most vulnerable buildings.

An *upgrading* intervention can be defined as affecting one or more items of the form, and be assumed to bring the concerned item(s) into the best condition, i.e., to reduce to zero the contribution of that item to the vulnerability index, thus decreasing its value. In the example presented in Section 3.2 below, following previous suggestions, three possible intervention types have been defined, namely: (i) in *L* (*light* intervention) the horizontal connections between orthogonal walls are secured, thus the contribution of item 1 to the vulnerability index vanishes; (ii) *M* (*medium* intervention) includes also the strengthening of the horizontal diaphragms and brings to zero also the contribution of item 5; (iii) finally, *H* (*heavy* intervention) includes also an increase in the overall strength against horizontal actions and brings to zero the contribution of item 3.

Table 1
Scheme of the survey form used lately by G.N.D.T.

No.	*Item*	*Item condition (a)*				*Weight (b)*	*(a) × (b)*
1	Connection of walls	0	5	20	45	1	....
2	Type of walls	0	5	25	45	0.25	....
3	Total shear resistance of walls	0	5	25	45	1.5	....
4	Soil condition	0	5	25	45	0.75	....
5	Horizontal diaphragms	0	5	15	45	var.	....
6	Plan regularity	0	5	25	45	0.5	....
7	Elevation regularity	0	5	25	45	var.	....
8	Transverse walls: spacing/ thickness	0	5	25	45	0.25	....
9	Roof	0	15	25	45	var.	....
10	Non structural elements	0	0	25	45	0.25	....
11	General maintenance conditions	0	5	25	45	1	....
		Vulnerability index V					....

Of course, to be significative for prevision of damages and evaluation of the effectiveness of loss-reduction campaigns, the values of the vulnerability index must be calibrated versus actual damages. Such a calibration requires the definition of a measure of earthquake intensity (usually referred to a *macroseismic scale*) and of the *degree* of damage. Much work is in progress on the subject: however, the vulnerability-intensity-damage relationships are still very much affected by uncertainties, some due to incomplete calibration, some due to their inherently random nature. For simplicity's sake, in the examples presented in Section 3.2 (and in previous papers) the vulnerability index V (defined in the range 0-282, according to an earlier version of the survey form), the MSK earthquake intensity I and the degree of damage D have been assumed to be related by the deterministic curves shown in Fig. 3, which were obtained from a statistical analysis of the damages caused by some recent Italian earthquakes [5]: D = 0 corresponds by definition to no damage, and D = 1 to total collapse.

1.3 Damage Probability Matrices

Another definition of vulnerability assumes that all relevant buildings can be subdivided into a limited (say, 3 to 5) number of *vulnerability classes,* and associates each class X with a damage probability matrix (*DPM*). By definition, each element $P_{ji}(X)$ of the *DPM* pertaining to the vulnerability class X, is the probability that a building of that class undergoes a damage of level j, if subjected to an earthquake of intensity i. The damage of the buildings and the intensity of the earthquakes must therefore be described according to discrete scales.

DPM's can be obtained from statistical analyses of the damages due to one or more earthquake, when many buildings of a similar nature are affected, in areas of different intensities. For instance, the DPM's shown in Table 2, which are used in the example presented in Section 3.1, originated from the statistics of the damages to masonry buildings caused by the 1980 Irpinia earthquake [6]: they are based on an eight-level scale of damage (ranging from no apparent damage to complete collapse) and define three *vulnerability classes* A, B and C (A being the least safe, C the most). In Section 3.1, also a fourth ideal class D of earthquake-resistant structures, including buildings which belonged to A, B, or C and have been fully upgraded, is considered: P_{ji} (D) = 0 by assumption.

Table 2
Damage probability matrices, elaborated from data on damages subsequent to the 1980 Irpinia earthquake [6]

Class X		A *(worst)*			B *(medium)*			C *(best)*		
MSK int. i		*6*	*7*	*8*	*6*	*7*	*8*	*6*	*7*	*8*
Damage level j	*1*	0.15	0.07	0.01	0.33	0.20	0.04	0.64	0.52	0.06
	2	0.19	0.12	0.03	0.25	0.26	0.11	0.24	0.24	0.24
	3	0.25	0.16	0.05	0.25	0.26	0.20	0.08	0.15	0.20
	4	0.19	0.20	0.06	0.10	0.13	0.16	0.03	0.05	0.17
	5	0.12	0.21	0.07	0.05	0.08	0.14	0.01	0.03	0.11
	6	0.07	0.17	0.12	0.02	0.05	0.13	0.00	0.01	0.10
	7	0.03	0.05	0.32	0.00	0.02	0.12	0.00	0.00	0.09
	8	0.00	0.02	0.34	0.00	0.00	0.10	0.00	0.00	0.03

All significant modifications to the vulnerability of a building can be indicated by its initial and final class: e.g., AB, AC, ... BD, ... correspond to *upgrading* interventions, while CB, BA, CA, ... would be examples of *degradation* of the structure.

2. OBJECTIVES OF OPTIMIZATION

As hinted in the Introduction, any structural design and any programme of seismic loss reduction should take many aspects into account, like, e.g., economic losses, casualties and deaths, damages to the artistic and cultural heritage, environmental damages, deterioration of the quality of life. Many of these quantities are incommensurable with each other, and therefore cannot be combined into a single *objective function*, not even by means of weighting factors (how to *weigh* and compare economic costs versus human lives, or versus the destruction of a historical village?): the right approach to a rational strategy for seismic risk reduction appears to formulate and solve a problem of *multi-objective optimal resource allocation*.

Fortunately, the objectives of the optimizations usually do not conflict with each other (a preventive intervention aimed at reducing the expected economic losses would also reduce the expected number of victims), but the respective optimal solutions - in general - do not coincide, as examples will show in the following.

As discussed in Section 1 above, much research and statistical investigations are in progress on the seismic *vulnerability* of existing buildings, so that the expected damage after an earthquake can be estimated. Also, many retrofitting techniques, aimed at *upgrading* buildings (i.e., at reducing their expected damage after an earthquake), are being developed.

However, comparatively little attention has been paid - at least to the authors' knowledge - to the several possible consequences of the damages, other than the *direct economic costs*: therefore, cost-benefit analyses of a campaign of preventive interventions seem possible only with reference to this aspect, and the question remains very open on how to account for the other, non-monetary aspects (often denoted *intangibles*) that have been quoted above.

A possibility would be to correlate directly the earthquake intensity (but the same could be applied to any other environmental or man-made hazard) and each of the consequences: e.g., casualties. Not much significant work is available along these lines, but some now begins to be published (cf. [7]). This approach, in principle the most correct, would require specific and independent statistics for each type of consequences: and, for instance, damage statistics, elaborated with reference to economic costs only, would be useless with regard to *intangibles*. On the contrary, the possibility of using the vulnerability statistics in all cases requires that *damage* be defined and measured independently from the specific consequence. Other statistical relationships should relate the damage to each relevant consequence: indeed, this approach can be applied only if reliable damage-consequence relationships of this type are available [8].

In a ideal world of perfect mathematics and complete knowledge the two approaches would not differ one from the other. In the real word, they do.

The great asset of the *vulnerability* approach lies in the unified treatment of the damage and its statistics, and in the possibility of studying the results of preventive interventions as a decrease of vulnerability, also independent of the specific consequences. Its greatest liability might appear the necessity of formulating other and separate relationships between damages and consequences, thus introducing an extra step in the calculations. But if one considers that in any case a reliable relationship between action and consequences is necessary but in many instances not (or not yet) available, it should be clear that such an approach allows to obtain al least approximate results through extrapolations of known relationships (e.g., assume that the expected earthquake casualties in wooden buildings are sought, and that direct statistics are not available, because the specific problem was never posed before; assume also that the structural damages of timber can be forecast, and that statistics relating damages and casualties for all buildings in the area are available: this latter statistics could be assumed valid for the wooden buildings, and introduced in the calculation of the expected casualties). If the relationship between damage and consequence is deterministic and immediate (as is implicitly assumed when no distinction is made between damage and its, say, economic cost), then the introduction of the extra relationship does not pose any problem whatsoever.

Thus, the great liability of this approach remains in the unified quantitative definition of damage, be it made in linguistic terms (e.g., slight, significant, heavy, etc.) or in fractions or percentages (usually, 0 corresponds to no damage, and 100% to complete collapse; but also intermediate values, e.g. 50% or 70%, must be defined in an unequivocal way) or, perhaps better, according to a small number of *damage levels*.

However, the vulnerability approach appears indeed essential in an optimal allocation procedure, which looks for the *best* distribution of the upgrading interventions, whose costs are assumed known, under a constraint on the total expenditure. In fact, it allows to calculate and introduce unified relationships between the costs of the interventions and the reduction of the vulnerability, evaluate the reduction of expected damage for each distribution of interventions, and make use of the relationships between damages and the consequence chosen as the objective of the optimization in order to choose the *most efficient* one. In the following, it will be seen that such alternative optimizations are possible by simplified procedures or by sophisticated mathematical instruments.

3. OPTIMAL ALLOCATION OF RESOURCES: BUILDINGS

3.1 Allocation to vulnerability classes

Let us start from the last (and simplest) description of vulnerability, i.e., through DPMs. For the sake of clarity, the procedure is illustrated with direct reference to an example pertaining to masonry buildings [9], assuming the DPMs of Table 2 to hold. Realistic costs have been estimated (as percentages of the construction cost), both for restoring a building of each class to its original condition after a level j damage, and for each type of upgrading intervention. For simplicity's sake, all these percentages (and the construction cost per unit building volume) have been assumed to be constant irrespective of the building volumes actually involved in the operations.

The restoration costs after an earthquake of any relevant intensity can be forecast - for each class A, B and C - by multiplying the probabilities of Table 2 by the unit restoration costs estimated for each damage level, and summing up the columns: the results (again in percent of the construction cost) are shown in Table 3; small but non-zero costs, corresponding to minor (non structural) damages, have been assumed also for the ideal class D. A preventive upgrading intervention changes the class of the building, and therefore the forecast cost to be read in Table 3: columns (2)-(4) of Table 4 show the *unit gains* δr_i due to each type of intervention, forecast for each given intensity i, that is, the differences between the forecast restoration costs without and with the interventions indicated in column (1).

Table 3
Forecast restoration costs of masonry buildings (in percent of the construction cost)

MSK int. i		*6*	*7*	*8*
Class X	A	56.4	73.3	98.3
	B	41.0	48.8	75.8
	C	30.3	34.7	64.0
	D	0.00	3.30	8.30

Multiplying these forecast gains by the probabilities of occurrence π_i of the relevant earthquake during the design life of the building, and summing up, the *expected unit gains* δr_p can be calculated. The values shown in column (5) of Table 4 have been calculated by introducing the probabilities: $\pi_6 = 0.5$; $\pi_7 = 0.2$; $\pi_8 = 0.1$; which, assuming a 100 years lifetime, correspond approximately to the seismicity of many areas in Central Italy.

Table 4
Forecast (δr_i) *and expected* (δr_p) *unit economic gains; cost* C_I *and efficiency* G_c *of interventions*

Intervention	δr_6	δr_7	δr_8	δr_p	C_I	G_c
AB	15.4	24.5	22.5	14.8	23.3	0.64
AC	26.1	38.6	34.3	24.2	33.3	0.73
AD	56.4	70.0	90.0	51.2	56.6	0.90
BC	10.7	14.1	11.8	9.3	28.3	0.33
BD	41.0	45.0	67.5	36.2	43.3	0.84
CD	30.3	31.4	55.7	27.0	26.6	1.01

Finally, column (6) of the same Table 4 shows the assumed (deterministic) costs C_I of each intervention (once more, estimated as percentages of the construction cost), and column (7) the ratio G_c between the values in columns (5) and (6), i.e., the *expected efficiency* of each type of intervention. It can be noted that, with the used numerical values (realistic, even if derived from rough estimates) most values of G_c are smaller than one, i.e., no economic advantage should be expected from preventive interventions. However, as already discussed, a number of considerations invalidates such a conclusion: it is therefore assumed that preventive interventions are actually performed and only their optimal choice is sought. (Note that, once the buildings of the examined ensemble have been assigned to a class, the procedure does not distinguish between individual buildings but can only refer to fractions of the volume of each class.)

No formal optimization procedure is necessary for the optimal choice, considering that the larger or smaller efficiency of an intervention depends on the relative values of the ratio G_c: such a comparison is easily achieved by drawing (as it has been done in Fig. 1a) straight lines with slopes equal to the values of G_c. The choice of the interventions to be performed in this specific case does not present any difficulty: in fact, Fig. 1a shows immediately that the most convenient interventions are, in the order, CD, AD and BD (while in Fig. 2 a more complicated case will be found). Therefore, if the amount of available resources is comparatively small, they are used to bring into class D the largest possible volume of buildings belonging to class C; if more money is available than necessary to upgrade all buildings of class C, the extra resources can be employed for intervention AD; then, if also class A can be fully upgraded, further resources can be employed for intervention BD.

It is thus possible to calculate the total gain $\delta R_p = \Sigma_l \, \delta r_p \cdot V_l$, where δr_p and V_l are the unit gain and the volume of each intervention, and the total expenditure $H = C_c \cdot \Sigma_l (C_l \cdot V_l)$. Examples of plots of the total gain δR_p and of the volume V_l of each intervention versus the amount of money H available for preventive upgrading are shown in Figs. 1b (solid line) and 1c; these plots have been calculated introducing the unit construction cost: C_c = 300,000 Lire/m³, and the following volumes of buildings of each vulnerability class (that have been estimated for the historic centre of Priverno, a small medieval town approximately 100 km south of Rome [9]): V_A = 441,854 m³; V_B = 197,169 m³; V_C = 223,543 m³.

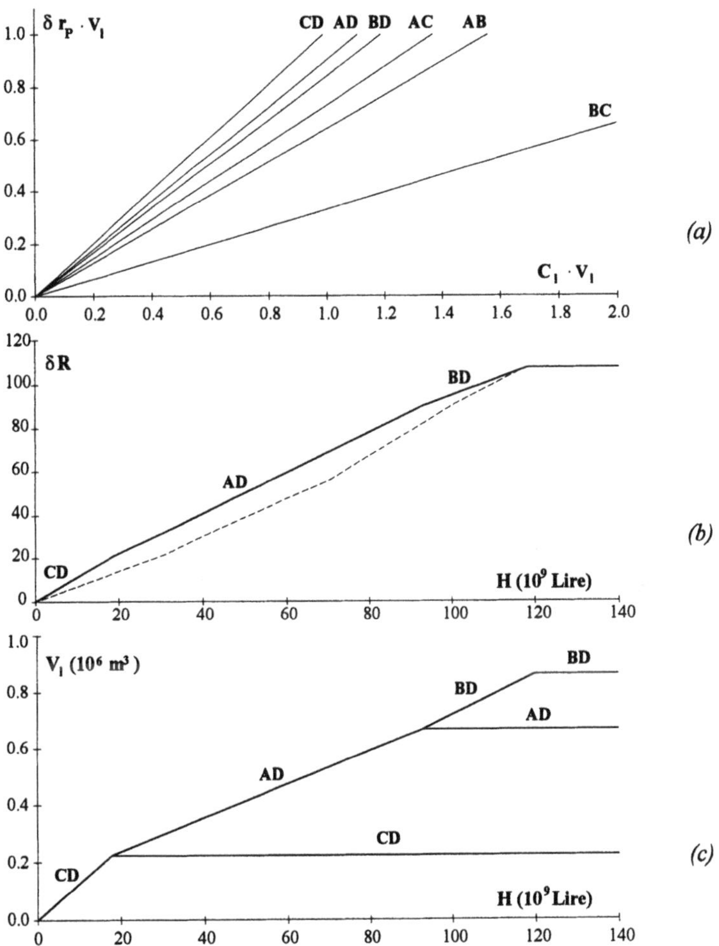

Fig. 1. *Interventions distributed among building vulnerability classes, optimized with respect to direct economic costs: (a) efficiency of interventions; (b) Expected economic gain (solid line) and comparison with the economic gain expected from the solution optimized with respect to saved lives (dotted lines); (c) interested volumes.*

In this way, the *intervention diagram*, optimized with respect to the direct economic losses, has been constructed in function of the available resources.

Table 5
Assumed ratio η_j between endangered and present people

Damage level j	*1*	*2*	*3*	*4*	*5*	*6*	*7*	*8*
η_j	0.00	0.00	0.00	0.04	0.12	0.24	0.40	0.80

The interventions can be optimized with respect to other objectives, for instance with respect to the decrease of the number of persons endangered by an earthquake. In want of reliable models for the number of persons present and endangered by an earthquake, the calculations for this optimization have been developed assuming that (i) 0.017 persons/m^3 inhabit the buildings (this value corresponds to the average density given by Italian statistics), (ii) 60% of the inhabitants are present in the buildings at the time of the earthquake, and (iii) the ratios between the number of endangered and present persons are given by Table 5 for each level of damage.

Table 6
Forecast (δn_j) and expected (δn_p) number of "saved" people; cost C_l *and efficiency* G_v *of interventions*

Intervention	δn_6	δn_7	δn_8	δn_p	C_l	G_v
AB	0.045	0.100	0.321	0.075	23.3	0.00320
AC	0.060	0.133	0.419	0.098	33.3	0.00296
AD	0.063	0.143	0.549	0.115	56.6	0.00203
BC	0.015	0.033	0.098	0.024	28.3	0.00084
BD	0.018	0.043	0.228	0.040	43.3	0.00093
CD	0.003	0.010	0.130	0.016	26.6	0.00062
Interv. substn.				$\Delta\delta n_p$	ΔC_l	G_v
AB → AC				0.023	10.0	0.00230
AC → AD				0.017	23.3	0.00073

Table 6 has been calculated in perfect analogy to Table 4. Namely, columns (2)-(4) show, for each intervention, the corresponding number δn_j of *saved* people (i.e. the reduction of endangered people) per unit volume, forecast for each earthquake intensity, while column (5) shows the expected unit number δn_p of saved people, assuming the already reported probabilities of occurrence. Finally, column (7) shows the efficiency of each intervention in terms of saved people, i.e. the ratio G_v between the expected unit gain of column (5) and the intervention cost of column (6): note that in the present case G_v is a ratio between two incommensurable quantities, which can be used only for comparative purposes. The last two rows of Table 6 show the differences in gains and costs between different interventions on class *A*, and the corresponding ratios G_v, which will be necessary to construct the optimal intervention diagram in the present case. The most convenient intervention is AB (Fig. 2a). However, if more resources are available than necessary to upgrade to class B the whole class A, it becomes next convenient not to intervene on more volumes, but to substitute intervention AC to AB: the intervention diagram is constructed as indicated in Figs. 2b and 2c, taking into account that the efficiency of the substitution AB → AC is $G_v = 0.0023$ (Table 6, col. 7). If the whole class A can be upgraded to class C, the next convenient intervention is BD ($G_v = 0.00093$), then the substitution of AD to AC ($G_v = 0.00073$), and finally CD ($G_v = 0.00062$). The optimal intervention diagram is thus completed (Figs. 2b and 2c).

Thus, two optimal allocations have been performed, but their objectives are not commensurable; hence, as already discussed, an overall multi-objective optimum cannot be defined.

However, the final choice of the solution should take into account the results of both calculations. To give some indications to this purpose, Figs. 1b and 2b show also, in dotted lines, respectively (Fig. 1b) the total economic gain of the solution optimized in terms of saved people, and (Fig. 2b) the people saved by the optimal economic solution. Although no general conclusions can be drawn, it can be noted that in this case the solution optimized with reference to saved people is close to optimal with respect to economic costs (Fig. 1b), while the reverse is not true (Fig. 2b).

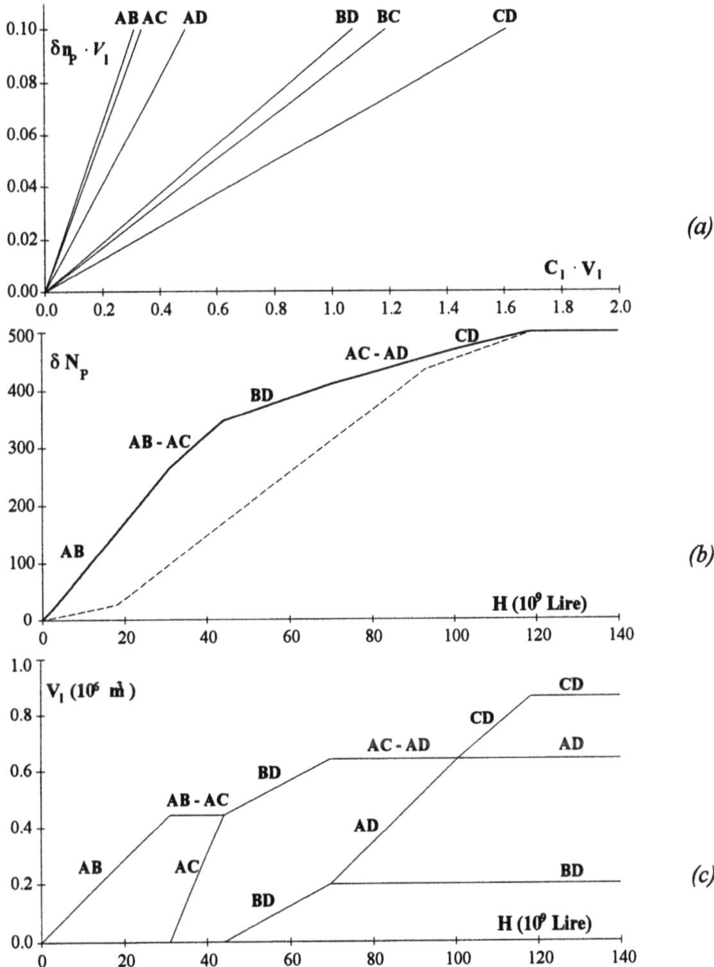

Fig. 2. *Interventions distributed among building vulnerability classes, optimized with respect to the decrease in the number of persons endangered by an earthquake (saved people): (a) efficiency of interventions; (b) expected number of saved people (solid line) and comparison with number of saved people expected from the solution optimized with respect to economic costs (dotted lines); (c) interested volumes.*

3.2 Allocation to individual buildings

In this Section, the seismic vulnerability of each building is measured by a number V (the *vulnerability index*) that, as anticipated in Sec.1.2, is related to the degree of damage D and the MSK earthquake intensity by the curves shown in Fig. 3. In the same Sec.1.2, three possible types of interventions (L, M and H) are defined: their assumed (deterministic) costs, together with the cost of construction, are shown in Table 7.

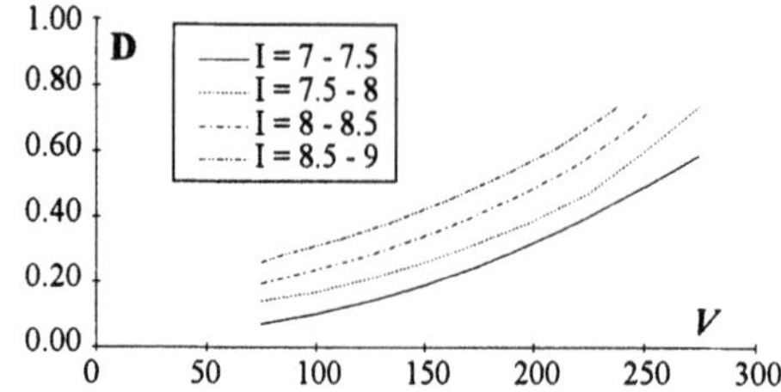

Fig. 3. *Degree of damage* D *vs. vulnerability index* V [5]

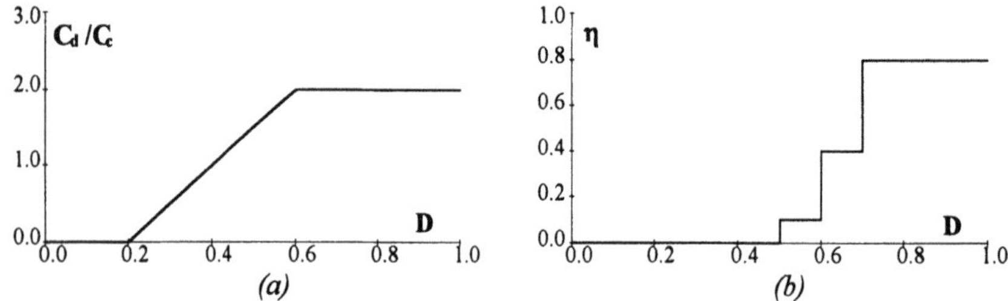

Fig. 4. *Ratios damage/construction costs (a) and endangered/present people vs. degree of damage* D *(b)* [10]

As in Sec. 3.1, direct economic costs and number of endangered persons are taken as alternative objective functions. The assumed relationships between the degree of damage D and respectively the monetary losses and the number of endangered persons n are shown in Figs. 4a and 4b. The number of people present at the moment of the earthquake is again assumed equal to 0.6×0.017 persons/m³.

Table 7
Assumed construction and intervention costs per unit volume of buildings (Lire/m³)

Construction	Intervention		
C_c	C_L	C_M	C_H
200,000	20,000	40,000	80,000

In order to present formally the optimization problem, define *gain* or *return* $g_i^k(C_l)_m$ of an intervention of cost C_l performed on the *m*-th building, the decrease in the expected damages when an earthquake of intensity *i* occurs (the index k = c will indicate economic returns, k = v returns in terms of saved people). In other words, the return is equal to the difference between the damages that would occur without any intervention and after having performed the intervention of cost C_l. For discrete types of interventions (like the quoted three types L, M, H), the return functions are multiple step functions.

Then, with reference to the *forecast* return under an earthquake of given intensity i, the optimization problem can be formulated as follows:

- maximize the total return (for either k = c or k = v):

$$F_{k|i}\{(C_1)_1,(C_1)_2,\ldots,(C_1)_m\} = \Sigma_m g_i^k (C_1)_m$$

- subject to:

$$\sum_{1}^{N} {}_m (C_1)_m \le C_{ava}$$

where the index m = 1, 2, ..., N indicates the building, the index i = 1, 2, 3, 4 the relevant interval of seismic intensity, and C_{ava} is the maximum amount that can be spent in preventive interventions (available resources).

The maximum can also be sought of the *expected* total return:

$$F_k = \Sigma_i \Sigma_m \pi_{im}\, g_i^k (C_1)_m$$

where π_{im} is the probability of occurrence of an earthquake in the intensity interval i at the site of building m.

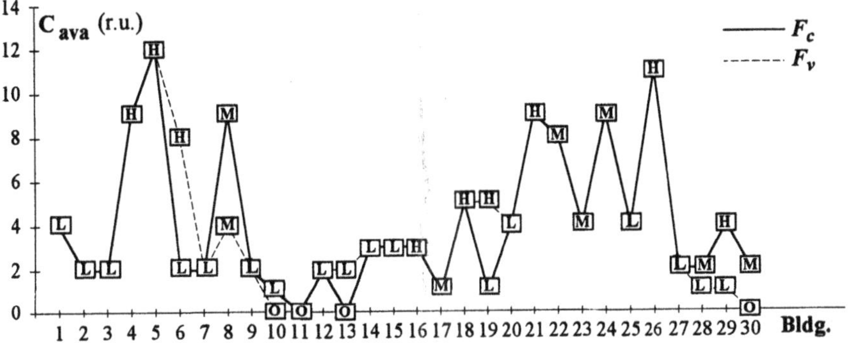

Fig. 5. *Interventions on 30 buildings, allowed by a given total amount of available resources, and optimized with respect to economic damages (F_c) and to saved lives (F_v)* [10]

Table 8
Assumed occurrence probabilities of earthquakes in 100 years

	i	*1*	*2*	*3*	*4*
	MSK *intensity*	*7 - 7.5*	*7.5 - 8*	*8 - 8.5*	*> 8.5*
Buildings	*Site (town)*	π_{1m}	π_{2m}	π_{3m}	π_{4m}
m = 1 - 10	Bastia Umbra	0.18	0.11	0.08	0.11
m = 11 - 20	Città di Castello	0.19	0.12	0.13	0.12
m = 21 - 30	Cascia	0.39	0.23	0.13	0.17

As already stated, the non-linearity and discontinuity of the relevant relationships do not allow the use of differential maximization procedures. But the objective functions $F_{k\,i}$ or F_k are the sum of as many quantities as the buildings, each in turn a function of the resources assigned to the *m*-th building only: hence, the optimization process is a *multi-stage decisional process*, which can be tackled with a comparatively small number of operations by dynamic programming [10][11]. Fig. 5 shows an example of optimal allocation of 120 *resource units* (r.u.), with respect to F_c and to F_y, obtained by dynamic programming among 30 buildings located in three different areas of Umbria, a region of Central Italy, where the 100-year probabilities of earthquake occurrence shown in Table 8 had been approximately estimated: details on the volume and vulnerability of the buildings are given in [10].

4. OPTIMAL ALLOCATION OF RESOURCES: NETWORKS AND LIFELINES

4.1 General considerations

At first sight, no significant difference appears whether the optimal allocation problems presented in Section 3 refer to buildings or other facilities (e.g., bridges). But in the case of buildings, dealt with so far, the initial vulnerability, the consequences of failures and the benefits derived from an intervention on any element of the ensemble can be assumed - at least as a first approximation - to be independent from each other and then summable, which simplifies much the problem. On the contrary, if the facilities are elements of a *system*, this is no more possible: the consequences of their failure, hence the effectiveness of any preventive measure, depend not only on the vulnerabilites of the single facilities, but also in an essential way on the logical diagram of the system, the critical condition considered and the collocation *(role)* of each element; therefore the vulnerability of the *system* must be evaluated on its own account.

On the other hand, it is now a well recognized fact (as very recent examples have confirmed) that the disruption of communication networks and other *lifeline* systems are among the most damaging effects of earthquakes. Indeed, as recent examples have confirmed, damages of this type can not only have immediate dramatic effects in the aftermath of an earthquake, but also consequences lasting for months and years on the economy, as well as on the conditions of life, of the whole area affected by an earthquake (or by any other disaster). And the increasing relevance of communications and services in modern life makes these effects all the more important. It becomes thus essential to develop the optimal allocation methodology, not only with regard to single buildings, but also to *systems*, and in particular lifeline networks, as first pointed out in [12].

A lifeline system can be in general modelled as a redundant network, comprising a number of vulnerable (or *critical*) elements, that may themselves be complex redundant structural or mechanical systems. The network topology is usually described by its minimal cut sets or its minimal path sets, and depends on the connections between the elements and on the assumed functionality condition. From the network topology and the element vulnerabilities, it is possible to derive the reliability of the network as a whole. To elaborate a strategy for its improvement, it is also necessary to estimate the costs and the benefits of possible preventive measures in terms of their effects on the vulnerability of critical elements and of the whole system.

4.2 Vulnerability of r.c. bridges

As a specific, but typical, case, Section 4.3 will deal with highway networks in which, by assumption, the only vulnerable elements are the bridges. It is also assumed that the seismic vulnerability of the bridges is described by *fragility curves*, known before and after some well defined *upgrading intervention.*

More specifically, the example bridges [2][14] are r.c. girder bridges: the decks are simply supported on piers of hollow rectangular section (of two different types). Five structural diagrams have been considered (Fig. 6) in four different conditions, i.e., either as originally designed (O) for a peak ground acceleration a_g = 0.10g (in accord with the Italian Regulations), or upgraded in one of three ways, which follow two different techniques, namely: jacketing of the piers with shotcrete cover and addition of longitudinal reinforcement to improve the pier flexural capacity and shear strength (the reinforcement is increased by 50% in intervention I; by 100% in intervention II); elimination of expansion joints between the decks and introduction of isolation/dissipation devices on piers to replace the existing bearings (intervention III).

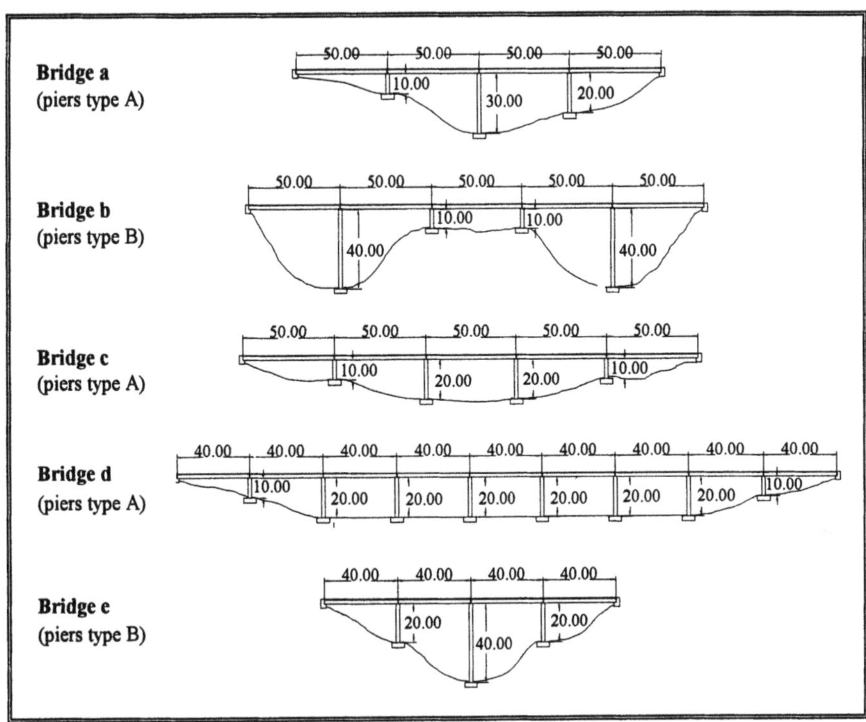

Fig. 6. *Structural diagrams of example bridges (measures in* m*)*

Table 9
Assumed costs of construction and of upgrading of bridges; conditional probabilities of failure of original (O) *and retrofitted bridges*

Bridge diagram		*a*	*b*	*c*	*d*	*e*
Construction cost		56	72	66	100	48
Upgrading cost	I	3	6	3	7	5
	II	4	8	4	9	6
	III	7	9	9	14	7
$P_f \mid a_g = 0.25g$	O	$3.15 \cdot 10^{-1}$	$2.82 \cdot 10^{-1}$	$5.60 \cdot 10^{-1}$	$6.29 \cdot 10^{-1}$	$4.43 \cdot 10^{-3}$
	I	$2.77 \cdot 10^{-1}$	$9.62 \cdot 10^{-2}$	$4.71 \cdot 10^{-1}$	$4.96 \cdot 10^{-1}$	$2.30 \cdot 10^{-3}$
	II	$1.94 \cdot 10^{-1}$	$2.71 \cdot 10^{-2}$	$3.49 \cdot 10^{-1}$	$3.59 \cdot 10^{-1}$	$3.69 \cdot 10^{-3}$
	III	$7.29 \cdot 10^{-3}$	$2.33 \cdot 10^{-3}$	$2.66 \cdot 10^{-3}$	$3.10 \cdot 10^{-3}$	$3.40 \cdot 10^{-4}$
$P_f \mid a_g = 0.35g$	O	1.00	1.00	1.00	1.00	$2.42 \cdot 10^{-1}$
	I	1.00	$8.72 \cdot 10^{-1}$	1.00	1.00	$1.15 \cdot 10^{-1}$
	II	1.00	$4.94 \cdot 10^{-1}$	1.00	1.00	$1.22 \cdot 10^{-1}$
	III	$3.02 \cdot 10^{-2}$	$1.54 \cdot 10^{-2}$	$1.14 \cdot 10^{-2}$	$2.50 \cdot 10^{-2}$	$7.57 \cdot 10^{-3}$

The costs of construction and intervention shown in Table 9 have been assumed in the numerical calculations: they are referred to the construction cost of bridge *d*, taken equal to 100 *resource units* (r.u.).

The failure condition of the bridges has been identified with the attainment of an appropriate threshold value of an indicator of the damage level in the critical sections of the piers (as reported in detail in [14]). The fragilities of each bridge in the four conditions have been evaluated by a MonteCarlo procedure, improved by Importance Sampling and Directional Simulation [13], using as inputs simulated seismic accelerograms compatible with the Eurocode spectrum S2, scaled to several values of the peak ground acceleration a_g (taken as the measure of the earthquake intensity). In this way, fragility curves were plotted as functions of a_g; the probabilities of failure of the five bridges corresponding to two values of the peak ground acceleration a_g are shown in Table 9.

4.3 Optimal allocation to the critical elements of a lifeline network

The aim of the network has been identified with ensuring the connection between a *source* node S and a *destination* node D. Thus, the network fails when this connection is severed: this definition has obvious limitations, because many factors are not taken into account (e.g., the capacity of traffic in the emergency that follows an earthquake), but it has been considered satisfactory for a first approach to the problem. Since the bridges are the only vulnerable elements, the network can fail only because one or more bridges fail.

Only the main results of some applications are presented in the following, while for a description of the procedure and other details the reader is referred to [2].

The five example networks diagrammaticaly represented in Fig. 7 have been considered. Each bridge is labelled by a serial number (1-5 or 1-10) and a letter indicating the structural diagram (Fig. 6).

The first network, denoted SE, is an elementary chain of elements in series, and may correspond to bridges located along a single highway stretch. It fails if any one of the bridges fail: therefore, assuming that bridge failures under a given earthquake are stochastically independent of each other, the (conditional) probability of network survival ($1 - P_{SE}$) is equal to the product of the probabilities of survival of all elements, whence:

$$P_{SE} = 1 - \Pi_i \{1 - P_i\}$$

where P_i is the probability of failure of element i subjected to a given earthquake.

The second network, denoted PA, is an elementary bundle of elements in parallel, and may represent the situation of a city cut by a river. The connection between the two banks fails if all bridges fail, whence:

$$P_{PA} = \Pi_i P_i$$

The analogous, but more complicated, laws yielding the (conditional) probability of failure of the other networks are presented in [2] and [3].

Note that the first four networks [2] can be represented as a combination of independent subsystems in series and/or in parallel, while this is not possible for the network CO [3]. Therefore, notwithstanding the small number of nodes, this is a *complex* network, according to the definition given in Ref. [11], Chap. 6.

Table 10 shows the failure probabilities of the five networks, in the original design condition (O) and after interventions of the same type on all bridges, for two values of a_g (namely 0.25 and 0.35 g, that correspond respectively to medium and high seismicity zones in Eurocode No. 8); the corresponding costs are also reported in the same table.

As described in [2], resorting to dynamic programming, it is possible instead to distribute preventive upgrading interventions on the bridges in such a way that, for a given total amount of employed resources, the increase in the expected reliability after an earthquake of given intensity is maximized. The distributions of the interventions, optimized in this way for two values of a_g, are shown in Table 11, while the conditional failure probabilities of the networks are plotted in Fig. 8 for three values of a_g versus the amount of resources C_{ava}.

The whole range of values of C_{ava} has been investigated from nil up to the value that would allow the most efficient intervention (III) on all bridges, i.e., 46 r.u. for the five-bridges networks, and 92 for the ten-bridges network SP2; calculations have been limited to $a_g = 0.35g$

for the parallel network PA, because its reliability under weaker earthquakes is already very large in the original condition.

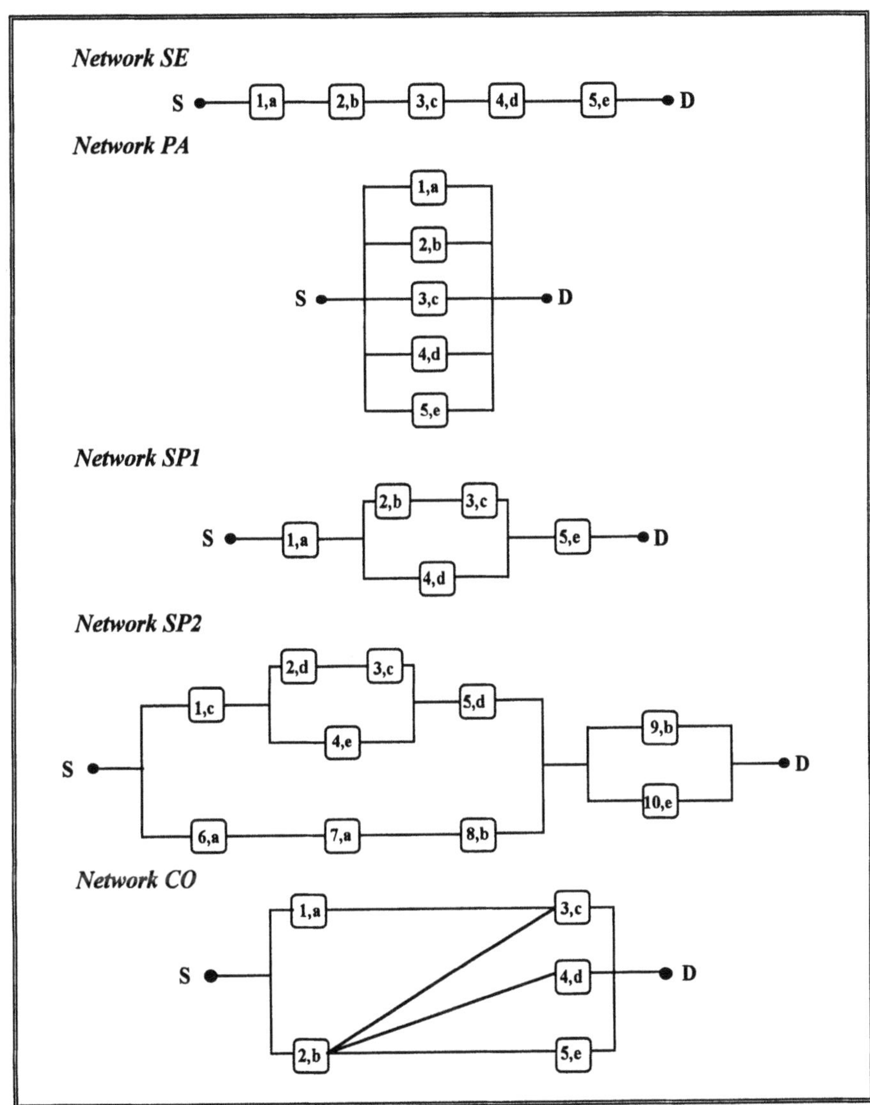

Fig. 7. *Diagrams of five example networks; locations and identification of critical elements*

With regard to the complex network CO, the optimal intervention distribution has been also obtained through an *exhaustive search* (in this example such a search is possible with a reasonable computational effort, because of the small number of elements). As a matter of fact, in the case of a complex network the results of the two procedures may not coincide, because in dynamic programming the problem is analyzed by successive steps that, in this specific instance, do not correspond to independent minimal cut sets. However, the two solutions have been found identical for all practical purposes, being different only in the range C_{ava} = 16 - 17 r.u. for a_g = 0.25g and a_g = 0.35g: this result seems to indicate the possibility of applying the procedure based on dynamic programming also to complex networks.

Table 10
Assumed costs of retrofitting of bridges and conditional probabilities of failure of original and retrofitted networks

Network		SE	PA	SP1	SP2	CO
Upgrading cost	I	26	26	26	52	26
	II	31	31	31	62	31
	III	46	46	46	92	46
$P_f \mid a_g = 0.25g$	O	$9.20\cdot10^{-1}$	$1.39\cdot10^{-4}$	$6.32\cdot10^{-1}$	$5.85\cdot10^{-1}$	$3.52\cdot10^{-1}$
	I	$8.26\cdot10^{-1}$	$1.43\cdot10^{-5}$	$4.65\cdot10^{-1}$	$3.87\cdot10^{-1}$	$7.12\cdot10^{-2}$
	II	$6.74\cdot10^{-1}$	$2.43\cdot10^{-6}$	$3.03\cdot10^{-1}$	$2.15\cdot10^{-1}$	$1.51\cdot10^{-2}$
	III	$1.56\cdot10^{-2}$	$\cong 0$	$7.64\cdot10^{-3}$	$1.10\cdot10^{-3}$	$2.32\cdot10^{-5}$
$P_f \mid a_g = 0.35g$	O	1.00	$2.42\cdot10^{-1}$	1.00	1.00	1.00
	I	1.00	$1.00\cdot10^{-1}$	1.00	1.00	$9.86\cdot10^{-1}$
	II	1.00	$6.03\cdot10^{-1}$	1.00	1.00	$7.75\cdot10^{-1}$
	III	$8.66\cdot10^{-2}$	$1.00\cdot10^{-9}$	$3.82\cdot10^{-2}$	$2.81\cdot10^{-3}$	$6.43\cdot10^{-4}$

Inspection of Fig. 8 and Table 11 can suggest many considerations.

For instance, it is interesting to note how the distribution of the optimized interventions sometimes changes drastically when the amount of the resources varies.

The convenience of an optimal versus a rule-of-thumb allocation of resources can also be put in evidence. Let for instance refer to the 10-bridge network SP2: if a_g = 0.25g and intervention II is performed on all bridges, 62 r.u. are employed and P_f is reduced from 0.58 to 0.21 (Table 10); if the same 62 r.u. are distributed in the optimal way, P_f becomes as low as $0.11\cdot10^{-3}$ (Fig. 8d).

In the same Fig. 8d, it can be also noted that, when the resources are optimally allocated, the reduction of P_f with C_{ava} is very slow beyond 68 r.u.: therefore, a sensible general policy of good exploitation of resources would allocate no more than 68 r.u. to the upgrading of bridges in the considered network.

It may be also of some interest to distinguish the preferential paths automatically chosen by the optimization procedure: in the already quoted network SP2, ⟨6-7-8⟩ if C_{ava} is rather small, ⟨1-4-5⟩ if it is larger.

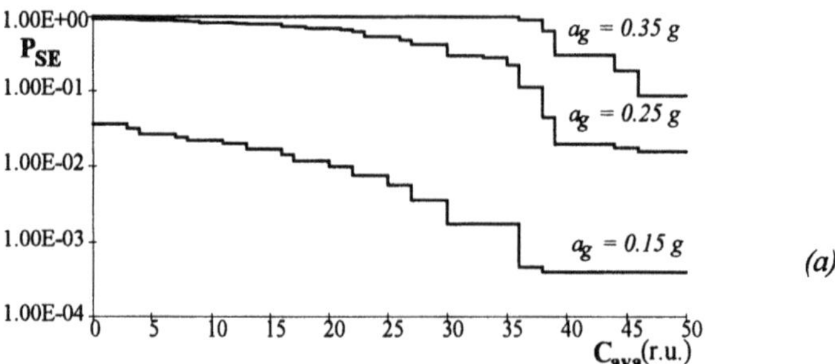

Fig. 8. *Probability of failure versus employed resources (optimized): a) network SE*

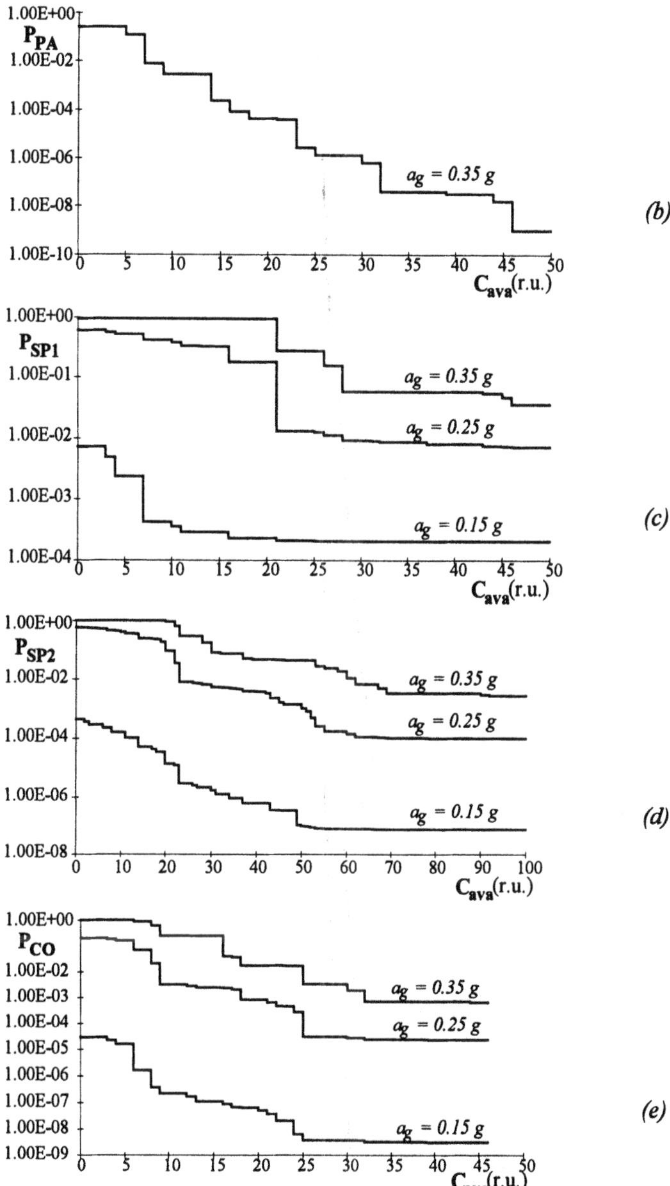

Fig. 8 (cont.d). *Probability of failure versus employed resources (optimized); peak ground acceleration* a_g = *0.15g, 0.25g and 0.35g: b) network* PA; *c) network* SP1; *d) network* SP2; *e) network* CO.

Table 11

Optimized interventions on each bridge of the five example networks vs. employed resources for $a_g = 0.35g$.

Network SE																
C_{ava}	3	6	9	12	15	18	21	24	27	30	33	36	39	42	45	46
1	-	-	-	-	-	-	-	-	-	-	-	III	III	III	III	III
2	-	-	-	-	-	-	-	-	-	-	-	I	III	III	III	III
3	-	-	-	-	-	-	-	-	-	-	-	III	III	III	III	III
4	-	-	-	-	-	-	-	-	-	-	-	III	III	III	III	III
5	-	-	-	-	-	-	-	-	-	-	-	-	-	-	I	III

Network PA																
C_{ava}	3	6	9	12	15	18	21	24	27	30	33	36	39	42	45	46
1	-	-	-	I	III	-	III	III	III	III	III	III	III	III	III	III
2	-	-	-	-	-	III	-	-	III	III	III	III	III	III	III	III
3	-	-	III	III	-	III	III	III	III	III	III	III	III	III	III	III
4	-	-	-	-	-	-	-	-	-	-	-	-	III	III	III	III
5	-	I	-	-	III	-	I	III	-	I	III	III	-	-	I	III

Network SP1																
C_{ava}	3	6	9	12	15	18	21	24	27	30	33	36	39	42	45	46
1	-	-	-	-	-	-	III	III	III	III	III	III	III	III	III	III
2	-	-	-	-	-	-	-	-	-	-	-	-	-	-	II	III
3	-	-	-	-	-	-	-	-	-	-	-	-	-	-	III	III
4	-	-	-	-	-	-	III	III	III	III	III	III	III	III	III	III
5	-	-	-	-	-	-	-	-	I	III	III	III	III	III	III	III

Network SP2																			
C_{ava}	5	10	15	20	25	30	35	40	45	50	55	60	65	70	75	80	85	90	92
1	-	-	-	-	-	-	-	III	III	III	III	III	III	III	III	III	III	III	III
2	-	-	-	-	-	-	-	-	-	-	-	-	-	-	-	-	-	III	III
3	-	-	-	-	-	-	-	-	-	-	-	-	-	-	-	-	-	III	III
4	-	-	-	-	-	-	-	III	III	III	-	III	III	III	III	III	III	II	III
5	-	-	-	-	-	-	-	III	III	III	III	III	III	III	III	III	III	III	III
6	-	-	-	III	III	III	III	-	-	-	III	III	III	III	III	III	III	III	III
7	-	-	-	III	III	III	III	-	-	-	III	III	III	III	III	III	III	III	III
8	-	-	-	II	III	III	III	-	-	-	III	III	III	III	III	III	III	III	III
9	-	-	-	-	-	-	III	III	III	III	III	-	III	III	III	III	III	III	III
10	-	-	-	-	-	III	-	-	II	III	-	III	-	III	III	III	III	III	III

Network CO																
C_{ava}	3	6	9	12	15	18	21	24	27	30	33	36	39	42	45	46
1	-	-	-	-	-	-	-	-	III	III	III	III	III	III	III	III
2	-	I	III	III	III	III	III	III	III	III	III	III	III	III	III	III
3	-	-	-	-	-	III	III	III	III	III	III	III	III	III	III	III
4	-	-	-	-	-	-	-	-	-	-	-	-	III	III	III	III
5	-	-	-	-	I	-	-	I	-	I	III	III	-	-	I	III

4.4 Alternative objectives of the optimization

So far, the allocation of the resources has been optimized exclusively with respect to the probability of failure of the network, i.e., by definition, with respect to the probability of severing the S-D connection. However, other factors should also be considered in the optimization process: among these, the length of the time in which the network remains out of service, either after an earthquake or during the upgrading works.

A first attempt at taking into consideration the time factor has been made in [3] assuming that the most efficient set of interventions is the set that yields the largest increase of reliability in the shortest time; hence, an example has been developed of an alternative optimization of network CO, with respect to a new objective function denoted *time-efficiency* and equal to the ratio

$$\eta = \frac{\Delta R}{T^*}$$

between the variation ΔR of the network reliability yielded by a set of interventions and the time T* they require.

The resources have then been allocated in the following way: for each of the paths connecting S and D, the distribution of interventions that maximizes the reliability of the connection in the shortest time is determined; then, the path corresponding to the largest time-efficiency is selected. If the available amount of resources is larger than the amount necessary to ensure the functionality of the most efficient path, the allocation procedure is iterated on the remaining paths. In the developed example, in this iteration both objective functions defined above have been tried: namely, the additional interventions have been planned either on the most time-efficient path among the alternative ones, or to maximize the further increase in the system reliability. In these cases, two alternative interventions shall appear in the relevant boxes of Table 12 (but the differences are very small).

In these operations, a variant has been used of the algorithm first presented by Horn [15] and applied in [16] to the restoration of lifelines damaged by a seismic event. This algorithm searches in a graph the path of *shortest length* (of *largest time-efficiency*, in the present case) between S and D: this path yields the largest rate of the restoration curve, that is of the plot of the level of efficiency attained by the system (which in [16] is given by the ratio between the number of the restored elements and the total number of elements) versus the time needed to attain it. This path corresponds to a global optimum and does not coincide, in general, with the path obtained by choosing the optimal solution at each node of the graph.

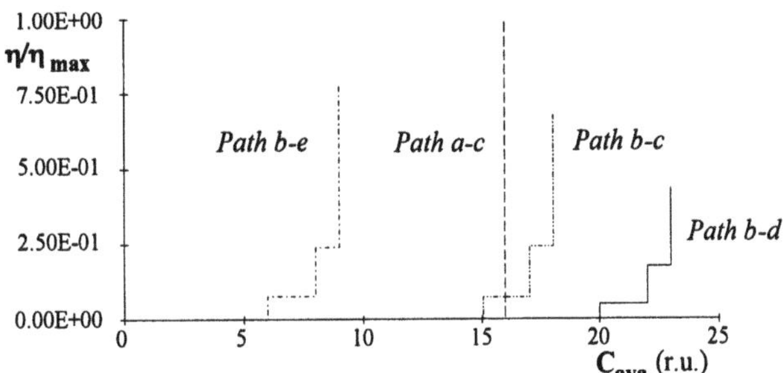

Fig. 9. *Network CO: Time-efficiency functions η versus available resources;* a_g = *0.35g.*

In Fig. 9 the time-efficiency functions η are plotted for $a_g = 0.35g$ and the different paths, versus the required resources. Inspection of these plots shows that *a-c* is the overall most efficient path; however, at least 16 r.u. are needed for its functionality. Therefore, if a smaller amount of resources is available (but at least 6 r.u., i.e., those needed for intervention I on bridge *b*) the S-D connection is assured by path *b-e*.

In deriving these plots, the time T* has been evaluated as the sum of the times needed to implement each intervention. However, similar results are obtained for the other limit case of T* equal to the longest time needed for one intervention, as if all interventions were applied at the same time. The only differences appear with reference to the path *b-e*, because if 14 r.u. are available, an additional upgrading intervention on bridge *e* yields a larger increase of the efficiency than indicated by Fig. 8e.

The distributions of interventions, thus optimized with respect to time-efficiency, are reported in Table 12 for a_g = 0.35g: *a-c* appears as the most efficient path. Note however that the probability of collapse of bridge *a* is very close to one both in the original state and after interventions I and II: hence, at least 16 r.u. are needed to ensure the functionality of the *a-c* path. If smaller amounts of resources are available, the optimal solution is to upgrade path *b-e*.

Table 12.
Interventions on each bridge (1-5) of network CO, *optimized with respect to time-efficiency, vs. employed resources (3-46 r.u.) for* a_g = *0.35g.*

Cava	*3*	*6*	*9*	*12*	*15*	*18*	*21*	*24*	*27*	*30*	*33*	*36*	*39*	*42*	*45*	*46*
a	-	-	-	-	-	III	III	III	III	III	III	III	III	III	III	III
b	-	-	-	-	-	-	-	II	III	III	III	III	III	III	III	III
c	-	-	-	-	-	-	-	III	III	III	III	III	III	III	III	III
d	-	-	-	-	-	-	-	-	-	-	-	-	-	-	-	III
e	-	-	-	-	-/I	-	-	-	-	I	III	III	III	III	III	III

The thick lines in Fig. 10 show, versus the available amount of resources, the variations of the failure probability P_f of the network for three values of a_g: it can be noted that the decrease of P_f is significant only beyond 16 r.u. The thin lines in the same Fig. 10 reproduce the analogous lines of Fig. 8e, that correspond to a resource allocation optimized with respect to P_f: the differences between the two plots (especially significant for a_g = 0.15g and 0.25g in the range 6-16 r.u.) show that optimization with respect to reliability and optimization with respect to time-efficiency lead, in general, to different sets of interventions. However, it can be noted that there are no significant differences between the two optimized allocations of an amount of resources larger than the amount necessary to secure the functionality of at least one S-D path.

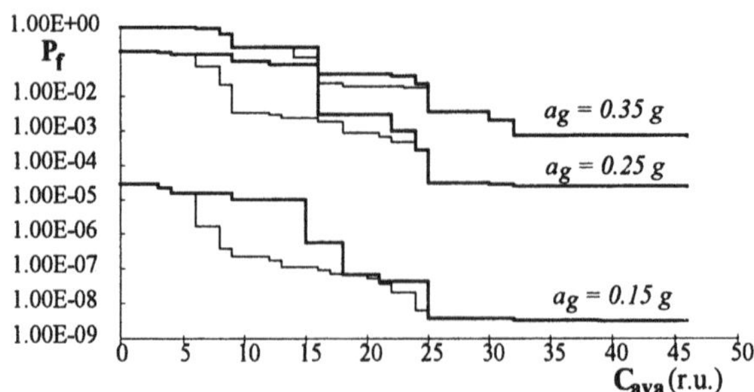

Fig. 10. *Probability of failure of network* CO *versus employed resources, optimized with respect to the time efficiency of the interventions (thick lines) and to the network reliability (thin lines: cf. Fig. 8e); peak ground acceleration* a_g = *0.15g, 0.25g and 0.35g.*

5. CONCLUDING REMARKS

In Section 3, a procedure has been illustrated for the optimal allocation of the resources available for seismic upgrading of existing buildings, whose damages are assumed independent of each other. Then, Section 4 illustrates an analogous procedure for the bridges of a highway network. In this case, the reliability of the network as a *system* had to be considered: it has been defined as the probability that, when an earthquake of given intensity hits the area, a connection is maintained between a source and a destination node.

It appears fair to say that, notwithstanding the many simplifying assumptions that have been assumed, the procedures could already be applied to real problems of resource allocation. This is true also for network optimization, at least when the reliability of the connection is the main concern. In fact, although so far only simple networks have been studied and rough quantitative estimates have been introduced, the interesting and significant results obtained appear worth further investigation.

Further research is also in progress for introducing alternative objective functions of the optimization process of the network (like, say, the largest expected traffic capacity between source and destination after the earthquake has occurred, or the minimum repair time necessary to restore the S-D connection when it is severed as a consequence of the earthquake), and for taking account of several degrees of damage (of the elements and/or the network), implying a reduction of load capacity of bridges and consequently of traffic capacity of the network.

As a first tentative in this direction, an efficiency function defined as the ratio between the obtained increase of reliability and the time necessary to perform the upgrading interventions has been introduced in Section 4.4 as an alternative objective function. Comparing the results with those of the optimization with respect to reliability, it has been found that the two optimal solutions do not coincide, i.e., consideration of time modifies the set of interventions to be performed: this indicates the interest of further studies on this problem.

ACKNOWLEDGEMENTS

This research has been partially supported by grants from the Ministry of University and Research (M.U.R.S.T.) and from the G.N.D.T.

REFERENCES

1. G. Augusti, A. Borri and M. Ciampoli, Seismic protection of constructed facilities: optimal use of resources, *Structural Safety*, 16(1-2) (1994) 91-102.
2. G. Augusti, A. Borri and M. Ciampoli, Optimal allocation of resources in the reduction of the seismic risk of highway networks, *Engineering Structures*, 16(7) (1994) 485-497.
3. G. Augusti and M. Ciampoli, On the seismic risk of highway networks and its reduction, *Risk Analysis, Proceedings of a Symposium*, Ann Arbor, Michigan (1994) 25-36.
4. G. Augusti, D. Benedetti and A. Corsanego, Investigations on seismic risk and seismic vulnerability in Italy, *Proc. 4th Int. Conf. on Structural Safety and Reliability* ICOSSAR '85, Kobe, Japan (1985) Vol. 2, 267-276.
5. D. Benedetti and G.M. Benzoni, Seismic vulnerability index versus damage for unreinforced masonry buildings, *Proc. Int. Conf. on Reconstruction, Restoration and Urban Planning in Seismic prone Areas*, Skopje (1985) 333-347.
6. F. Braga, M. Dolce and D. Liberatore, Southern Italy November 23, 1980 earthquake: a statistical study on damaged buildings and an ensuing review of the MSK-76 scale, *7th European Conf. on Earthquake Engineering*, Athens (1982).
7. R.M. Wagner, N.P. Jones and G.S. Smith, Risk factors for casualties in earthquakes: The application of epidemiologic principles to structural engineering, *Structural Safety*, 13(3) (1994) 177-200.
8. G. Augusti: Discussion of Ref.[7], *ibid.* [in print]

9. G. Augusti and A. Mantuano, Sulla determinazione della strategia ottimale per la prevenzione del rischio sismico nei centri abitati: un nuovo approccio, *L'Ingegneria Sismica in Italia 1991, Proc. 5th Italian Nat. Conf. Earthquake Engrg.*, Palermo (1991) 117-128.
10. G. Augusti, A. Borri and E. Speranzini, Optimum allocation of available resources to improve the reliability of building systems, *Reliability and Optimization of Structural Systems '90, Proc. 3rd IFIP WG 7.5 Conference,* Berkeley, California (1991) 23-32.
11. S.S. Rao, *Reliability-Based Design*, McGraw-Hill, Inc. (1992).
12. G. Augusti, A. Borri and M. Ciampoli, Optimal allocation of resources in repair and maintenance of bridge structures, *Probabilistic Mechanics and Structural and Geotechnical Reliability, Proc. 6th ASCE Specialty Conf.*, Denver, Colorado (1992) 1-4.
13. M. Ciampoli, R. Giannini, C. Nuti and P.E. Pinto, Seismic reliability of non-linear structures with stochastic parameters by directional simulation, *Proc. 5th Intern. Conf. on Structural Safety and Reliability* ICOSSAR '89, S.Francisco (1989) 1121-1126.
14. M. Ciampoli and G. Augusti, Seismic reliability assessment of retrofitted bridges, *Structural Dynamics, Proc. 2d European Conference on Structural Dynamics* Eurodyn '93, Trondheim, Norway (1993) Vol. 1, 193-200.
15. W.A. Horn, Single-machine job sequencing with treelike precedence ordering and linear delay penalties, *SIAM, Journal of Applied Mathematics*, 23(2) (1972) 189-202.
16. N. Nojima and H. Kameda, Optimal strategy by use of tree structures for post-earthquake restoration of lifeline network systems, *Proc. Tenth World Conference on Earthquake Engineering*, Madrid (1992) Vol. 9, 5541-5546.

2

CODIFIED RELIABILITY OF STRUCTURES

O. Ditlevsen, Department of Structural Engineering, Technical University of Denmark, Build. 118, DK 2800 Lyngby, Denmark

ABSTRACT For the practical applications of probabilistic reliability methods it is important to make decisions about the target reliability level. Presently calibration to existing design practice seems to be the only practicable and politically reasonable solution to this decision problem. However, several difficulties of ambiguity and definition show up when attempting to make the transition from a given authorized partial safety factor code to a superior probabilistic code. For any chosen probabilistic code format there is a considerable variation of the reliability level over the set of structures defined by the partial safety factor code. Thus there is a problem about which of these levels to choose as target level. Moreover, if two different probabilistic code formats are considered, then a constant reliability level in the one code does not go together with a constant reliability level in the other code. The last problem must be accepted as the state of the matter, and it seems that it can only be solved pragmatically by standardizing a specific code format as reference format for constant reliability.

By an example this paper illustrates that a presently valid partial safety factor code imposes a quite considerable variation of the reliability measure as defined by a specific probabilistic code format. Decision theoretical principles are applied to get guidance about which of these different reliability levels of existing practice to choose as target reliability level. Moreover it is shown that the chosen probabilistic code format not only has strong influence on the formal reliability measure but also on the formal cost of failure to be associated if a design made to the target reliability level is considered to be optimal. In fact, the formal cost of failure can be different by several orders of size for two different by and large equally justifiable probabilistic code formats. Thus the consequence is that a decision theoretical code format formulated as an extension of a probabilistic code format must specify formal values to be used as costs of failure.

A principle of prudency is suggested for guiding the choice of the reference probabilistic code format for constant reliability. To the author's opinion there is an urgent need for establishing a standard probabilistic reliability code. This paper presents some considerations that may be debatable but nevertheless point at a systematic way to choose such a code.

1 Introduction

It may prevent confusion about the meaning of the concept of limit state, if the concept is formally introduced without a specific physical interpretation. Then the limit state concept is just a mathematical criterion that categorizes any value set of the relevant structural variables (load-, strength-, geometrical variables) into one of two categories, the "desirable" category ("safe" set) and the "adverse" category ("failure" set). The word failure then means "failure of satisfying the limit state criterion" rather than a failure in the sense of a dramatic physical event.

The purpose of introducing the limit state is to represent any chosen kind of less desirable physical behavior (adverse behavior) in a sufficiently idealized way for setting up a set of practically operational rules that ensure that such adverse behavior rarely occurs, given that the rules are followed.

Codified probabilistic reliability theory contains a definition of a measure of rareness of violation of any considered limit state. For example, this measure may be defined as the reliability index β with respect to the limit state (technically: $\beta = -\Phi^{-1}(p_f)$, $\Phi(\cdot) =$ standardized normal distribution function, $p_f =$ probability of adverse behavior calculated by use of codified probability distribution types for the basic variables).

The problem is, however, to decide what "rarely" means. It seems obvious that the reliability index corresponding to a given limit state should depend on the physical consequence (and thus the socio-economical consequence) of the occurrence of the adverse behavior. This leads directly to the rationale of cost-benefit analysis (decision theory). While such an analysis in a rational way can lead to a determination of the relative variation of the reliability index over a set of different limit states, there remains the problem of choosing the absolute reliability level for the structure as it will be physically represented when built. Since the costs of consequences of structural damage or collapse embrace intangible socio-economic costs such as injury or loss of human lives and possibly irremediable damages on the environment, there seems to be no other way than to establish some fix-points in the most frequent existing practice and calibrate the reliability level to these fix-points.

This point of view, however, presents a problem of what is most frequent existing practice. An illustrative example concerns the design of reinforced concrete structures. It is current practice in some countries and among some consulting engineers that the ultimate (the most adverse) limit state is defined in terms of load effects and stresses obtained from the theory of elasticity. In other countries it has become practice to define the ultimate limit state as corresponding to a plasticity-theoretical collapse. In spite of these differences of defining the ultimate limit state, existing practice has through codification of a fixed set of partial safety factors by and large assigned the same formal reliability level to the two different types of limit states. However, except if the elasticity theoretical formulas have been corrected on the basis of carrying capacity experiments to represent the collapse situation, the consequences of violating the "elastic" ultimate limit state are in several important examples less adverse with respect to the physical consequences than the consequences of collapse as idealized by the "plastic" ultimate limit state. This means that reinforced concrete designs made according to the elastic ultimate limit state have larger reliability with respect to real physical collapse than designs made according to the plastic ultimate limit state.

For code calibration purposes the problem is now which of the two reliability levels with respect to collapse should be codified in a probabilistic code. It is by certain parties claimed that only the current design practice based on the elastic ultimate limit state provides sufficient structural reliability; if this is so, then several existing reinforced concrete structures are underdesigned. On the other hand, since such structures designed according to the (effectivity factor adapted) plasticity theory with usual partial safety factors have been accepted by the society (the authorities), and since such existing concrete structures show no particular indications of weaknesses, it can be claimed that their general reliability level is sufficiently high. There is also a stronger argument for accepting the plastic ultimate limit state as the ultimate limit state for which the codified reliability level should be valid. This stronger argument is that the designs made according to the elastic ultimate limit state have nonuniform reliability with respect to similar consequences. In some cases of current practice the elastic and the plastic ultimate limit states are made practically coincident (by use of empirical correction factors, e.g. in the case of bending failures of statically determinate shallow beams); in other cases there are very large differences (for example for shear walls). It is, however, a well established experience that the violation of the plastic ultimate limit state implies that a situation close to structural collapse occurs. By and large this experience is valid uniformly over the different types of reinforced concrete structures (plates and shells excepted, perhaps). The occurrence of the adverse event defined by the elastic ultimate limit state may or may not indicate such a near collapse situation.

Let us make the supposition that there in a given country (or union of cooperating countries) is a decision maker (a committee, say) that possesses superior authority with respect to deciding what is optimal structural design. This decision maker is imagined to be asked to point out a specific structure type (structural element type, say) that will get optimal dimensions if it is designed to the limit according to a given partial safety factor code and a specific ultimate limit state model both specified by the authority itself. This means that the superior authority by answering this question implicitly evaluates the intangible socio-economic cost of adverse behavior of the considered structure type such that any stronger or weaker design will have a larger total cost than that of the optimal design. In fact, if the decision theory of von Neumann and Morgenstern [9] is accepted as a valid principle of decision making the evaluation of the socio-economic costs of the different possible consequences of the decision is precisely the task of the decision maker. The minimization of the total expected cost then leads to the optimal decision. With the optimal decision given the intangible cost of adverse behavior can therefore be backcalculated under the adoption of a specific probabilistic code. In this way having calibrated the intangible cost of the design decision problem to the authoritative declaration of optimality, the future design decisions for the considered structure type can be based on rational decision theory implemented within the given probabilistic code.

Even though the superior authority might admit that its partial safety factor code may be too simplified to have the potential of giving optimal designs for all the structure types for which the code is used, there should be at least one structure type within the domain of the code that by the authority is considered to be optimally designed. Otherwise the authority is not self-consistent and as the proper authority it should therefore change the value settings of the code [1]. Therefore the outlined strategy for setting up a probabilistic code that can be approved by the superior authority seems reasonable. However, more detailed analysis reveals several problems that require authoritative decisions.

The first problem to be illustrated in the next section is that the authority is not asked which specific structure of the considered type is the optimal structure. Of course, such a question can be asked, but as it will become evident in the following it is not necessary and hardly wise to require such a specific and difficult question to be directly answered by the authority. By only declaring optimality for the specified structure type class as such it turns out that for a given probabilistic code there is a considerable uncertainty about the value of the optimal reliability index. It will be shown that this problem can be solved in a rational way by a decision theoretical argument.

If the authority declares that two or more classes of structure types within the domain of the partial safety factor code possessoptimality then the corresponding reliability indices and intangible costs of adverse behavior can be calculated for each structure type by use of the given probabilistic code. Of course, it is not necessarily so that these costs should be the same for all structure classes. However, it may be a surprise to the authority if the ordering of the structure classes according to increasing cost of adverse behavior turns out not to be consistent with the ordering obtained from a realistic engineering evaluation of the consequences of the adverse behavior. Such experience should force the authority to do a reexamination of its decisions about optimality. Having consistence in the ordering it may also become clear to the authority that the calculated costs of adverse behavior should be adjusted relative to each other to better reflect the engineering judgement of the consequences of the adverse behavior. Of course, at this state the authority should totally abandone the original partial safety factor code and replace it by the probabilistic code as the superior reference for codified structural reliability. Future simplified codes of type as partial safety factor formats or other simplified code formats should then be calibrated to this superior probabilistic reference code.

The philosophical problems of establishing a standardized probabilistic code are not over by the above strategy. There remains the problem of which probabilistic code format to standardize, that is, which types of probability distributions to assign to the specified list of basic variables of the code. The often discussed tail sensitivity problem cannot be ruled out solely by supporting the choice on empirical evidence and physical modeling. By a simple example the ambiguity problem is shown to be serious. Calculations made by Sørensen and Enevoldsen [2] may be taken as an indication that there may even be a problem of whether two different equally possible probabilistic code formats define the same ordering with respect to the reliability index. To come to a decision about which format to choose among a finite set of probabilistic code formats a prudency principle is suggested that ensures as limited deviation from the authoritatively instituted partial safety factor design practice as possible.

All these different aspects are discussed in the following. The paper advocates a rational and operational procedure to transfer present codified practice into a more structured and rational probabilistic code. An attempt to formulate an example of how such a code could look like (without any calibrations to existing codes) is given by Ditlevsen and Madsen [6] as a part of the activity within the Joint Committee on Structural Safety (JCSS).

As probabilistic reliability methods gain increasing interest in engineering practice as judged from the growing community of reliability engineers and from the increasing number of conferences on the subject, it becomes increasingly more urgent to reach general concensus about the standardization on a specific probabilistic code format.

This paper is meant as a debate paper on this topic. The first draft of the paper was presented to JCSS for discussion leading to extensive revisions. It is emphasized that the statements of this paper are solely expressing the opinion of the author and they cannot be taken as approved by JCSS.

If should finally be mentioned that several code calibration applications of ad hoc probabilistic reliability models without reference to a standardized probabilistic code format have been reported in the literature [1,3,4,5].

2 Illustrative example

As an illustration let us make the supposition that the superior authority points at the structure type class consisting of the tubular joints of an offshore structure as being optimal if the joints are designed according to the Danish offshore code (DS 449) [7] as explained in the following.

For a standard type plane tubular joint in an offshore structure the limit state with respect to failure of the joint is assumed to be modeled by an empirically based equation of the form

$$1 - \frac{P}{P_u} - \left(\frac{M}{M_u}\right)^{\alpha_1} - \left(\frac{N}{N_u}\right)^{\alpha_2} = 0 \tag{1}$$

in which P is an internal normal force, M is an internal moment in the plane of the joint, and N is an internal moment orthogonal to the plane. The corresponding strength variables are the random variables P_u, M_u, and N_u, respectively. Further details about the geometry and mechanical behavior of the joint are not needed in this example. For each joint there is an influence matrix $\boldsymbol{C}$ such that the internal forces of the joint are obtained as the linear function $[P\ M\ N]' = \boldsymbol{C}[G\ Q\ S\ V\ W]'$ of the five random load variables G (self weight), Q (operational load), S (snow and ice load), V (wind load), and W (wave and current load).

All eight random variables are assumed to be mutually independent even though this assumption may not be fully realistic. In particular the three strength variables could be mutually dependent as also the nature loads S, V, and W could be dependent. However, for the purpose of this illustration these possible dependencies are not essential. Moreover the problem is simplified by assuming that the exponents α_1, and α_2 are deterministic: $\alpha_1 = 2.1$, $\alpha_2 = 1.2$.

The probabilistic code is defined by specifying that the three resistance random variables P_u, M_u, N_u have lognormal distributions of common expectation θ and coefficients of variation 0.20, 0.25, 0.25, respectively, and that the load variables G, Q, S, V, W have normal distributions of common mean 1 (with a suitably physical unit) and coefficients of variation 0.05, 0.15, 0.20, 0.25, 0.30, respectively. The common expectation θ of the strength variables is taken as the design variable ideally to be determined such that the joint gets a prespecified reliability with respect to the limit state (1).

We now make the supposition that the joint is chosen at random among all the joints of a large offshore structure. This means that the influence matrix $\boldsymbol{C}$ is an outcome of a random matrix. It is assumed that the value of the design variable θ for the chosen joint is obtained according to the Danish offshore code (DS 449) (the format of which is similar to the general format of the Euro-codes) [7]. Thus θ is determined as the largest of the five values obtained according to the five design load combinations

$$\begin{bmatrix} p_{d1}\cdots p_{d5} \\ m_{d1}\cdots m_{d5} \\ n_{d1}\cdots n_{d5} \end{bmatrix} = \boldsymbol{C} \begin{bmatrix} \gamma_g g_k & \gamma_g g_k & \gamma_g g_k & \gamma_g g_k & 1.15 g_k \\ \gamma_f q_k & \psi_q q_k & \psi_q q_k & \psi_q q_k & 0 \\ \psi_s s_k & \gamma_f s_k & \psi_s s_k & \psi_s s_k & 0 \\ \psi_v v_k & \psi_v v_k & \gamma_f v_k & \psi_v v_k & 0 \\ \psi_w w_k & \psi_w w_k & \psi_w w_k & \gamma_f w_k & 0 \end{bmatrix} \quad (2)$$

substituted together with the design values $p_{ud} = p_{uk}/\gamma_m$, $m_{ud} = m_{uk}/\gamma_m$, $n_{ud} = n_{uk}/\gamma_m$ of the resistance variables into the limit-state equation (1). The characteristic values p_{uk}, m_{uk}, n_{uk} are defined as the 5%-fractiles of the distributions of P_u, M_u, N_u, respectively. For the assumed data of the probabilistic code these characteristic values are $[p_{uk} \;\; m_{uk} \;\; n_{uk}] = [0.708 \;\; 0.647 \;\; 0.647]\theta$. The material strength partial safety factor is put to $\gamma_m = 1.2$. The characteristic values q_k, s_k, v_k, w_k of the loads Q, S, V, W are defined as the 98%-fractile in the distribution of the yearly extreme, while the characteristic value g_k of the self-weight G is defined as the 50% fractile. Thus $[g_k \;\; q_k \;\; s_k \;\; v_k \;\; w_k] = [1.000 \; 1.308 \; 1.411 \; 1.514 \; 1.616]$. The last matrix in (2) contains the design values obtained from the characteristic values by application of the partial coefficients γ_g and γ_f on self-weight and variable load, respectively, and the load combination factors $\psi_q, \psi_s, \psi_v, \psi_w$. These factors are put to $\gamma_g = 1.0$, $\gamma_f = 1.3$, $\psi_q = 1.0$, $\psi_s = \psi_v = \psi_w = 0.5$.

In order to assess how much the reliability varies among the joints of the considered offshore structure given that all joints are designed according to the partial safety factor code defined here, it is assumed as an example that the 15 influence numbers in $\boldsymbol{C}$ are mutually independent random variables that are uniformly distributed between 0 and 1. Bjerager [8] considered this example and simulated 1000 outcomes of $\boldsymbol{C}$, and for each of these the design parameter value θ and thereafter the geometric reliability index β (Hasofer-Lind) were calculated. The obtained sample of 1000 values of β turned out to be reasonably well described by a normal distribution of mean $\mu \simeq 4.40$ and a standard deviation $\sigma \simeq 0.30$. Thus there is a considerable variability of the reliability index resulting from constant partial safety factor design. This variability is comparable in size with the difference between two neighbour safety classes as they are defined by The Nordic Committee on Building Regulations [4]. The difference between high safety class and normal safety class is expressed by a difference in the β-level of 0.5, Fig. 1.

On the basis of this result it is seen that the probability p_f of occurrence of failure within one year and the corresponding reliability index β_{class} of a randomly chosen joint is

$$p_f = \frac{1}{\sigma}\int_{-\infty}^{\infty} \Phi(-\beta)\varphi(\frac{\beta-\mu}{\sigma})d\beta \simeq 1.25 \cdot 10^{-5}, \beta_{class} = -\Phi^{-1}(p_f) = 4.21 < E[\beta] = 4.40 \quad (3)$$

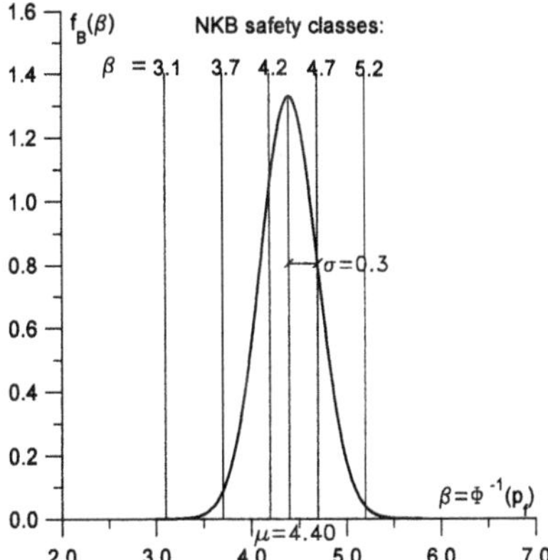

Fig. 1. Distribution of the reliability index as obtained by simulating deterministic designs of tubular joints according to a presently valid partial safety factor code for offshore structures (DS 449) when compared to a specific probabilistic code for the joint resistance and the external loads.

The problem is now whether the optimal target reliability index β_t should be chosen as β_{class} or whether it should be some other value of β. One argument could be that since the authority has accepted that the fraction x of the joints according to the present code has a reliability index smaller than $\beta(x) = \mu + \sigma\Phi^{-1}(x)$, then $\beta(x)$ should in principle be acceptable as target reliability index if x is not too small ($x = 25\%$, for example). However, the problem is not solved before some rational principle of choosing x is established.

3 Optimal reliability

The discussion of the problem about the choice of the reliability level can conveniently be supported on the following often considered simplified decision theoretical model.

The anticipated capital investment in a structure under design is obviously a function $c(p)$ of the required failure probability p. This function can be assumed to be decreasing. Assume that the loss of the structure by a failure will imply an investment equal to $c(p)+d$ where d is the direct damage cost of the failure event (compensations for losses, clearing, etc.) measured in monetary units. For each value of p we thus have a lottery (a game) between two consequences, namely a consequence with the utility $-c(p)$ corresponding to the event of no failure, and a consequence with the utility $-2c(p) - d$ corresponding to the event of failure. These two consequences refer to a specific period of time. In order not to complicate the discussion with considerations on capitalizations (interest rate considerations) that require an extension of the model to keep account of the point in time at which the failure possibly occurs, the reference period is taken to be short like 1 year, say. The total cost then is

$$c(p) + [c(p) + d]\mathbf{1}_{failure}, \quad \mathbf{1}_{failure} = \begin{cases} 1 & \text{if failure occurs} \\ 0 & \text{otherwise} \end{cases} \tag{4}$$

and it has the expected value $c(p) + [c(p) + d]p$. Following von Neumann's and Morgenstern's decision rule [9] we should among these infinitely many lotteries between two consequences choose the lottery for which the expected total cost is smallest. By differentiation we get the condition $c'(p)(1 + p) + c(p) + d = 0$ for the determination of the optimal failure probability. For real structures the value of d will generally be sufficiently large for p to be very small. Assuming that $p << 1$ the last equation reduces to

$$d \simeq -c(p) - c'(p) \tag{5}$$

from which d is directly obtained as a function of p, or from which p can be determined as a function of d.

The variation of the capital investment within the relevant interval of small failure probabilities can in many cases be approximated well by a function of p which is linear in the generalized reliability index $c(p) - c(p_0) = a(\beta - \beta_0)$, where a is the increase of cost per unit increase of β, and β_0 is a suitable representative reliability index value (e.g., $\beta_0 = 4.76$ corresponding to the probability $p_0 = \Phi(-\beta_0) \simeq 10^{-6}$). The equation (5) then becomes

$$\frac{d + c(p_0)}{a} \simeq \sqrt{2\pi} e^{\frac{1}{2}\beta^2} - (\beta - \beta_0) \simeq \sqrt{2\pi} e^{\frac{1}{2}\beta^2} \simeq \begin{cases} 2.3 \cdot 10^2 & \text{for} \quad \beta = 3 \\ 0.75 \cdot 10^4 & \text{for} \quad \beta = 4 \\ 0.67 \cdot 10^6 & \text{for} \quad \beta = 5 \end{cases} \tag{6}$$

Since $c(p_0)/a$ presumably is of order of size as 1, $c(p_0)$ can obviously be neglected in the ratio $[d + c(p_0)]/a$ on the left side of (6). In order to get optimal reliabilities of the size required in present codes the direct damage costs therefore must be several orders of magnitude larger than the investment $c(p)$.

For given d and for values of β that are relevant in structural reliability we thus find from (6) that the optimal reliability index is approximately

$$\beta_{opt} \simeq \sqrt{2 \log \left(\frac{d}{a\sqrt{2\pi}} \right)} \tag{7}$$

For an optimal value of β equal to 4.5 a change of the ratio d/a by a factor of 10 will imply a change of the optimal value of β by about 10%.

Let us as a vision assume that engineering practice for the choice of the target reliability in structural design develops in the direction of determination of the optimal reliability according to decision theoretical principles. Then the engineering profession must formulate regulating codes for certain utility value assessments that ensure a generally acceptable and authoritatively approved preference ordering of the consequences of any building activity. Codes for utility value specifications should in an initial phase be calibrated to such values that the code by and large leads to the reliability levels approved by the superior authority and pointed out by this authority on the basis of the existing codes. As shown this calibration will lead to very large values of the direct failure costs d.

Often d will be orders of magnitude larger than the compensations ordinarily paid out by the insurance companies. That the values of d are so large may be seen as a consequence of the natural aversion of the engineering profession against experiencing failures. Such aversion can partly be due to a fear of loosing prestige and goodwill in the society but may also be seen as an ethical attitude that command the engineering activity to be practiced such that it will cause no harm to human health and life.

A practical consequence of the high reliability levels is that the serviceability requirements to the structures in many cases need not be checked. Thus the motive for the high reliability levels may be to free the designer for problems about the serviceability requirements. Definitely this motive has nothing ethical to it but it might be defendable from a cost benefit point of view.

If the optimality postulate stated by the superior authority is accepted, that is, if it is taken for grated that the reliability levels defined implicitely for certain structure type classes by a current authorized code is optimal [1] it turns out that the damage costs d do not vary proportionally with realistic monetary compensation values. Thus the damage cost d is a socio-economic value.

Besides specifying critical utility values in suitable units that can be transformed into monetary units a decision theoretical code just like a probabilistic code with specified target reliability levels should standardize a set of probability distribution types for the input variables.

4 Uncertain gain and utility functions in the decision analysis

Construction activities imply direct monetary costs that at the state of the design decision making are not known with certainty. The costs are assessed by considering present price levels and perhaps by setting up prognosis models for the future economical environments of the project. Such considerations may lead to an assessment of a probability distribution of the actual monetary costs. With respect to monetary costs the realization of the project is thus considered as a draw from this distribution, that is, as a run of a lottery. This lottery is a component of a composite lottery in the set of lotteries of relevance for the actual decision problem. When taking the expected value of the utility of the composite lottery the monetary cost component will contribute merely with its expected value. Thus it follows that the future monetary costs need only be assessed up to the expected value.

The decision maker may also be uncertain about his or hers preferences. This is reflected in uncertainty about the choice of the utility of the consequences K, that is, of the utility function $u(K)$. In the particular case where the utility is identified with the gain, the problem has been clarified above. If the considered consequence K is not directly measurable on a monetary scale implying that a socio-economic utility value must be assiged to K, the problem may be dealt with in the same way. The uncertainty in the mind of the decision maker is expressed by his or hers assessment of a probability distribution of utility values rather than by a choice of a unique value. Thus the consequence K is with respect to utility dissolved into a set of socio-economic values. A lottery between these values is included as a component of the relevant composite lottery. The lottery probabilities are given by the assessed probability distribution. As for the monetary gain it follows that it is sufficient to use the expected value in the utility distribution as the input utility value of the consequence K in the decision model.

The simple conclusion of these considerations is that if von Neumann's and Morgenstern's decision axioms are accepted [9], then it should also be accepted that *in presence of uncertain gain or utility functions it is only the expected gain or utility function which is relevant for the final decision problem.*

As an example consider the variability of the reliability index of the tubular joints obtained by partial safety factor design as reported in Section 2. This variability can be transformed into variability of d by considering the ratio $d/(a\sqrt{2\pi})$ obtained from (6) as a random variable $D/(a\sqrt{2\pi})$. Thus the partial safety factor code as applied to the tubular joint class of structures gives uncertain information about the damage cost d. If all the variability is taken as an expression of uncertainty in the assessment of the value of d (or better expressed, perhaps, the uncertainty left over by the superior authority to the code committee that faces the problem of calibrating a probabilistic code to an existing partial safety factor code), it follows from the decision theory that d should be put to the expectation $E[D]$. Since

$$\frac{E[D]}{a\sqrt{2\pi}} \simeq \frac{1}{\sigma}\int_{-\infty}^{\infty} e^{\frac{1}{2}x^2}\varphi(\frac{x-\mu}{\sigma})dx = \frac{1}{\sqrt{1-\sigma^2}}\exp\left[\frac{\mu^2}{2(1-\sigma^2)}\right] \tag{8}$$

in which $\mu = 4.40$ and $\sigma = 0.30$, the optimal reliability index according to (7) becomes

$$\beta_{opt} \simeq \sqrt{\frac{\mu^2}{1-\sigma^2} - \log(1-\sigma^2)} \simeq 4.62 > E[\beta] \tag{9}$$

This value is about as much larger than the expectation $E[\beta] = 4.40$ as $\beta_{class} = 4.21$ is smaller than the expectation, (3). It is seen that $E[D]$ increases with σ and that $E[D] \to \infty$ ($\beta_{opt} \to \infty$) as $\sigma \to 1$. Of course, these specific conclusions can only be claimed to be valid for the example considered here, where the uncertainty distribution of β is normal.

More refined analyses may be made in specific realistic examples of practice. For example, the coefficient a in the expansion $c(p) - c(p_0) = a(\beta - \beta_0)$ can be a random variable when considering the variation over structures.

Remark If β_{opt} as given by (9) is taken as the target reliability index β_t for calibration of a design value code format the penalty function on the reliability index deviations from the target reliability index should ideally be such that it preserves the damage cost expectation $E[D]$. This suggests that the penalty function should be chosen such that it is a function of β solely through $\exp[\frac{1}{2}(\beta^2 - \beta_t^2)] - 1$ (and not through $\beta - \beta_t$ as it is most often suggested).

5 Probabilistic code ambiguity

For simplicity of analysis in this section consider a structure type for which a resistance R and a load effect S can be defined as two mutually independent random variable. The failure event is then the event $R - S < 0$.

Let a structure with small uncertainty of the resistance R relative to the uncertainty of the load effect S be designed to the limit according to a current partial safety factor code. The result is a structure with given resistance R. Assume now that we are asked to formulate a probabilistic code that leads to exactly the same resistance R of the considered structure. To do that we first need to specify the distribution types on which the probabilistic code should be based.

The examples in the appendix suggest three different probabilistic code formats, in the following denoted as Code (N,N), Code (LN,N), and Code (LN,LN) where N $\sim$ normal distribution and LN $\sim$ lognormal distribution. Code (N,N) is the format where both R and S are assigned normal distributions. This format is inconvenient except if the probability $P(R < 0)$ is negligible as compared to the failure probability $P(R < S)$. In fact, Code (N,N) can by elementary calculations be shown to give

$$\frac{E[R]}{E[S]} = \frac{1 + \beta\sqrt{V_R^2 + V_S^2 - \beta^2 V_R^2 V_S^2}}{1 - \beta^2 V_R^2} \tag{10}$$

which for $V_R = 0$ is the same as (A.2). It is seen that $E[R] \to \infty$ as $V_R \to 1/\beta$, which is an unreasonable behavior caused by the erraneous assignment of positive probability to the event $R < 0$.

Ruling out Code (N,N) we are left with the two other formats: Code (LN,N) and Code (LN,LN). Since Code (LN,N) and Code (N,N) coincides in the limit $V_R \to 0$ we may calibrate Code (LN,N) by use of (A.2). To give the same design as determined by the partial safety factor code we thus should put β in (A.2) to

$$\beta_N = \left(\frac{R}{E[S]} - 1\right)\frac{1}{V_S} \tag{11}$$

However, for Code (LN,LN) we obviously should calibrate by use of (A.6) with $V_R = 0$. Thus β in (A.6) should be put to

$$\beta_{LN} = \frac{1}{\sqrt{\log(1 + V_S^2)}} \log\left[\frac{R}{E[S]}\sqrt{1 + V_S^2}\right] \tag{12}$$

It is seen that $\beta_{LN} \to \beta_N$ asymptotically as $V_S \to 0$. Otherwise $\beta_{LN} < \beta_N$. By substituting $R/E[S]$ from (11) into (12) we get the relation

$$\beta_{LN} = \frac{\log[(1 + \beta_N V_S)\sqrt{1 + V_S^2}]}{\sqrt{\log(1 + V_S^2)}} \tag{13}$$

In case $V_R > 0$ the relation between β_N and β_{LN} cannot be given in explicit analytical form. Fig. 2 shows the relation for different values of V_R and V_S. For $V_R > 0$ it is calculated approximatively by a standard first order reliability analysis method (FORM, Section 7) using the DNV program PROBAN.

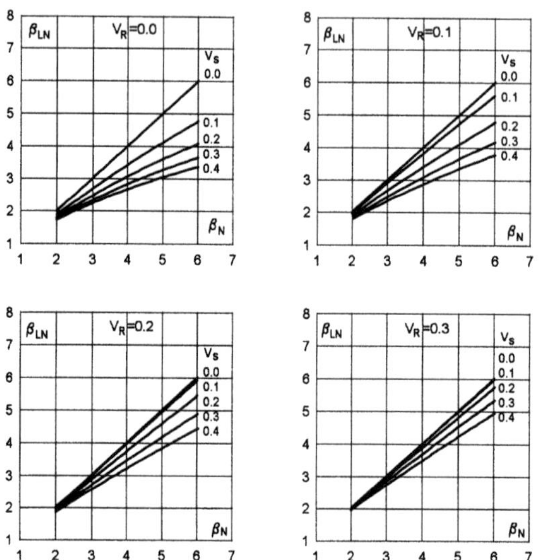

Fig. 2. Equivalent reliability indices β_N and β_{LN} corresponding to Code (LN,N) and Code (LN,LN), respectively.

It is seen that design with respect to constant reliability index of the one code implies a substantial variation of the reliability index of the other code. This is just an illustration of the old tail-sensitivity problem that since the early days of structural reliability theory repeatedly has been an obstacle of concern but also by many contributors to the field has been conveniently neglected or swept under the carpet. Under the optimality postulate the variation of the reliability index is by (6) directly transformed into a very large variation of the damage cost d. This emphasizes the role of d as a formal calibration quantity that cannot be separated from the considered probabilistic reliability model. Also the large variability emphasizes the necessity of codifying a specific probabilistic reliability model as a standard reference code.

The two probabilistic code formats compared here are grossly simplified and serve only as an illustration of the code definition problem. Realistic probabilistic code formats will specify a set of different basic input variables that define geometrical quantities, material resistances and loads. Each code format specifies the joint probability distribution of the basic variables. The simple (R, S) examples discussed here show the importance of making similar comparisons between different suggested realistic probabilistic code formats, and to choose one of these code formats as the basis for the standard reference code that should be calibrated to the optimal existing design practice as declared to be so by the superior authority.

6 Prudency principle for choice of probabilistic code

A principle of prudency can be applied to guide the choice of the standard reference probabilistic code format. Clearly it is a reasonable principle first to rule out all those code formats that specify distribution types that are strongly inconsistent with available empirical distributions and general engineering knowledge. The choice among the remaining code formats is arbitrary, as pointed out long ago by Freudenthal [10]. A requirement of reasonable calculational operability may rule out some further formats.

Left over with a finite set of different probabilistic code formats we need a principle for making a unique choice among these formats. Referring first to the tubular joint example with the limit state (1) the prudency principle is to choose that code format for which there is minimum of the coefficient of variation of the reliability index for the population of designs obtained according to the given partial safety factor code with fixed factor values. Thus the prudency principle advocates an attitude of minimal change of current design practice.

This prudency principle may lead to different probabilistic code formats in dependence of the considered structure type class with associated type of ultimate limit state. To overcome this ambiguity problem the coefficient of variation of the reliability index can be treated as a random variable defined on the population of relevant structure types. The prudency principle may then be to choose that probabilistic code format for which the expectation of the coefficient of variation is minimal. To be operational, this principle requires that probabilities are assigned to the different considered ultimate limit states. These probabilities could reasonably be chosen to reflect the relative frequency of occurrence of the structure type in practice.

For each structure type with associated ultimate limit state there is an infinity of realizations of the limit state surface. In the tubular joint example the population of these realizations was defined by randomly generating the influence coefficients by which the external loads are transformed into the internal forces of the joint. Clearly this way of obtaining the population of limit state surfaces is used in the example just for illustration. A realistic code calibration excercise should be based on countings and structural analysis investigations to assess the relative frequency of occurrence of the different influence coefficients in $\boldsymbol{C}$.

7 Partial safety factor code calibration

After the choice of a reference probabilistic code format a specific probabilistic code should be established through assignment of parameter values, certain cost values (intangibles), and/or reliability levels such that the code at the time of its authorization possesses socio-economic optimality as assessed by the superior authority. The probabilistic code may then be used directly in engineering practice. However, for the design of standard types of structures it may be more convenient and operational to base the routine design work on a partial safety factor code format. Therefore it is relevant to determine partial safety factor values by calibration to designs obtained by use of the reference probabilistic code. It goes without saying that this partial safety factor code only approximates the probabilistic code reasonably well if it is allowed to contain several sets of partial safety factor values to be used in dependence of structure type and ultimate limit state type.

To appreciate this statement it is illustrative to study the relation between the partial safety factor concept and the reliability index concept. To make the text directly accessible to non-specialists of probabilistic reliability calculus the standard approximative procedure known as FORM (First Order Reliability Method) is explained here in its simplest possible form.

Consider a reliability problem with only two random variables X_1 and X_2. Let the limit state be a curve in the (x_1, x_2)-plane defined by the equation $g(x_1, x_2) = 0$, where $g(x_1, x_2)$ is a function defined on the basis of structural analysis or experiments. Moreover assume that X_1 and X_2 are mutually independent random variables with distribution functions $F_{X_1}(x)$ and $F_{X_2}(x)$ ($\equiv$ the probability that $X_2 \leq x$, an input information), respectively. Then define the mapping

$$u_1 = \Phi^{-1}[F_{X_1}(x_1)], \quad u_2 = \Phi^{-1}[F_{X_2}(x_2)] \tag{14}$$

of the (x_1, x_2)-plane onto the (u_1, u_2)-plane, where $\Phi^{-1}(\cdot)$ is the inverse function to the standard normal distribution function $\Phi(\cdot)$. This mapping has the property that the corresponding random variable pair (U_1, U_2) obtained from (X_1, X_2) has the rotational symmetric 2-dimensional standard normal density function

$$f_{U_1,U_2}(u_1, u_2) = \varphi(u_1)\varphi(u_2) = \frac{1}{2\pi}\exp[-\frac{1}{2}r^2] \tag{15}$$

where $r^2 = u_1^2 + u_2^2$. In the (u_1, u_2)-plane the limit state becomes represented by a curve defined by the equation

$$g(F_{X_1}^{-1}[\Phi(u_1)],\ F_{X_2}^{-1}[\Phi(u_2)]) = 0 \tag{16}$$

For any reasonable reliability problem this curve divides the (u_1, u_2)-plane in two parts such that the origin is in the image part of the safe set in the (x_1, x_2)-plane. Thus the probability p_f of the failure event is the same as the probability obtained from the normal density (15) integrated up over that part of the (u_1, u_2)-plane that does not contain the origin. For simplicity assume that the curve (16) is a straight line. Then the rotational symmetry of the normal density about the origin ensures that p_f is determined solely by the distance β from the origin to the straight line. In fact, we have $p_f = \Phi(-\beta)$ or $\beta = -\Phi^{-1}(p_f)$.

Even though the limit state curve in the (u_1, u_2)-plane is rarely a straight line the curvature at the point closest to the origin is most often so small that the curve for reliability evaluation purposes can be replaced by the tangent at this point. Thus the geometric reliability index (Hasofer-Lind index [11]) defined as the smallest distance from the origin to the limit state curve is most often a very good approximation to the reliability index defined by $\beta = -\Phi^{-1}(p_f)$.

The situation is illustrated in Fig. 3 with two different limit states 1 and 2. The radius β of the circle defines the common reliability level. If this β is the target reliability level the value of the design variable (reinforcement area, say) in the limit state problem 1 is determined such that the limit state curve 1 becomes tangential to the circle. In its most elementary form this is design according to the reliability index method. For this particular design there are an infinity of partial safety factor pairs applied to given characteristic values for X_1 and X_2. Clearly all what is needed is arbitrarily to define a single design point Q on the limit state curve, given that the family of limit state curves for the design situation 1 is a one-parameter family with the design variable as parameter. If D_1 is chosen as design point, where D_1 is the point on the limit state curve 1 closest to the origin (D_1 is in the literature conveniently called the "most central" or "most likely" limit state point, but less conveniently it also appears with the name "design point"), the partial safety factors that produce the same design become

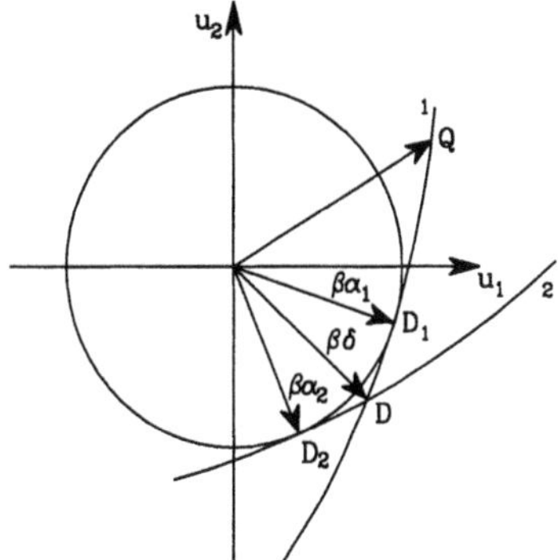

Fig. 3. Illustration of design points corresponding to a given geometric reliability index β (Q: ambiguous for one limit state; D: unique for two limit states, single design points non-existent for more than two limit states).

$$\gamma_{11} = \frac{F_{X_1}^{-1}[\Phi(\beta\alpha_{11})]}{x_{1k}}, \quad \gamma_{12} = \frac{x_{2k}}{F_{X_2}^{-1}[\Phi(\beta\alpha_{12})]} \tag{17}$$

where (in accordance with Fig. 3) x_{1k} is an upper fractile characteristic value of X_1 (load type variable) and x_{2k} is a lower fractile characteristic value of X_2 (resistance type variable), e.g. $x_{1k} = F_{X_1}^{-1}(p_1)$, $x_{2k} = F_{X_2}^{-1}(p_2)$ with $p_1 = 0.98$, say, and $p_2 = 0.05$, say. For the other limit state 2 the point D_2 may be taken as design point and partial safety factors γ_{21}, γ_{22} be calculated.

Needless to say, since the partial safety factor method only serves a purpose if it can be used directly for design without first making a design according to the reliability index method, it is necessary that the partial safety factors be fixed at least within a reasonably wide class of design problems. For the case with the two limit states in Fig. 3

it is immediately seen that if the intersection point D between the two limit state curves is taken as design point, then the partial safety factors are common for the two limit states. The vector $\boldsymbol{\delta}$ acts as a replacement vector for $\boldsymbol{\alpha}_1$ and $\boldsymbol{\alpha}_2$ without introducing any error of approximation. Clearly, with more than two limit states in the class of design problems containing solely the two random variables X_1 and X_2, there is no replacement vector $\boldsymbol{\delta}$ that exactly reproduces all the designs. By a given superior requirement there will only be allowed one design point $\beta\boldsymbol{\delta}$ for the class, and all the limit state curves are therefore adjusted to contain this point. Thus the reliability indices will vary over the class. The strategy is then to determine the replacement vector $\boldsymbol{\delta}$ such that the expected loss (penalty) for the entire design class becomes minimal, [12]. Of course, the range of obtained reliability index values within the class should be controlled. If the range is judged to be too wide, the class should be divided into subclasses each with its own set of partial safety factors.

In this connection it is of interest to evaluate the sensitivity of the replacement vector $\boldsymbol{\delta}$ and the corresponding partial safety factors to changes of parameters. The parameters may have influence on the limit state before the mapping into the standard Gaussian space (the (u_1, u_2)-plane) or they may have influence on the probability distributions of the basic variables. If the reliability index is kept at the target value β all what essentially happens in the example of Fig. 3 by a minor parameter variation is that each limit state curve is changed to another curve close by the first curve. This new curve is also tangential to the circle of radius β. The averaging effect of the penalty optimization will tend to cause that the change of the replacement vector $\boldsymbol{\delta}$ will be smaller than the changes of the individual $\boldsymbol{\alpha}$-vectors.

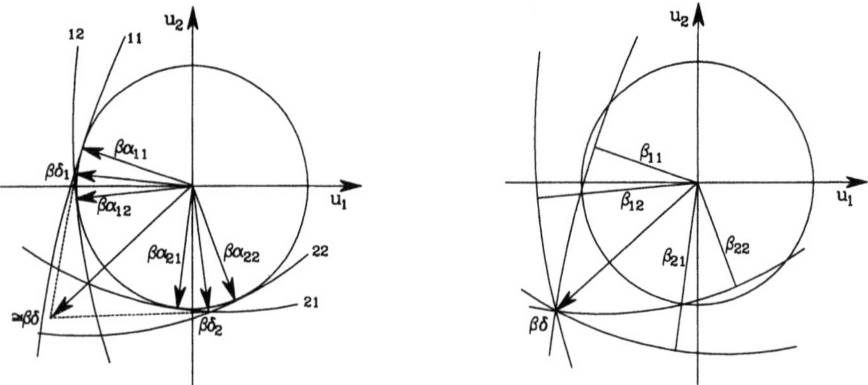

Fig. 4. Illustration of two clusters of most central limit state points separated by an angle of about 90° with well approximating separate replacement vectors $\boldsymbol{\delta}_1, \boldsymbol{\delta}_2$ *and the resulting less good approximating common replacement vector* $\boldsymbol{\delta}$.

There is a particularly important issue related to the classification of design problems into classes of constant partial safety factors. For limit states of the same mechanical type (same collapse mode) the $\boldsymbol{\alpha}$-vectors as defined in Fig. 3 tend to be close to each other

such that the corresponding most central points form a cluster on the sphere of radius β (in Fig. 3 the points D_1 and D_2 on the circle of radius β). As an example consider reinforced concrete, and in particular consider as one type the bending failure of a beam cross-section where the yield moment is mainly determined by the yield strength X_1 of the tension reinforcement and, as another type, the compression failure of a short column where the carrying capacity is mainly determined by the compression strength X_2 of the concrete. Then the situation in the (u_1, u_2)-plane is as scetched in Fig. 4. For the bending failure limit states the variability of X_2 has only small influence and the cluster of most central limit state points is therefore situated on the circle in the vicinity of $(-\beta, 0)$. For the limit states of compression failure type the cluster of most central limit state points is similarly situated on the circle in the vicinity of $(0, -\beta)$, about 90° apart from the first cluster of points. If the two limit state types are put in two different design case classes we get the two optimal replacement vectors $\boldsymbol{\delta}_1$ and $\boldsymbol{\delta}_2$ centrally placed within each cluster, as illustrated in Fig. 4 (left). If both types of limit states are required to be treated by the same partial safety factors, that is, if they are put together in one class, the optimal replacement vector $\boldsymbol{\delta}$ will have some direction between the two clusters depending on the relative weighting of the two types. Fig. 4 shows two limit states of each type. With equal weight on the two types a first crude assessment of an optimal $\boldsymbol{\delta}$ is obtained by letting $\beta\boldsymbol{\delta}$ be the vector to the crossing points of the two orthogonals to $\boldsymbol{\delta}_1$ and $\boldsymbol{\delta}_2$ at the points $\beta\boldsymbol{\delta}_1$ and $\beta\boldsymbol{\delta}_2$, respectively.
Fig. 4 (right) illustrates the possible consequence of having a single replacement vector $\boldsymbol{\delta}$ for the $\boldsymbol{\alpha}$-vectors of about 90° separated clusters of most central limit state points. All the limit states are now required to contain the point $\beta\boldsymbol{\delta}$. Obviously the approximation errors for the reliability indices increase when extending the class of design cases. What intuitively can be seen from Fig. 5 is that from having quite small approximation errors when keeping $\boldsymbol{\delta}_1$ and $\boldsymbol{\delta}_2$ for practical design, the error may blow up to considerable size when requiring that there should only be one set of partial safety factors.

Whether the minor complication of having some few classes of partial safety factors is of sufficient inconvenience to justify the resulting much larger reliability variability when there is only a single class, can in principle be decided in a rational way by a cost benefit analysis. At the very least it must be ensured that there within the class is no relevant structure for which the reliability index becomes much smaller than the target reliability index.

8 Conclusions

Some of the following conclusions are about facts while other conclusions are of a debatable philosophical or political nature expressing the opinion of the author. The author sees that there is an increasing need for establishing standard reference probabilistic code formats. Therefore a debate in the profession about this issue would be wellcome. The author's conclusions are:

1. The reliability index of a limit state formulated for a given structure carries information about the structural reliability with respect to the limit state only if the reliability index

is reported together with a reference to the specific probabilistic code format by which it is defined. Constant reliability in one code does not imply constant reliability in another code.

2. A specific probabilistic code format should be declared to be the authorized reference code for constant reliability.

3. The choice of the authorized probabilistic reference code format can be based on a principle of prudency after ruling out those formats that are significantly inconsistent with available generally valid empirical data and scientific evidence. By the principle of prudency that probabilistic format is chosen by which the reliability index variation as induced by a current authorized partial safety factor code in a certain sense becomes as small as possible.

4. A partial safety factor code with a fixed set of safety factor values induces a considerable variation of any reliability index over the set of structures designed to the limit according to the safety factor code.

5. Under the postulate that the intention of any code is that it should produce designs among which there are at least some designs that are optimal in a socio-economic sense, the reliability index uncertainty induced by the code can be transferred into damage cost uncertainty. The optimal designs should be pointed out by a superior authority.

6. The damage cost distribution induced by a given authorized code can for the calibration of a probabilistic code to the given code be taken as an expression of uncertainty about the damage cost assessment. Adopting the decision theory (von Neumann and Morgenstern) as a superior basis for rational decision making the damage cost uncertainty distribution can simply be replaced by the expected value of the distribution. The corresponding reliability index value is then the calibration result.

Acknowledgements

Ph.D. students Claus F. Christensen, Johannes M. Johannesen, and Søren Randrup-Thomsen are acknowledged for computer programming help. The Joint Committee on Structural Safety has discussed the topic and its members are acknowledged for valuable and quite critical comments to the first draft of the paper. This work has been financially supported by the Danish Technical Research Council.

References

1. N.C. Lind, i) Reliability-based structural codes. Optimization theory. ii) Reliability-based structural codes. Practical calibration, iii) Safety level decisions and socio-economic optimization. In *Safety of Structures under Dynamic Loading*, Holland, Kavlie, Moe, Sigbjørnson, eds., Vol. 1, Tapir, Trondheim, 1978, 135-175.

2. J.D. Sørensen and I. Enevoldsen, Sensitivity weaknesses in application of some statistical

distributions in First Order Reliability Methods, *Structural Safety*, 12 (1993) 315-325.

3. A.R. Flint, B.W. Smith, M.J. Baker, and W. Manners, *Proc. Conf. on the New Code for the Design of Steel Bridges*, Cardiff, 1980. Granada Publishing, 1981.

4. The Nordic Committee on Building Regulations (NKB), *Recommendation for Loading- and Safety Regulations for Structural Design*, NKB-Report No 36, Ministry of Housing, Copenhagen, 1978.

5. P. Thoft-Christensen and M.J. Baker, *Structural Reliability Theory and its Applications.* Springer Verlag, Berlin, Heidelberg, New York, 1982.

6. O. Ditlevsen and H.O. Madsen, Proposal for a code for the direct use of reliability methods in structural design. *Working Document of the Joint Committee on Structural Safety* (JCSS), IABSE-AIPC-IUBH, ETH-Hönggerberg, Zürich, 1989.

7. DS 449, Dansk Ingeniørforening's Code of Practice for Offshore Steel Structures on Pile Foundations (in Danish: Norm for pælefunderede offshore stålkonstruktioner), 1. edition. Teknisk Forlag, Copenhagen, 1983.

8. P. Bjerager, *Offshore Konstruktioner 5947, Notat 8, Pålidelighedsanalyse.* Teaching note in Danish. Department of Structural Engineering, Technical University of Denmark, 1985.

9. Von Neumann and Morgenstern, *Theory of Games and Economical Behavior.* Princeton University Press, 1943.

10. A. Freudenthal, Introductory Remarks, *Proc. of ICOSSAR '72* (International Conference on Structural Safety and Reliability), A. Freudenthal, ed., Pergamon Press, Oxford, 1972, 5-6.

11. A.M. Hasofer and N.C. Lind, An exact and invariant first-order reliability format, *Journal of Eng. Mech.*, ASCE, 100 (1974) 111-121.

12. K. Skov and R. Rackwitz, Note on the evaluation of partial safety factors for unknown and partially known shape of the limit state problem. *DIALOG 77*, Danish Engineering Academy, Lyngby, 1977, 99-115.

APPENDIX

As in the section on probabilistic code ambiguity consider a structure type for which resistance R and load effect S can be defined as two mutually independent random variables. The failure event is then the event that $R - S < 0$.

Assume that the cost $c(p)$ within the domain of practical interest can be written as $c(p) = h + kE[R]$, where h and k are given constants while R is the random resistance of the structure. Then (5) can be written as

$$\begin{aligned} &d + c(p_0) \simeq \\ &-[c(p) - c(p_0)] - \frac{d}{dp}[c(p) - c(p_0)] = k\{(E[R] - E[R_0]) + \frac{1}{\varphi(\beta)}\frac{d}{d\beta}(E[R] - R[R_0])\} \end{aligned} \tag{A.1}$$

where $\beta_0 = -\Phi^{-1}(p_0)$ is a suitable representative reliability index and R_0 is the random resistance when the structure is designed to have the reliability index β_0.

First consider the simple example where the randomness of R is so small that it can be neglected as compared to the randomness of the load effect S, and where S has a Gaussian distribution. Then we simply have

$$E[R] = E[S](1 + \beta V_S) \tag{A.2}$$

where $V_S = D[S]/E[S]$ is the coefficient of variation of S. Thus (A.1) can be written as

$$\frac{d + c(p_0)}{a} \simeq \frac{1}{\varphi(\beta)} - (\beta - \beta_o) \simeq \sqrt{2\pi}\, e^{\beta^2/2} \tag{A.3}$$

where $a = kE[S]V_S$. Next consider the example where R and S both have lognormal distributions. Then

$$E[\log R] = E[\log S] + \beta\sqrt{\mathrm{Var}[\log R] + \mathrm{Var}[\log S]} \tag{A.4}$$

or

$$\log E[R] - \frac{1}{2}\log(1 + V_R^2) = \log E[S] - \frac{1}{2}\log(1 + V_S^2) + \beta\sqrt{\log(1 + V_R^2) + \log(1 + V_S^2)} \tag{A.5}$$

so that

$$\frac{E[R]}{E[S]} = \sqrt{\frac{1 + V_R^2}{1 + V_S^2}} \exp\left\{\beta\sqrt{\log[(1 + V_R^2)(1 + V_S^2)]}\right\} \tag{A.6}$$

Thus

$$\frac{E[R]}{E[R_0]} = \exp\left\{(\beta - \beta_0)\sqrt{\log[(1 + V_R^2)(1 + V_S^2)]}\right\} \tag{A.7}$$

and (A.1) reads

$$\frac{d + c(p_0)}{a} = \frac{1}{b}\left\{\exp[b(\beta - \beta_o)][1 + \frac{b}{\varphi(\beta)}] - 1\right\} \simeq \sqrt{2\pi} e^{\beta^2/2} \tag{A.8}$$

where $a = kE[R_0]b$ and $b = \sqrt{\log[(1 + V_R^2)(1 + V_S^2)]}$.

3

Towards Consistent-Reliability Structural Design for Earthquakes

Luis Esteva

Institute of Engineering, National University of Mexico,
Ciudad Universitaria, 04510, Mexico, D. F., Mexico

1. INTRODUCTORY NOTE

This is an abridged version of Ref. 1. It eliminates most of the references, as well as the discussion concerning the probabilistic modelling of seismic hazard and ground motion, and concentrates on the system reliability concepts.

2. STATE OF THE ART: AN OVERVIEW

Establishing reliability-based criteria for earthquake resistant design of structural systems is not an easy task, nor one free of conflicting viewpoints, wide knowledge gaps and significant implementation difficulties. The troubles are linked to the complexities inherent in the phenomena under study --ranging from earthquake generation and propagation to structural response and behavior-- and, of course, to the process of making safety-related decisions.

Take, for instance, the modeling of the seismic excitation. One school of thought advocates the formulation of probabilistic models of the process of earthquake occurrence at a site and of the detailed characteristcs of ground motion time histories during each event. The followers of this approach have to confront the problems arising from the lack of sufficient statistical information on which to base those probabilistic models. Therefore, they have to make simple assumptions concerning the forms of the models, and to estimate the corresponding parameters by means of bayesian statistical analysis. For this purpose, they have to adopt largely subjective prior probability distributions to describe their degrees of belief on sets of alternate assumptions based on geological and geophysical features, as well as on statistical information about events occurred in other seismic regions, considered to be similar to those for which the estimates of activity are to be obtained. The set of mathematical models used to represent the joint probability distributions of earthquake magnitudes and interoccurrence times includes models as simple as the homogeneous Poisson time process with independent random selection of magnitudes and spatial coordinates, and as refined as those that take into account correlation among time, magnitude, spatial coordinates and history of energy liberation by previous activity.

Opponents to the foregoing approach are found mainly among those responsible for the formulation of design requirements for very important structures. To support their position, they argue against the subjective component of the bayesian estimation approach mentioned above, and against the difficulties associated with extrapolating to the very low hazard interval the time-magnitude-intensity probability distributions derived on the exclusive basis of the direct statistical evidence. They choose, instead, to base their design recommendations on an assumed worst credible event, consisting in the largest earthquake magnitude at the shortest possible distance from the site of interest. This approach is called "deterministic", not because it is devoid of uncertainties, but because the decisions arising from it are supposed to be conservative enough as to make irrelevant any formal evaluation of uncertainty. Although much smaller, uncertainties associated with intensity attenuation, ground motion time history and response spectra for a given magnitude and distance, are covered by the formal probability-based criterion of adopting for design the response values corresponding to a specified number of standard deviations above the mean of the conditional distribution of the maximum response, given the assumed occurrence of the so called worst credible combination of magnitude and source-to-site distance. The rest of this paper will focus on the formal probabilistic approach.

The second link in the seismic reliability analysis chain was mentioned in connection with the deterministic approach (abandoned for the rest of this paper), but it appears as well in the formal probabilistic analysis of seismic hazard. It deals with the use of intensity-attenuation expressions; i.e., algorithms to predict ground motion characteristics for given magnitudes and distances. The first such expressions were used to predict peak amplitudes of ground acceleration, velocity and displacement on firm ground, or ordinates of the response spectra for those local conditions. The information on which they were based included very few records at short hypocentral distances, and therefore their predictions of intensities near the source area are very poor. At moderate distances, typical prediction errors are characterized by variation coefficients of the order of 0.5.

Our capability for making probabilistic predictions of ground motion intensities and spectral amplitudes has improved substantially during the last few years, mainly as a result of the use of Fourier spectra and random vibration theory, and the introduction of the empirical-Green's function concept. The latter has been particularly valuable for the prediction of high-intensity ground motion characteristics at soft-soil sites. Considerable effort has been devoted to the development of mathematical models and algorithms to study the influence of local soil conditions and some geological and morphological features, but they are not discussed here, as their contribution to the uncertainties about ground motion characteristics will be considered as included in the attenuation expressions specifically applicable to each site of interest.

The study of structural reliability for earthquake actions may be divided into two steps: a) reliability and expected-damage analysis for single earthquakes, of given intensities, and b) integration with respect to the probabilities of occurrence of earthquakes with different characteristics during given time intervals. And there remains the problem of taking into account the influence on structural properties and system reliability of damage accumulation resulting from previous excitations (other earthquakes, differential settlements, etc.).

Research efforts on the reliability analysis of structural systems subjected to stochastic models of earthquakes have been concentrated on the probabilistic prediction of response, either by means of random vibration theory or by Monte Carlo simulation. Typical results of the former group of methods are the superposition criteria of modal responses of linear multi-degree-of-freedom systems, and the estimation of first passage probabilities of linear-system responses to segments of stationary and nonstationary gaussian ground motion. Other significant result applicable to the same types of excitation is the development of the method of stochastic equivalent linearization to obtain time-varying covariance matrices of the response vectors of nonlinear hysteretic or deteriorating multi-degree-of-freedom systems with deterministic properties.

Much less attention has been given to the influence of uncertainties about structural properties and to the definition of failure criteria for nonlinear multi-degree-of-freedom systems and to the analysis of failure probabilities of those systems when subjected to earthquakes. References 2 and 3 describe applications of Monte Carlo simulation to evaluate those probabilities for multistory building frames and for simple models of structure-foundation systems with uncertain properties. In order to solve the implied system reliability problem, the set of relevant failure modes is identified, and the maximum value reached of the capacity/demand ratio for any of the failure modes is used as the indicator of system damage. Probability distributions of that ratio are obtained by repeated evaluations of structural responses of systems with Monte Carlo-simulated properties subjected to artificial ground motion records with prescribed statistical properties.

A significant limitation of the few studies that try to evaluate seismic failure probabilities is the arbitrary way in which system failure criteria are defined. However, one of their main conclusions is incontestable: seismic failure probabilities of structural systems designed for the same earthquake spectra and the same general requirements may differ from each other by orders of magnitude. Discussing possible ways to reduce these differences is the main objective of this paper.

3. GENERAL FRAMEWORK

For the purpose of developing consistent-reliability seismic design criteria an integrated analysis is required of the concepts described in the following paragraphs (Ref. 1 includes a detailed analysis of these concepts.)

3.1. Seismic hazard: This should be expressed by the joint probability distributions of times of occurrence, intensities and evolutionary spectral functions (ESF), of the earthquakes that may be experienced at a given site.

3.2. System models: These include mass, stiffness, strength and, in general, constitutive laws of structural members subjected to large-alternating strains. The largest contributions to uncertainty about model properties and behavior rules are probably those related with failure criteria and damage indexes. Therefore, they will receive special attention in this paper.

3.3. System-reliability: Having in mind its expected use in the making of safety-related engineering decisions, this concept is best expressed in terms of its complement, that is, a set of system failure probabilities for different time intervals or, equivalently, by the probability density function of the waiting time to failure.

3.4. Safety-related decisions: Strategies for the selection of acceptable (or tolerable) failure probabilities for a given system on the basis of trade-off relations among the construction-and maintenance cost increments required to increase safety of that system and the expected reductions in the values of the failure consequences.

3.5. Design formats: The concepts and guidelines presented in the sequel may be formally and explicitly applied by individual designers or groups to very important specific projects. For typical structures to be covered by building codes, however, the best feasible option consists in specifying earthquake design spectra corresponding to different return intervals, to be chosen in accordance with the importance of a structure and the potential consequences of its failure, and a set of rules to determine the safety factors in accordance with the system-reliability characteristics of each structure.

4. SYSTEM RELIABILITY

4.1. Influence of model on dynamic response

From the discussions above it is concluded that two of the main difficulties for the determination of the seismic reliability of multi-degree-of-freedom structural systems are the possibility of occurrence of several correlated failure modes and the lack of clearly defined mathematical models expressing the conditions required for system-failure.

Another significant difficulty is the high sensitivity of the computed failure probabilities to the types of models adopted to estimate the response of nonlinear systems. For instance, the results presented in Ref. 4 show that the distributions of story ductility demands produced by earthquakes on multi-story buildings are much more irregular for buildings idealized as shear systems than for those represented by continuous, ductile frames. In the former systems, ductility demands are concentrated at the bottom stories, where they attain values several times higher than those appearing at the upper stories of the same buildings or anywhere along the height of the frame-building models.

Ref. 4 also shows that for frames with stiff and strong beams ductility demands may be very sensitive to slight deviations of column strengths as compared with those produced by standard design procedures. This is another argument in favor of the generally recomended strong-column-weak-beam design approach.

4.2. Reliability of elastoplastic systems

Several single-bay, single-story and multistory frames with uncertain properties were studied in Ref. 2. A set of possible random realizations of each frame was generated by Monte Carlo simulation. The members of each set were then subjected to artificial accelerograms belonging to either of two families: one with the same statistical properties as the El Centro 1940 NS component; the other, similar to the Mexico City 1985 SCT EW component. In order to study the influence of earthquake intensity on the computed failure probabilities, both the original and the artificial accelerograms were scaled up and down to several intensities (measured by peak ground acceleration), without changing the time scale.

The determination of system reliability for a given intensity was based on the assumption that under the action of an earthquake, the structure may fail in n different modes; for instance, each failure mode may correspond to exceedance of the capacity for ductile deformation at a given story. If S_i represents the maximum amplitude of the response variable governing the occurrence of the *i*-th failure mode (here, story deformation) and R_i is the value of S_i causing failure, the ratio $Q_i = S_i / R_i$ is the reciprocal of a random safety factor such that $Q_i \geq 1$ means failure in the *i*-th mode. It is also assumed that failure occurs precisely in the *i*-th mode and not in any other if $Q_i \geq Q_j$ for all $j = 1,\ldots,n$. This is a simplifying assumption that ignores the possibility that during the response process the condition $Q_j \geq 1$ may be reached before the condition $Q_i \geq 1$. From these assumptions, the probability of failure for a given intensity (y) equals the probability that the maximum of all the values Q_i exceeds unity. Thus, if that maximum is called Q, then $p_F(y) = P(Q \geq 1 \mid y)$.

Figures 1 to 4 were taken from Ref. 2. In them, V_μ is the variation coefficient of the available story ductility, μ, and μ^* its nominal value, assumed for the purpose of selecting a reduced design spectrum. μ^* and V_μ are related with the mean value $\bar{\mu}$ through the equation $\bar{\mu} = \mu^* \exp(0.55 \times 3 \times V_\mu)$. The terms HC and LC designate high and low spatial correlation coefficients. The values of these coefficients as well as of other relevant variables can be found in Ref. 2. Figures 1 and 2 show values of $Q = S/R$ obtained for different earthquake intensities by means of Monte Carlo simulation. They also show empirical expressions for the mean and standard deviation of the natural logarithm of Q in terms of the earthquake intensity, measured by the peak ground acceleration. The curves in figures 3 and 4 were obtained using these expressions, assuming that Q has a lognormal distribution. From these figures, and others similar to them, the following conclusions were reached (2):

1. The number of degrees of freedom has a great influence on the probability of failure.

2. Due to the form of the assumed relation between the expected and the nominal values of the available ductility, as a fuction of the variation coefficient of that variable, the probabilities of failure for a given intensity are greater for those cases for which that variation coefficient is lower.

3. Because the systems considered in the study are assumed to develop significant local yielding at several critical sections before a failure limit state is reached, the results reported are valid only if the safety factors with respect to local brittle failure modes are sufficiently higher than those associated to ductile modes as to prevent the occurrence of the former.

4. The variability of the failure probabilites obtained for the few cases studied justifies the delevoment of new studies, with the aim of widening the ranges of the variables covered and exploring the use of better representations of the mechanical behavior of structural members and systems.

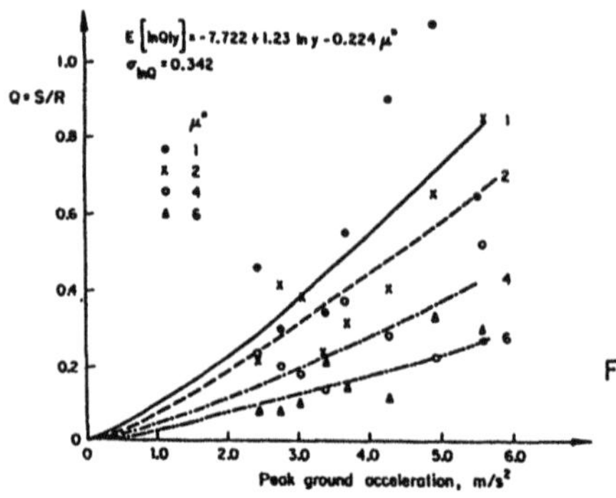

Fig 1 Normalized response of single-story frames; T = 0.36 s, V_μ = 0.3 (From ref. 2)

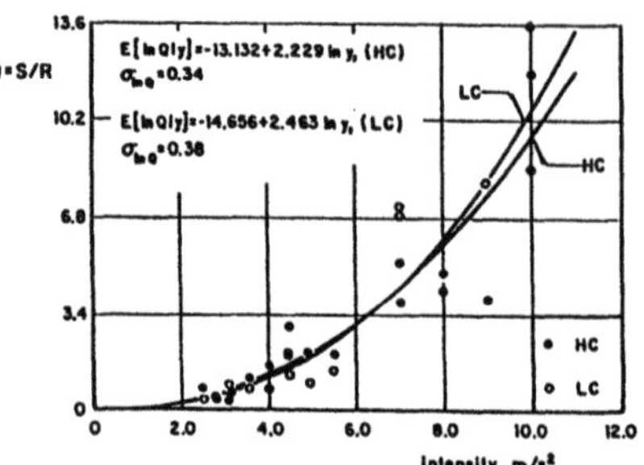

Fig 2 Normalized response of three-story frames; T=0.36 s, μ^*=4, V_μ=0.3. High (HC) and low (LC) spatial correlation (From ref. 2)

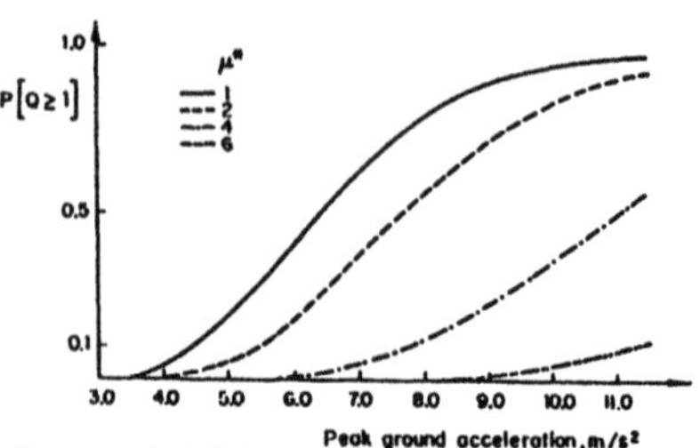

Fig 3 Failure probabilities of single-story frames designed for different ductility factors; V_μ = 0.3 (From ref. 2)

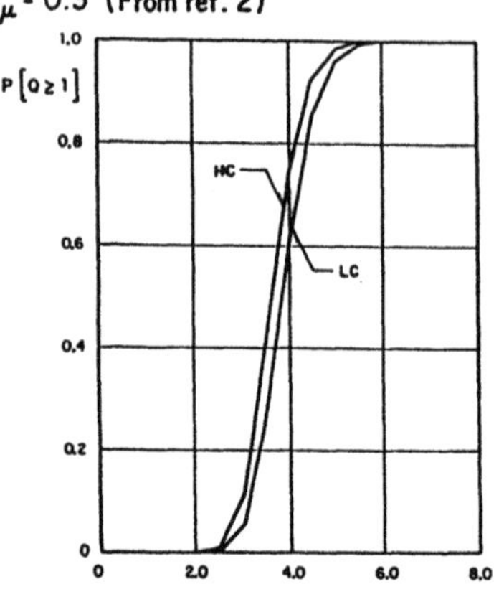

Fig 4 Failure probabilities of three-story frames; T = 0.36 s, μ^* = 4, V_μ = 0.3. High (HC) and low (LC) spatial correlation (From ref. 2)

4.3. Failure rates

Following Ref. 2, the rate of failure (per unit time) of a structure with uncertain properties (vector **R**) is

$$\nu_F = \int_0^\infty -\frac{\partial \nu\,(y)}{\partial y}\, p_F(y)dy \tag{1}$$

where $\nu(y)$ is the rate of occurrence of earthquakes with intensity greater than y and $p_F(y)$ is the failure probability of the structure under the action of that intensity. For the examples described in the foregoing paragraphs, $p_F(y)$ is given by curves as shown in figures 3 and 4.

The resulting failure rates show wide variability with respect to a trend to decrease when the design ductility factors increase. This trend can be explained because the higher the capacity of a structure to resist ductile deformations, the lower the lateral strength (additional to that available in a continuous frame designed for gravitational loads) required to resist a specified set of lateral forces and, therefore, the higher in proportion is the contribution of the member resistances needed for vertical loads to the lateral strength required to take an earthquake of a given intensity.

The wide variability of the normalized failure rates reflects that shown by the failure probability functions similar to those shown in Figs 3 and 4. This variability shows how far present design criteria are from our consistent-reliability goal. As important as to accomplish consistency among computed failure probabilities it is to improve the criteria used to define system failure and to calibrate response, damage and failure models with the observed behavior of actual systems.

4.4. Improved response models and failure criteria

One possible way of defining the failure conditions in multistory buildings subjected to earthquake ground motion is to express those conditions in terms of the damage accumulated throughout the members which participate in each failure mechanism. For instance, for failure associated with excessive damage accumulation at the ends of the columns and beams of a given story, damage can be measured by a sum of terms proportional to x_i^q, where x_i is the absolute value of the story deformation amplitude during the *i*-th cycle of its response time-history. Still, it is not easy to establish the value of the accumulated damage at which the story fails. A more practical indicator of damage, strongly correlated with the formerly proposed measure, is the decrease in the story-stiffness. If failure is expressed by the condition that the secant story stiffness (K_s) corresponding to the maximum story deformation during a given cycle of the response time history is equal to zero, then the accumulated damage can be measured by $\delta = (K_o - K_m)/K_o$, where K_o is the small-deformation tangent initial stiffness for the undamaged structure, and K_m is the minimum value of K_s.

The damage accumulation and the associated stiffness reduction for any given story will be determined by the corresponding local damage processes at the ends of the structural members of the story of interest and of those adjacent to it. Important efforts have been made during the last few years to develop realistic stiffness- and strength-degrading models of the moment-curvature (or moment-rotation) constitutive laws of the plastic hinges when subjected to repeated load applications (5, 6). The model proposed in Ref. 6 has been applied in introductory studies about the influence of cumulative damage on the seismic reliability of multistory frames and about the formulation of optimum decisions related to design and replacement of seismic energy dissipators. Those applications brought to the surface both the limitations of the model and the sensitivity of the computed failure probabilities to its form and parameters. However, the writer considers that the use of damaged-based stiffness-degrading models (DBSD, models relating stiffness degrading with accumulated damage) constitute a promising possibility for the modeling of system failure in the study of seismic structural reliability. This has stimulated the development of a new DBSD model, inspired on that of Ref. 6, and described in the sequel.

The model proposed is presented in Fig. 5. The upper portion shows an idealized moment-rotation behavior at one end of a simply supported bending member, with the other end free to rotate. Following the model adopted --linear elastic member with plastic hinges at the ends-- the tangent initial stiffness is that of the linear elastic member, while the deformations in excess of those represented by line 0'-0" are the rotations concentrated at the plastic hinge. The behavior of the latter is shown in the lower portion of the figure. Damage accumulation is assumed to take place independently in each of the two opposite loading directions. In each direction, damage is measured by the sum of the plastic deformations in that direction divided by the yield deformation of the virgin curve. The algorithm which transforms accumulated damage into stiffness reduction can be better explained following the cyclic response curves in Fig. 5. Suppose that when point b in Figs 5a and b is reached, the damage already accumulated in the positive direction of θ is D_b. A consequence of this damage is that the next reloading cycle in the positive direction of the θ_p axis in Fig. 5b will be directed from point f to point b_1, and may eventually reach the envelope curve at point g. If y_b and y_{b1} are respectively the ordinates of points b and b_1, corresponding to the moments acting at the beam end for a plastic hinge rotation equal to θ_{pb} (segment oc in Fig. 5b), then the level of degradation is given by the ratio $\varepsilon_b = \dfrac{y_b - y_{b1}}{y_b}$. This ratio is a function of the damage D_b accumulated up to point b: $\varepsilon_b = \varepsilon(D_b)$. A similar relation exists between $\varepsilon_h = \dfrac{y_h - y_{h1}}{y_h}$ and D_h, and between $\varepsilon_m = \dfrac{y_m - y_{m1}}{y_m}$ and $D_{m'}$. As figures 5a and b also show, the reloading branches may be directed towards a point in segment AB of the envelope, or may have such a small stiffness that they or their projections intersect the envelope at a point in segment BC.

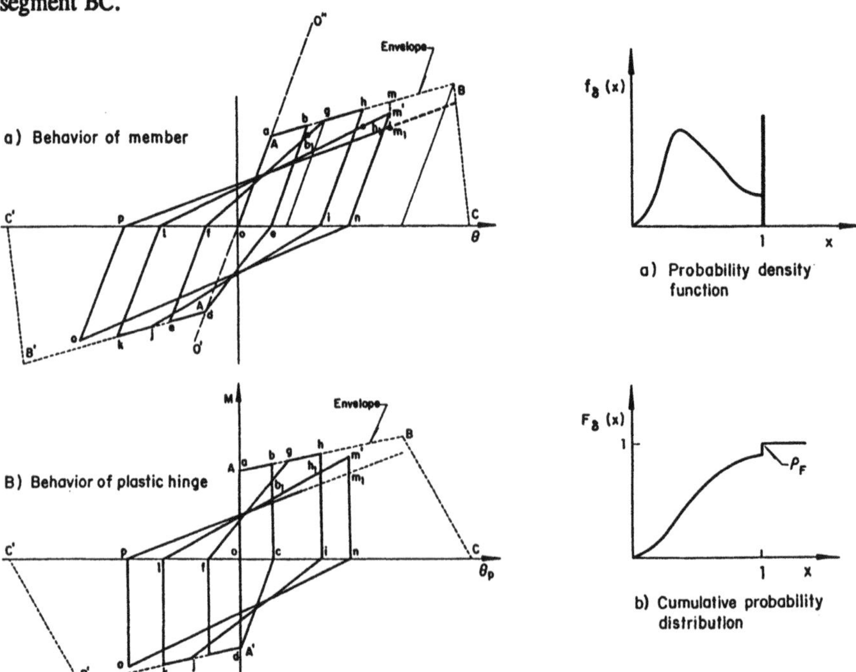

Fig 5 Stiffness and strength-degrading models of bending member and plastic hinge

Fig 6 Probability distribution of δ_{max}

4.5. A new approach to seismic reliability analysis

It is not easy to compute an indicator similar to D for a given story or subassemblage of a structure when obtaining the time-history of its dynamic response to a ground motion accelerogram, but it is reasonably simple to obtain values of K_t of any story throughout the time when the response of the system remains sufficiently high as to be significant for engineering purposes. This statement can be extended to the stiffnesses related to failure mechanisms other than excessive story deformation. Thus, the problem of determining the probability of failure of a system under the action of earthquakes randomly selected from a given population (which, of course, includes the case of earthquakes with a given intensity) is transformed into one of a) identifying a set of potential failure modes, each associated to the occurrence of excessive deformations of a certain portion or subassemblage of the system; b) expressing the response in each mode in terms of an effective force and an effective deformation, such that failure in that mode takes place when the value of K_t expressed in terms of them becomes null; and c) obtaining the probability that any of the K_m becomes null.

If δ_i, i = 1, N_m is the ratio $(K_{oi} - K_{mi})/K_{oi}$ for the *i*-th mode and N_m is the number of potential failure modes, then the latter probability is equal to that of the event that any of the δ'_is reaches a value of 1, or of the equivalent event that the maximum δ_i, called here δ_{max}, equals 1. The probability density function of δ_{max} has a shape similar to Fig. 6a. The spike at 1 has an area equal to the failure probability, equal to the discontinuity shown by the cumulative distribution depicted in Fig. 6b. If the value of this probability is low, the number of response analyses required to estimate it by counting the number of failures predicted by Monte Carlo simulation will be large. However, the information about values of δ_{max} smaller than 1 obtained by Monte Carlo simulation is also significant for the estimation of p_F.

If the form of $f_{\delta max}$ (x) for $x < 1$ is assumed known except for a set $\{\alpha\}$ of undetermined parameters, the estimation of these parameters may be formulated, for instance, by a maximum likelihood criterion or, more generally, resorting to bayesian statistical analysis. Regardless of which of these approaches is adopted, the solution rests on the calculation of the likelihood function for a sample consisting of n values of δ_{max} smaller than 1 (x_i , i=1, n) and n_1 values equal to 1. Formally, let

$$f_{\delta_{max}}(x) = f(x) + p_F \delta(x - 1) \tag{2}$$

where $\delta(.)$ is Dirac's delta function and the integral of f(x) between the extremes 0 and 1 equals $1-p_F$. This means that, if f(x) is chosen as a probability density function such that its integral from zero to infinity equals 1, p_F is equal to the integral of f(x) from 1 to infinity; therefore, it is a function of $\{\alpha\}$. Given the sample described above, the likelihood function is

$$L(\alpha \mid x_i , i = 1, n; n_1) = (1 - F(1 \mid \alpha))^{n_1} \prod_{i=1}^{n} f(x_i \mid \alpha) \tag{3}$$

where $F(x) = \int_0^x f(u)\,du$. Passing from here to maximum likelihood estimates or bayesian analysis is straightforward.

A drawback of the foregoing approach is that the results may depend excessively on the choice of f(x). This may be a valid argument when the interest is centered on the study of failure probabilities for given intensities, and those probabilities are very low. However, in most earthquake engineering problems the most important contribution to the uncertainty about the ratio of excitation to capacity comes from the uncertainty about the ground motion intensity. For those cases, the sensitivity of marginal failure probabilities to the form of f(x) is not so pronounced. But the study of this problem must accompany the implementation of the proposed methodology.

Finally, this approach opens the door to the possibility of studying the influence of cummulative damage on the reliability of structural systems to sequences of earthquakes (7).

4.6. Model uncertainty

The constitutive laws adopted to represent the behavior of structural members are in general simplified idealizations of complex, imperfectly known phenomena. They are based on very limited experimental information, most of it obtained under laboratory conditions, and therefore requiring extrapolation to natural conditions. Predicted responses to high-intensity earthquakes are very sensitive to model uncertainties --both model form and model parameters-- and so are the probabilistic predictions of response and the reliability estimates derived from them. Accounting for these uncertainties and reducing them through calibration of theoretical models with empirical evidence are key links in the excitation-response-reliability chain.

Calibrating structural-response and system-reliability models for high earthquake intensities implies contrasting observations of response and behavior with the predictions resulting from alternate model forms or parameters. Available field observations come in most cases from low- or moderate-intensity earthquakes, and they must be complemented with laboratory simulations of responses to high-intensity events. The nature of the information that can be obtained varies widely. Laboratory tests in most cases can only be performed on simple structures, but accurate instrumental and descriptive information can be obtained covering both general and local response functions and failure mechanisms; they can also provide qualitative (fuzzy) information about damage levels. Field observations can be made on complex systems, but the degree of complexity significantly affects our capability for interpreting the records. Instrumental information usually refers to global response time histories, but with adequate planning it can also include time histories of deformations strongly correlated with safety margins for specific failure mechanisms or modes. Another valuable outcome of these observation is, of course, the qualitative information about damage severity and space distribution.

Whatever the origin of the empirical information to be used in the calibration of mathematical models, the process includes evaluation and adjustment of the capacity to predict a) structural response, at various levels of detail (global, sub-assemblage, local); b) variation of the properties of simple, equivalent systems; and c) damage severity and distribution. Statistical information about numbers of failures of structures of given types and designed in accordance with given design criteria and quality control norms also provides useful calibration points, but due to the (fortunate) usual shortness of these numbers, our calibrations of failure-probability prediction models will rely more heavily on the information furnished by moderate numbers of severely damaged (but not collapsed) structures.

Among the challenges to be found in calibration studies, the following deserve special mention:

a) Identifying critical failure modes and deploying instrument arrays aimed at providing information about the corresponding safety margins.

b) Developing criteria and methods to relate information about global system response with safety margins for specific failure modes.

c) Relating moderate-intensity calibration results with those that should apply for higher intensities.

In spite of their limitations as predictors of system response and behavior, computer methods of response analysis can be used to develop probabilistic information about the relations among global, subassemblage and local responses, thus permitting the application of the methods proposed in point b) above. The suggested transformation can be achieved by means of Bayes' theorem, as explained in the following. Let X be the vector of responses of interest, which are used in the estimations of safety margins for specific failure modes. (In a multistory building, vector X would contain maximum instantaneous values of lateral story deformations or minimum instantaneous values of secant story stiffnesses.) Let Z be a vector of measured values of some elements of the (global) response time history. Suppose that, for a random ground motion with intensity and frequency content similar to those of the earthquake for which Z was measured, our knowledge about the model parameters is expressed by the *a priori* bayesian probability density function of the parameters of a probabilistic model of known form. (Accounting for the uncertainty about model form can easily be incorporated, at least in theory.) Let α be the vector of the mentioned parameters and $f_\alpha'(.)$ its *a priori* probability density function. Application of Bayes' theorem leads to the *a posteriori* probability density function:

$$f''_\alpha(u \mid z) \propto f'_\alpha(u)p(z \mid u) \tag{4}$$

where $p(z \mid u)$ is the likelihood function of α corresponding to the observation $Z = z$. The *a posteriori* probability density function of X is obtained by integrating with respect to the *a posteriori* probability density function of α :

$$f''_X(x \mid z) = \int f_X(x;u)\, f''_\alpha(u \mid z)du \tag{5}$$

Of course, the value of the information on Z to improve our knowledge about α would depend on the sensitivity of Z to α. This should be a basic consideration at the stage of planning the location of the instruments on the systems used for calibration.

5. CONSISTENT RELIABILITY DESIGN CRITERIA

The foregoing sections deal with uncertainty analysis and synthesis for the purpose of computing failure probabilities; they consider both single earthquake events and sequences that may occur during given time intervals. The rest of the paper deals with the selection of target reliability levels, and with the definition of structural design formats and values required to attain those levels.

5.1. Iso-reliability spectrum for given intensity and frequency content

This constitutes a simplified approach to the situation when the design requirements aim at a specified reliability for an earthquake intensity corresponding to a given return interval at a given site, or for the most unfavorable excitation that may affect that site. It is presented here to illustrate some features of the more general case when the intended reliability level refers to the worst condition that may affect the structure of interest during a given time interval.

Both the intensity and the evolutionary frequency content are assumed given, leaving the detailed ground motion time-history as random. The structures studied in this example are elasto-plastic single-degree-of-freedom systems with uncertain mass, stiffness, lateral strength, and failure deformation, the latter expressed as the product of an "available" ductility factor by the yield deformation. Each system analyzed was defined in terms of the mean values and coefficients of variation of the properties mentioned above. For each of them a nominal natural vibration period was defined, as that computed in terms of the mean values of mass and stiffness. The values assumed for this period range between 0.25s and 5s. The mean and the nominal values ($\bar{\mu}$, μ^*, respectively) of the available ductility (μ, with a variation coefficient V_μ), are related through the equation $\mu^* = \bar{\mu} \exp(-2 V_\mu)$. Several alternative values were assumed for the mean lateral strength. These were sequentially chosen during the study, with the aim of covering an interval of values of the safety index β comprised between 1.0 and 6.0. The following variation coefficients were assumed: mass, 0.08; stiffness, 0.10; available ductility, 0.25; lateral strength, 0.15 and 0.30. A correlation coefficient of 0.7 was assumed between strength and stiffness; the other variables were taken as stochastically independent.

For the application of Monte Carlo method, several samples of values of mass, stiffness, strength and available ductility were generated for each possible combination among the mean values and variation coefficients of these variables. Each sample determined a system which was subjected to several artificial accelerograms similar in intensity and frequency content to the EW component of the Mexico City SCT accelerogram of 19 September 1985. The simulated accelerograms were normalized to the same values of Arias' intensity. The maximum ductility demand of each system for each accelerogram was obtained from its calculated dynamic response time history. A safety factor was calculated for each case analyzed, as the ratio of available ductility to ductility demand. The value of the safety index β was calculated for each combination of structural properties, as the ratio of the mean to the standard deviation of the safety factor. Typical results are shown in Figs. 7-10, corresponding respectively to nominal design ductilities, μ^*, of 4 and 2, and variation coefficients of lateral strength, V_R, of 0.15 and 0.30. The ordinates, expressed as spectral accelerations, correspond to expected values of lateral strength leading to the specified β

values for the family of accelerograms considered. The average linear spectral ordinates for 0.05 damping for this family are also shown in the figures, for comparison with the iso-reliability spectra. It is clearly seen that the latter may adopt forms which differ substantially from those corresponding to the average linear spectra, and even from those adopted by nonlinear systems with deterministically known properties. The deviations are very sensitive to the uncertainty associated to the natural period of the structure, at least for the narrow-band ground motion accelerograms considered in the study reported.

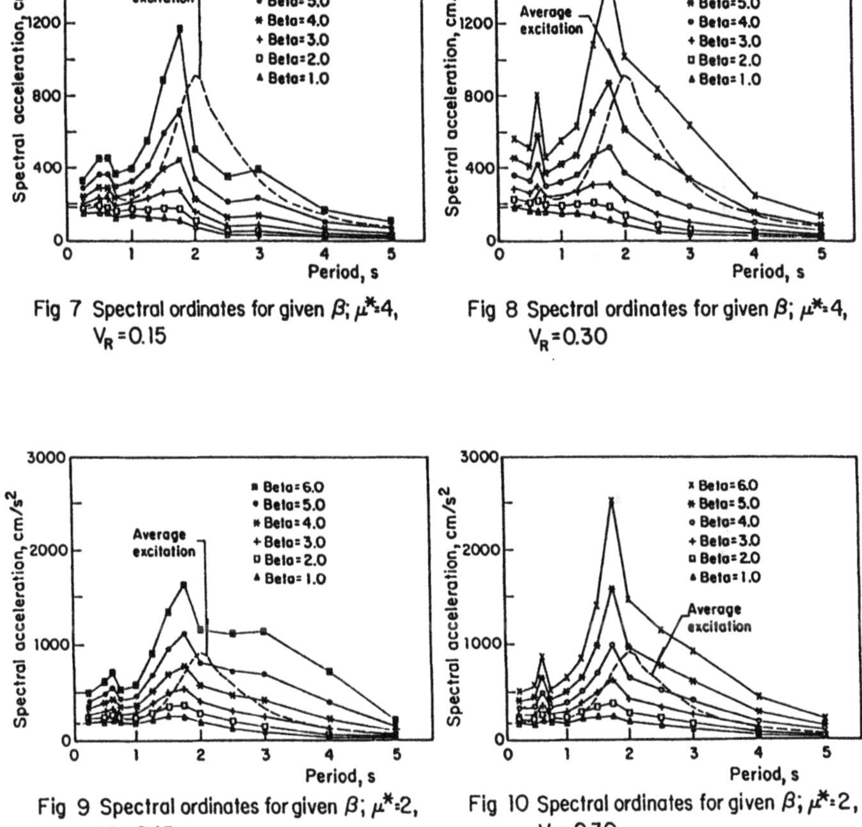

Fig 7 Spectral ordinates for given β; μ^*=4, V_R=0.15

Fig 8 Spectral ordinates for given β; μ^*=4, V_R=0.30

Fig 9 Spectral ordinates for given β; μ^*=2, V_R=0.15

Fig 10 Spectral ordinates for given β; μ^*=2, V_R=0.30

The differences in required strength that may arise from alternate assumptions regarding uncertainties about mass, stiffness, strength and deformation capacity have not been systematically studied, even at an exploratory level. The wide variabilities which characterize those uncertainties in practical cases point at the convenience of undertaking deeper and more extensive studies on this topic, including multi-degree-of-freedom systems.

Finally, when the determination of the target safety level is formulated as an optimization problem, it may be convenient to refer its value to an expected failure rate per unit time, rather than to a failure probability for an earthquake with specified intensity and frequency content. The forms of the corresponding iso-reliability design spectra do not necessarily coincide with those of the cases studied above.

5.2. Design format

From the above it is concluded that conventional earthquake-resistant design for specified response spectra does not lead to consistent reliability levels for structures of different types, and not even of the same type but with different numbers of degrees of freedom. It is clear that we have to learn much about failure probabilities of different systems and understand the influence of uncertainties about several variables before we can significantly improve the consistency of codified design rules. And we will also have to transform our findings on easy-to-apply recommendations. It is very likely that the best option will include the following concepts.

1. Several sets of nonlinear response spectra for structures with different energy-dissipation characteristics. Each set should correspond to a given return interval (or to its reciprocal, an exceedance rate), to be selected in accordance with the importance of the system to be designed and/or the expected consequences of its failure. These spectra should lead to specified failure rates for single-degree-of-freedom systems with typical uncertainties about their mechanical properties.

2. Load factors, expressed as functions of the nominal natural period of vibration, for systems of different types. These load factor functions should take care of all relevant influences on system reliability.

The determination of load factor functions can be accomplished according to the following steps:

1. Response spectra for given exceedance rates are determined. The exceedance rates selected should be of the order of twenty times the target failure rates.

2. Typical structures belonging to each family being studied are designed for the spectra defined in step 1.

3. Failure probability functions in terms of intensity ($p_F(y)$) are obtained in accordance with criteria discussed in previous sections.

4. Expected failure rates ν_F are obtained by combining the failure probability functions $p_F(y)$ with the intensity-recurrence functions $\nu(y)$.

5. The cycle is repeated until the calculated ν_F values are sufficiently close to those aimed for.

The size of the effort required is not small, but the benefits expected from attaining consistent reliability levels justify it.

5.3. Consistent-safety decisions

Structural reliabilities for different structures need not (indeed, **must** not) be equal, nor do they for similar structures at different locations. They should be optimum, in the sense of maximizing a utility function consisting of a sum of positive (benefits) and negative terms (initial and maintenance costs, damage and failure costs and consequences). Because these contributions are uncertain, their values must be expressed as the expectations of the mentioned variables. Because they may occur at different instants in the future, we must work with their present values; that is, we must add up the expected values of the positive or negative contributions that may take place in the future, each multiplied by an actualization factor, which we will take as $\exp(-\gamma t)$, where t is the time of its occurrence and γ is an effective discount rate.

The difficulties that hinder formal applications of the optimization criterion sketched above are well recognized. Prominent among the causes for those limitations are some of philosophical nature: how to assign values to non-monetary losses, such as human deaths or injuries, social disruption, political instability, etc. The analysis of these problems constitutes a focus of interdisciplinary research of engineers and social scientists and is out the scope of this paper. Other difficulties arise from the embrionic stage of development of the methods for computing seismic failure probabilities, and have been dicussed above. Therefore, most of the usefulness of the formal optimization process advocated in this section does not lie at present on its direct application to obtaining optimum design spectra for specific systems but mainly on its use as a calibration tool for the formulation of mutually consistent sets of seismic design spectra for given familis of structures (8).

6. FINAL COMMENTS

The advances of the last few decades in the areas of seismic hazard analysis and random vibration theory have permitted the development of practical criteria and methods applicable to some concepts of great importance in the seismic reliability analysis of engineering systems and, consequently, in the corresponding structural design methods. The estimation of intensity-recurrence functions and the application of random vibration analysis to approximately represent the ground motion or to make probabilistic predictions of the responses of linear and (some) nonlinear systems are among the most important results. However, very little has been done to obtain estimates of the failure probabilities of complex nonlinear systems, closely representing real systems, under the action of random earthquakes and sequences of earthquakes, and to assess the validity of those estimates in the light of observations during moderate- or high-intensity earthquakes. Consequently, very little profit has been made of the progress in probabilistic response analysis methods to attain consistent, optimum reliability levels for structures of different types. This explains the interest of the writer about the needs, rather than about the accomplishments.

The intention of this paper has been to provide an integrated view of the problems of single-earthquake and long-term reliabilities of complex systems, of the decisions required to set the target safety levels, of the research required in the system-reliability-area, and of possible formats for implementing the knowledge to be developed into codified design rules. It also intends to contribute to the solution of some of the individual problems pointed at; in particular, of the analysis of seismic system-reliability. The efforts related with system-modeling are concentrated on building frames. This is just for illustrative purposes, but it is recognized that the modeling of other typical structures may be more complicated and is less developed. Also, damage other than that due to system collapse was not discussed, despite its importance in design criteria related with serviceability failure modes. These concepts should receive more attention in the near future.

Finally, several very important topics are either ignored or only cursorily mentioned in this paper: uncertainties about seismic hazard models and parameters, social consequences of catastrophic failures, and risk-related social attitudes, to name a few. The challenges we have to face are huge, but the benefits we may expect are worth the efforts.

REFERENCES

1. Esteva, L., "Towards consistent-reliability structural design for earthquakes", manuscrip prepared for a keynote lecture presented at the 6th IFIP WG7.5 Working Conference, Assissi, Italy (1994).

2. Esteva. L., and Ruiz, S. E., "Seismic failure rates of multistory frames", **ASCE Journal of Structural Engineering, 115,** 2 (1989), 268-284.

3. Esteva, L., Díaz, O., Mendoza, E., and Quiroz, N., "Reliability based design requirements for foundations of buildings subjected to earthquake", **Proc. ICOSSAR 89**, San Francisco (1989).

4. Díaz, O., Mendoza, E., and Esteva, L., "Seismic ductility demands predicted by alternate models of building frames", **Earthquake Spectra**, 10, 3 (1994), 465-488.

5. Park, Y. J., Reinhorn, A.M. and Kunnath, S.K., "IDARC": Inelastic Damage Analysis of Reinforced Concrete Frame-shear-wall Structures", Technical Report **NCEER-87-0008**, State University of New York at Buffalo, July 1987.

6. Shah, S. P., and Wang, M. L., "Reinforced concrete hysteresis model based on the damage concept", **Earthquake Engineering and Structural Dynamics, 15** (1987), 993-1003.

7. Esteva, L., and Díaz, O., "Optimum decisions related to design and replacement of seismic energy dissipators", **Structural Safety and Reliability,** Proc. of ICOSSAR 93, Balkema, Rotterdam (1993), 653-660.

8. Esteva, L., and Ordaz, M., "Riesgo sísmico y espectros de diseño en la República Mexicana", accepted for publication in **Revista de la Sociedad Mexicana de Ingeniería Sísmica** (1994).

4

Fuzzy logic and its contribution to reliability analysis

H. Furuta

Department of Informatics, Kansai University,
2-1-1, Ryozenji-cho, Takatsuki, Osaka 569, Japan

In this paper, outline of fuzzy logic and its historical development are first introduced briefly. From the standpoint of fuzzy logic, uncertainties and ambiguities can be classified into several kinds of uncertainties. Here, they are divided into two types of uncertainties; randomness and fuzziness. The reliability analysis based on fuzzy logic is described with emphasis on the probability of fuzzy events and fuzzy probability. Some comments are given on problems that arise in introducing the fuzzy logic in the structural reliability assessment. Several practical examples are presented to demonstrate the applicability of the fuzzy logic in structural engineering. Future problems and possible developments are summarized and discussed.

1. INTRODUCTION

Since the pioneer paper by Zadeh in 1965 (Zadeh 1965), fuzzy logic has been developed in various fields. As well as engineering fields, social sciences, medicine, economics, and linguistics have paid attention to the characteristics and advantages of fuzzy logic (Kandel 1991). In Japan, many attempts have been made to develop practical fuzzy controllers in various fields since the pioneer work of the train control of subway in Sendai City (Sugeno 1985). Recently, many Japanese have become familiar with so-called "fuzzy electric appliances" which use micro processors of fuzzy controller to provide flexible and subtle treatment. Especially, washing machine, vacuum cleaner and video camera have gained great success from both technological and commercial points of view.

In the field of civil engineering, many applications have been attempted, some of which have been applied for practical use. In structural engineering, the fuzzy logic has been introduced into the optimum design, reliability and damage assessment, fatigue analysis and cable tension adjustment. Dam control and water quality control are typical examples of applications in hydrology and water resource engineering. In geotechnical engineering, soil or rock classification, seepage analysis, soil improvement and slope stability analysis are promising areas. Traffic control and traffic flow estimation are considered in transportation engineering. In construction engineering, the fuzzy controller for tunneling is famous as a successful application (Furuta 1994a).

In the field of structural engineering, Brown and Leonards (Brown and Leonards 1971) first introduced and discussed the application of fuzzy in 1971, Blockley (Blockley 1975) published an excellent paper on the likelihood of structural accidents. In 1979, Brown introduced a fuzzy safety measure combining the application of classic theory of structural reliability with the interpretation of subjective evaluation to obtain more realistic failure rates. Later Shiraishi and Furuta (Shiraishi, Furuta and Kawamura 1982) discussed the application of fuzzy sets to the design of reinforced concrete beams. Furuta et al. (Furuta et. al. 1983) developed the optimum design of earthquake resistant structures using fuzzy mathematical programming. Yao (Yao 1980) and Yao et al. (Yao, Bresler and Hanson 1984) discussed the application of fuzzy sets to safety evaluation of existing structures. Ishizuka et al. (Ishizuka, Fu and Yao 1981) developed a scheme for the combination of evidence using fuzzy sets in their study on expert systems. Chameau et al. (Chameau et. al. 1983) discussed potential applications of fuzzy sets in various subject areas of the civil engineering profession. In 1985, Symposium on Application of Fuzzy Set in Civil Engineering was held at Purdue University, chaired by Brown and Yao (Brown and Yao eds. 1985), and International Symposium on Fuzzy Mathematics in Earthquake Researches was held at Beijing, China (Feng and Liu 1985).

So far, the fuzzy logic has been applied to various structural problems such as damage assessment of existing structures (Furuta et. al. 1991a, Furuta 1993), design planning of bridges (Furuta and Shiraishi 1987), analysis of structural accidents (Blockley 1977, Furuta and Shiraishi 1984), structural reliability assessment (Shiraishi and Furuta 1983, Yao and Furuta 1986), fatigue analysis (Shiraishi, Furuta and Ozaki 1988), and cable tension adjustment (Kaneyoshi et. al. 1990).

In this paper, outline of fuzzy logic and its historical development are first introduced briefly. From the standpoint of fuzzy logic, uncertainties and ambiguities can be classified into several kinds of uncertainties. Here, they are divided into two types of uncertainties; randomness and fuzziness. The reliability analysis based on fuzzy logic is described with emphasis on the probability of fuzzy events and fuzzy probability. Some comments are given on problems that arise in introducing the fuzzy logic in the structural reliability assessment. Several practical examples are presented to demonstrate the applicability of the fuzzy logic in structural engineering. Future problems and possible developments are summarized and discussed.

2. OUTLINE OF FUZZY LOGIC

Roughly speaking, fuzzy logic consists of fuzzy set theory and fuzzy measure theory. As well-known, fuzzy set theory was developed by Zadeh, while fuzzy measure was proposed by Sugeno in 1972 (Sugeno 1972). Although the fuzzy sets provides us with an intuitive pleasing method of representing one form of uncertainty, there can be another type of uncertainty which is related to the degree of evidence (Klir and Folger 1988). In order to treat this type of uncertainty, the fuzzy measure was developed.

A fuzzy set is defined as a subset of a universal (sample) space Ω by a membership function $\mu(x)$ that is a generalization or extension of the characteristic function defining the ordinal sets.

$$\mu(x) \rightarrow [0,1] \tag{1}$$

$$C_A(x) = \begin{cases} 1 & x \in A \\ 0 & x \notin A \end{cases} \tag{2}$$

where C_A is the characteristic function defined by only two elements, say 0 and 1, then A is said to be a non-fuzzy or crisp set. Eq. 1 implies that the boundary of the fuzzy set (subset) is not sharp. This property of the fuzzy set is useful in expressing the human subjectivity, because it provides a good interpretation of natural languages which have intrinsically vague or ambiguous and are practical tools for thinking way of human beings.

The arithmetic operations of fuzzy sets commonly used are given as follows:
The membership function for the union of two fuzzy sets A and B is

$$\mu_{A\cup B}(x) = \max\{\mu_A(x), \mu_B(x)\} \tag{3}$$

On the other hand, the membership function for the intersection of A and B is

$$\mu_{A\cap B}(x) = \min\{\mu_A(x), \mu_B(x)\} \tag{4}$$

The complement of A is denoted by A, and is given by

$$\mu_{\bar{A}}(x) = 1 - \mu_A(x) \tag{5}$$

For two sets X and Y, a binary relation R can be considered, which is important for many applications. Using a fuzzy relation R , a fuzzy set A is derived from another fuzzy set B as follows:

$$A = R \circ B \tag{6}$$

where the symbol $\circ$ denotes the composition of R and B. Eq. 6 can be executed by the extension principle.

$$\mu_A(x) = \max_y\{\min\{\mu_R(x,y), \mu_B(y)\}\} \tag{7}$$

If R is a fuzzy relation from X to Y, and S is a fuzzy relation from Y to Z, then the composition of R and S is a fuzzy relation which is described with the following membership function:

$$\mu_{R\circ S}(x,z) = \max_y\{\min\{\mu_R(x,y), \mu_S(y,z)\}\} \tag{8}$$

As mentioned above, the fuzzy measure assigns a value to each crisp sets of the universal set signifying the degree of evidence or belief that a particular element belongs in the set, whereas in the fuzzy set, a value is assigned to each element of the universal set signifying its degree of membership in a particular set with unsharp boundaries.

A fuzzy measure is defined by a function

$$g : P(X) \rightarrow [0,1]$$

which assigns to each crisp subset of X a number in the unit interval $[0,1]$. The axioms of fuzzy measures are as follows:

Axiom 1 (boundary condition) $g(\phi)=0,\ g(X)=1$

Axiom 2 (monotonicity) For every $A,B \in P(X)$, if $A \subseteq B$, then $g(A) \le g(B)$

Axiom 3 (continuity) For every sequence $A_i \in P(X)$ of subsets of X,

if $A_1 \subseteq A_2 \subseteq \cdots\cdots \subseteq A_n$, then $\lim_{n\to\infty} g(A_n) = g(\lim_{n\to\infty} A_n)$ (9)

Then, the probability measure can be derived by replacing the monotonicity condition with the additivity condition:

$$P(A \cup B) = P(A) + P(B) \qquad \text{whenever } A \cap B = \phi \tag{10}$$

This means that the probability measure is a special case of fuzzy measure, because the additivity axiom is evidently included in the monotonicity axiom.

As representative fuzzy measures, possibility measure, necessity measure and g-λ measure are considered. The possibility measure Π is defined as follows:

1) $\Pi(\phi)=0, \Pi(X)=1$

2) $\Pi(A \cup B) = \max\{\Pi(A), \Pi(B)\} \qquad A,B \in \Omega$ (11)

The necessity measure N, which is the dual one of the possibility measure, is as follows:

1) $N(\phi)=0, N(X)=1$

2) $N(A \cap B) = \min\{N(A), N(B)\} \qquad A,B \in \Omega$ (12)

Between these two measure, Π and N, the following relation holds

$$N(A) = 1 - \Pi(\bar{A}) \tag{13}$$

The g_λ measure is defined as

1) $g_\lambda(\phi)=0, g_\lambda(X)=1$

2) $g_\lambda(A \cup B) = \dfrac{g_\lambda(A)+g_\lambda(B)-g_\lambda(A \cap B)+\lambda g_\lambda(A) g_\lambda(B)}{1+\lambda g_\lambda(A \cap B)}$ (14)

Using the g_λ measure, the fuzzy integral (Sugeno 1972) can be defined as follows:

$$\int h(x) \circ g_\lambda = \max_{\alpha \in [0,1]} \{\min(\alpha, g_\lambda(H_\alpha(x))\} \tag{15}$$

where h(x) is the integrand and α is the α -level set expressed by

$$H_\alpha(x) = \{x | h(x) \geq \alpha\} \tag{16}$$

The fuzzy integral can provide a useful tool for evaluating the total effect of different kinds of factors.

3. APPLICATION OF FUZZY LOGIC TO RELIABILITY ANALYSIS

3.1 Uncertainties in Civil Engineering

In spite of the significance success of the probabilistic methods in structural reliability assessment, several investigators have indicated that there are other types of uncertainties in addition to that of randomness (Blockley 1980, Blockley 1983, Brown 1979). It is important and desirable for engineers to recognize several alternative ways of representing uncertainties, and to choose the most appropriate one(s) among them for a particular application (Yao and Furuta 1986).

Here, the application of fuzzy logic to structural reliability is outlined along with probabilistic methods from the viewpoint that a) they are two distinct concepts and b) their applications can be compatible and complementary to each other.

As one of implicit effects of fuzzy logic on structural reliability, one may recognize and distinguish various kinds of uncertainties involved in civil engineering. Generally, the term "uncertainty" may be associated with ambiguity, fuzziness, randomness, vagueness and imprecision of the events under consideration. These uncertainties may be delineated from one another as follows:

(1) Randomness is due to factors in the complex phenomena which are random in nature.
(2) Fuzziness results from the complexity of natural events, the knowledge of which is imprecise and/or incomplete, and/or subjective.
(3) Ambiguity results from the use of natural languages which can be meaningful but not clearly defined.
(4) Blur or vagueness is accompanied with inexact and/or ill-defined figures, pictures and scenes.
(5) Imprecision is due to the lack of information.

For the sake of simplicity, the last four items (i.e., (2) through (5)) are included in the expression of "fuzziness" in contrast to "randomness". Using these two concepts, namely randomness and fuzziness, one can categorize various problems in civil engineering as shown in Fig. 1. In Fig. 1, the upper right-hand corner refers to certain events with neither randomness nor fuzziness. For the analysis of these events, deterministic methods are appropriate in obtaining satisfactory solutions. On the other hand, the lower left-hand corner refers to events which cannot be described with classical mathematical models because of their highly complex and chaotic properties. Now, consider the remaining domain which is not shaded in Fig. 1. Although most common and significant problems in civil engineering belong to this domain, our capability to deal with these problems remains to be studied and developed further.

The probability theory is a useful tool, but it does not cover the whole domain. In the application of the probability theory, an event of interest is usually clearly defined. In other words, the application of the probability theory requires us to deal with events which are collections of outcomes of well-defined actions. These events must be subjected to repeatable testing and observations. It is realized that Bayesian statistics serves the purpose of extending the potential application of probabilistic methods. It is, however, noted that the Bayesian approach is not entirely free from the basic probability axioms. To avoid these axioms, Dempster & Shafer theory (Shafer 1976) has been developed. Including this approach and recognizing the existence of ignorance, the uncertainty diagram can be redrawn as Fig. 2 (Brown and Yao 1985).

Other statistical methods such as regression analysis, discrimination analysis, and multi-dimensional methods, may also be insufficient to cover all important problems. In any event, it is emphasized here that each method can become more powerful by using it combined with other available methods. The theory of fuzzy logic is one such new approach, with which more meaningful solutions may be obtained for complex problems in the state of nature.

Real-world problems are usually more complex than their corresponding mathematical models. To compensate the gaps between them, some verbal explanation occasionally adds to the results obtained through the models. The concept of fuzzy sets has been developed to deal with the verbal information which is usually meaningful but not clearly defined. Therefore, it is desirable to a) first model actual events with fuzzy sets and b) then analyze them using probability concepts.

To illustrate the need for an additional methodology, several simple examples are presented in the following. Consider an axially loaded plate specimen with butt welds as shown in Fig. 3. If all relevant information (e.g., weld quality, loading and environmental conditions, and statistical quantities) are completely and precisely known, various limit states such as those for ultimate failure yielding and fatigue may be plotted as shown in Fig. 4. In the real world, however, it is difficult to have complete and exact information even for such a simple axially-loaded structural member. For example, the fatigue behavior of butt welds with slag inclusion has been studied experimentally in detail by Bowman et al (Yao and Furuta 1986). Nevertheless, the behavior of actual welded joints depends on many kinds of defects, the effects of which may not be known in a precise manner. As another example along the same line, consider the 1978 AISC (American Institute of Steel Construction) specifications. The allowable stress ranges for various design lives are listed in Table 1. Test data as given by Munse and allowable stresses for 100 000-cycle life are plotted on a Modified Goodman

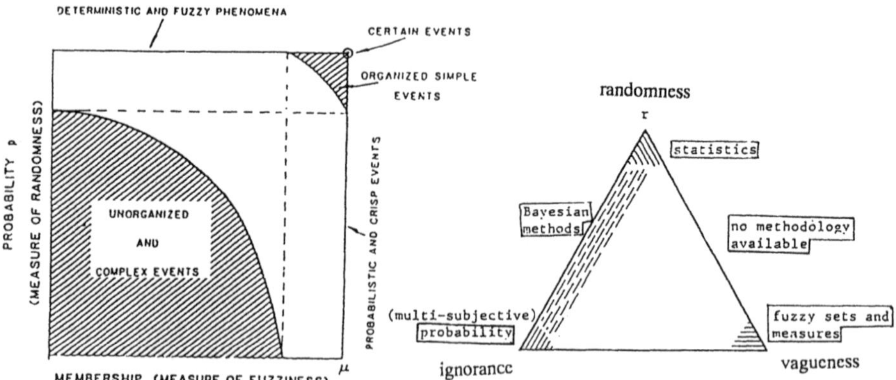

Fig. 1 Probability vs. membership relations.

Fig. 2 Uncertainty diagram

Fig. 3 Axially-loaded and butt-welded plate specimen

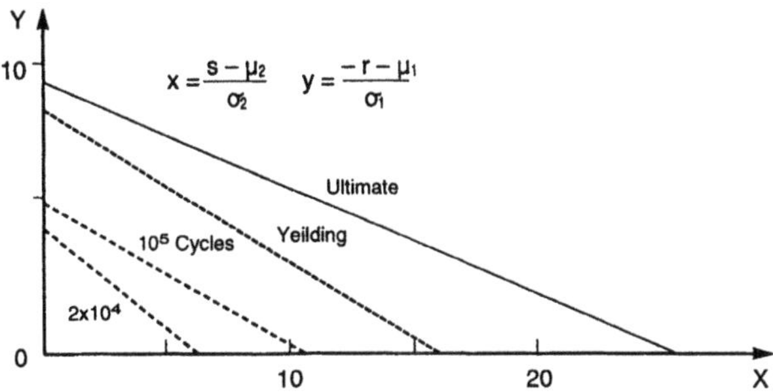

Fig. 4 Limit states for an axially-loaded and butt-welded plate specimen with complete and precise information

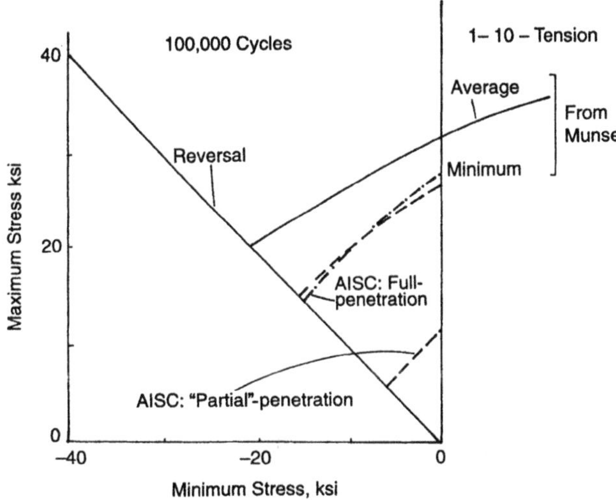

Fig. 5 Modified Goodman diagram for butt welds

Table 1 Allowable fatigue ranges for butt welded joints

Design life, cycles	Allowable stress range, ksi: Full-penetration	Allowable stress range, ksi: Partial-penetration
20,000–100,000	45	15
100,000–500,000	27.5	12
500,000–2,000,000	18	9
> 2,000,000	16	8

Diagram as shown in Fig. 5. It is seen that the allowable stresses for "partial penetration of butt welds are much lower than those for "full-penetration" welds. What constitutes "partial" penetration does not seem to be clearly defined.

3.2 Probability of Fuzzy Events

Most civil engineering problems involve uncertainties which may be classified into randomness and fuzziness. As an example consider the damage assessment of existing structures. The assessment of structural damage involves several types of uncertainties including randomness in applied loads as well as structural resistances, ambiguity and vagueness involved in observed data or inspection results, imprecision due to the lack of data and limitations in instrumentation, subjective interpretations, and approximation in modeling.

It is possible to evaluate the fuzzy uncertainties with the use of linguistic variables which may be expressed in terms of fuzzy sets. Such uncertainties in the damage assessment are evaluated for each source of available information. As examples, available information sources include (1) building documents, (2) results of a visual examination, (3) field testing, (4) laboratory testing, and (5) structural analysis. By using all available information separately in terms of verbal expressions such as no damage, slight damage, more or less damage, sever damage, and collapse.

Let E_i denote the evaluation of the information source. E_T is given by the fuzzy sets with membership functions as given in Fig. 6 and their supports are defined by the real number [0,1]. Then, the total assessment may be computed as

$$E_T = \sum_{i=1}^{n} W_i E_i \tag{17}$$

where the summation and product correspond to the union and intersection, respectively; W_i denotes a fuzzy set representing the weight of the i-th evaluation. Because Eq. 17 is an equation of fuzzy quantities, maximum and minimum operators are applied according to the properties of the possibility measure.

$$\mu_{ET}(u) = \bigvee_i (\mu_{Wi}(u) \wedge \mu_{Ei}(u)) \tag{18}$$

where μ_{ET}, μ_{Wi} , and μ_{Ei} denote the membership functions of E_T, W_i, E_i, respectively. The symbols $\vee$ and $\wedge$ denote maximum and minimum operations, and u is the support for the linguistic variables.

Assume that the damage of the structure is defined by the reserve strength Z with a probability density function f_Z. Damage occurs when Z is less than or equal to zero. Then, the probability of having a certain damage state may be calculated as follows:

$$P_f = \int_{-\infty}^{\infty} \mu_z(z) f_z(z) dz \tag{19}$$

where the membership function μ_z for the damage state is obtained by the following composition procedure.

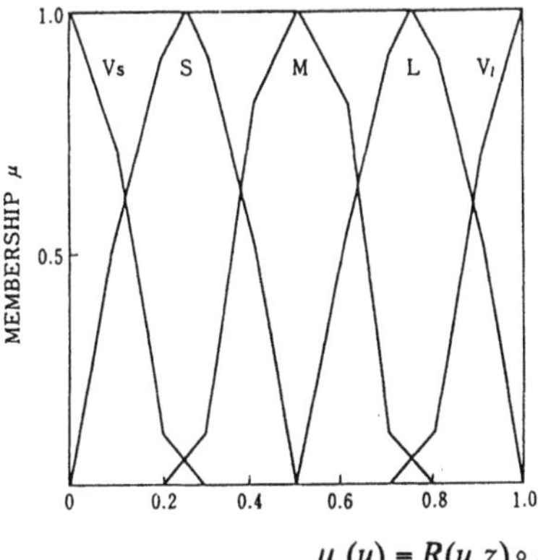

Fig. 6 Membership functions for linguistic variables

$$\mu_z(u) = R(u, z) \circ \mu_{ET}(u) \tag{20}$$

where R is a fuzzy relation between the total assessment E_T and the reserve strength Z.

To illustrate this method, consider a numerical example where the probability of having a certain damage state is calculated. The grade of a structural damage is evaluated in the linguistic forms such as "Very large (Vl)", "Large (L)", "Medium (M)", "Small (S)", and "Very small (Vs)", which are characterized by the membership functions as shown in Fig. 6. In Fig. 6, the support u is a number between zero and one such that u=1 indicates collapse and u=0 indicates no damage. Suppose that the evaluation of Vs, Vl, M, L and S are based respectively on the results of examination of building documents, the visual inspection, the field testing, the laboratory testing, and the structural analysis. Weighing factors are, for the sake of simplicity, assumed to be given by crisp numbers; say 0.3 for building documents, 0.9 for visual inspection, 1.0 for field testing, 0,7 for laboratory testing, and 0.5 for structural analysis. In general, both the linguistic evaluation and weights are determined based on the judgments of experts. Then the total evaluation is calculated as

$$\begin{aligned} E_T &= 0.3/0 + 0.5/0.1 + 0.5/0.2 + 0.5/0.3 + 0.7/0.4 \\ &\quad + 1/0.5 + 0.7/0.6 + 0.7/0.7 + 0.7/0.8 + 0.7/0.9 + 0.9/1.0 \end{aligned} \tag{21}$$

Eq. 21 is plotted in Fig. 7. Moreover, assume that the probability density function of reserve strength is given in a discrete form as listed in Table 2 and the fuzzy relation between the total evaluation and the reserve strength is given in Table 3. Using Eq. 21, Table 2 and Table 3, the probability of the damage state is calculated as follows:

$$\begin{aligned} P_f &= 0.9 \times 0.0001 + 0.5 \times 0.001 + 0.7 \times 0.005 + 0.5 \times 0.01 + 1 \times 0.015 \\ &\quad + 0.5 \times 0.02 + 0.5 \times 0.025 + 0.5 \times 0.03 + 0.5 \times 0.04 = 0.082 \end{aligned} \tag{22}$$

For the damage state which is defined by Z<0, its occurrence probability is

$$P_f = 0.0001 + 0.001 + 0.005 + 0.01 + 0.015 = 0.031 \tag{23}$$

Comparing with Eq. 22, the answer as given in Eq. 23 is 2.6 times smaller.

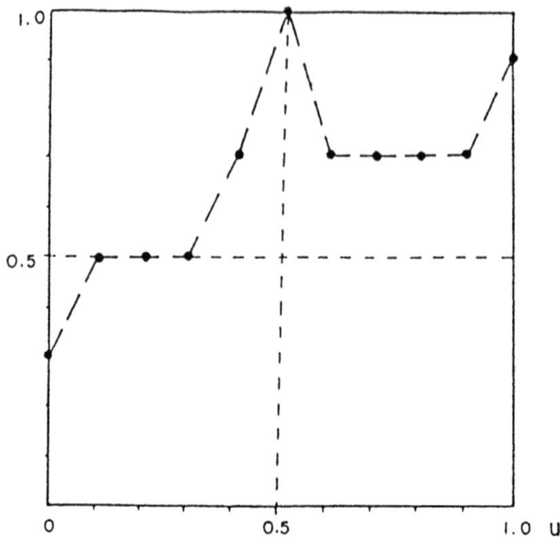

Fig. 7 Membership grades of E_T

Table 2 Probability density function of Z

Reserve strength (MPa)	Probability
−20	0.0001
−15	0.001
−10	0.005
− 5	0.01
0	0.015
5	0.020
10	0.025
15	0.030
20	0.04
25	0.01
30	0.7539

Table 3 Fuzzy relation matrix between u and z

	−20	−15	−10	−5	0	5	10	15	20
0								0.5	1.0
0.1						0.5	0.5	0.5	0.7
0.5						0.5	0.9	0.5	0.1
0.3				0.1	0.1	0.5	0.9	0.5	
0.4				0.5	0.7	0.5	0.5	0.5	
0.5				0.5	1.0	0.5			
0.6		0.5	0.5	0.5	0.7	0.5			
0.7		0.5	0.9	0.5	0.1				
0.8	0.1	0.5	0.9	0.5					
0.9	0.7	0.5	0.5	0.5					
1.0	1.0	0.5							

3.3 Fuzzy Probability

By definition, the probability of fuzzy events is deterministic and crisp number. Meanwhile, there exist situations where the probability is expressed in an ill-defined manner regardless of the fuzziness of the event involved. Such an example can be found in weather forecast. One may say that "the probability of having a cloudy day next Monday is approximately 0.6" or "it will likely be a cloudy day next Monday". As another example, say that the collapse of a specific bridge is likely to occur within the next 10 years. Or, say that the probability of the crushing of this particular reinforced concrete specimen is approximately 0.5 after a given compressive load is applied.

These ill-defined probabilities are called fuzzy or linguistic probability, which is characterized by a membership function on the probability measure. The fuzzy probability can be mathematically interpreted by considering the upper and lower limits of an integral as fuzzy quantities. By definition, the probability of the occurrence of an event A is given as

$$P_A = P_r[a \le x \le b] = \int_a^b f_A(x)dx \tag{24}$$

where $P_r[*]$ is the probability of the event * and f_A is its probability density function. On the other hand, the fuzzy probability of A can be defined as

$$P_A = P_r[\tilde{a} \le x \le \tilde{b}] = \int_{\tilde{a}}^{\tilde{b}} f_A(x)dx \tag{25}$$

In Eq. 25, the limits of the integral are fuzzy so that the resultant probability becomes fuzzy. Eq. 25 may be obtained using the extension principle, as will be explained later by an example. Using Eqs. 19 and 25, we can obtain the fuzzy probability of fuzzy events.

$$P = \int_{\tilde{a}}^{\tilde{b}} \mu_A(x) f_A(x)dx \tag{26}$$

As an illustrative application of fuzzy probability, consider an example in which the severely damaged state of a uni-axially loaded plate is investigated. Assume that the probability density function of the initial crack length is given by

$$f_Y(y) = \begin{cases} 10(1-5y) \cdots 0 \le y \le 0.2 \\ 0 \cdots\cdots\cdots otherwise \end{cases} \tag{27}$$

Thus, the severely damaged state is defined when the crack length is between approximately 10 and 20 % of the plate width. Using Eq. 25, the fuzzy probability of such a plate being in severely damaged state is given by

$$P_f = \int_{0.\tilde{1}}^{0.\tilde{2}} 10(1-5y)dy \tag{28}$$

Supposing that fuzzy limits 0.2 and 0.1 are given by

$$0.\tilde{2} = 0.7/0.15 + 1/0.2 + 0.2/0.25$$
$$0.\tilde{1} = 0.2/0.05 + 1/0.1 + 0.7/0.15 \tag{29}$$

the calculation of Eq. 28 can be executed by using the extension principle as follows:

$$\begin{aligned} P_f &= (0.7 \wedge 0.2)/(F(0.15) - F(0.05)) + (0.2 \wedge 1)/(F(0.2) - F(0.05)) \\ &+ (0.2 \wedge 0.2)/(F(0.25) - F(0.05)) + (0.7 \wedge 1)/(F(0.15) - F(0.1)) \\ &+ (1 \wedge 1)/(F(0.2) - F(0.1)) + (1 \wedge 0.2)/(F(0.25) - F(0.1)) \\ &+ (0.7 \wedge 0.7)/(F(0.15) - F(0.15)) + (1 \wedge 0.7)/(F(0.2) - F(0.15)) \\ &+ (0.2 \wedge 0.7)/(F(0.25) - F(0.15)) \\ &= 0.7/0 + 0.7/0.063 + 0.7/0.188 + 1/0.25 + 0.2/0.5 + 0.2/0.563 \end{aligned} \tag{30}$$

where Eq. 30 are plotted in Fig. 8. Although the fuzzy probability may appeal to our intuition, its calculation is complicated. The fuzzy probability was used in the reliability assessment of damaged structures (Shiraishi and Furuta 1988) and the seismic safety assessment (Furuta et. al. 1991b). However, it is difficult to interpret results of such calculations. In particular, the fuzzy probability of fuzzy events is too broad to be used for solving practical engineering problems at present.

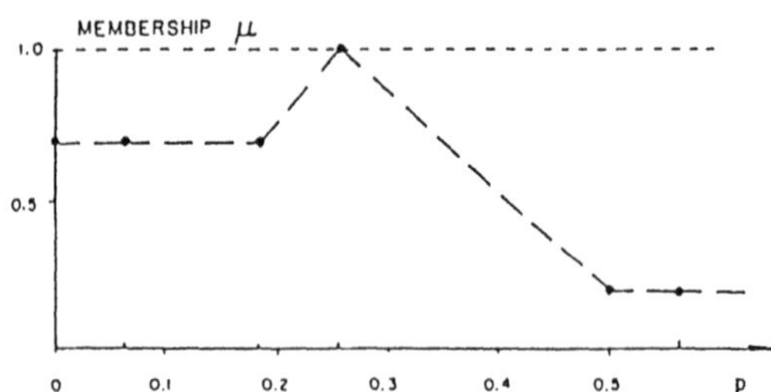

Fig. 8 Membership grades of fuzzy probability

4. APPLICATION EXAMPLES

4.1 Fuzzy Expert System for Damage Assessment

In order to establish an efficient repair and maintenance program, it is important to evaluate the durability of existing structures. However, the durability assessment of structures is not easy due to the lack of available information and the complex mechanism of structural deterioration. Therefore, the daily maintenance has been so far carried out on the basis of intuition and engineering judgment of experienced engineers. Therefore, attempts have been made to develop rule-based expert systems for assessing the durability of bridge structures.

A number of problems arise when an expert system is built for the practical use. How to treat uncertainties or ambiguities is one of problems which we face occasionally. Here, an expert system is introduced, which has such a remarkable feature that it includes a fuzzy-set manipulation system which can treat fuzzy sets in the process of data handling, rule representation and inference procedure (Furuta et al. 1991a). Using this system, it is possible to deal with various kinds of uncertainties and ambiguities involved inherently in the data, rules and inference process in a unified manner. Similar to the usual expert system (Shiraishi, Furuta, Umano & Kawakami 1987), this durability assessment system consists of interpreter, rule-base and working memory. In order to derive a meaningful conclusion from imprecise and ambiguous information or knowledge, a special inference procedure is necessary. In this system, a fuzzy reasoning method (Zadeh 1975) is employed for this purpose.

In this system, the past records and inspection results are used as the input data. When the inspection results regarding cracks are firstly input into the system, the matching processes for rules concerning their damage cause, damage degree and damage propagation speed are executed to provide a solution for the remaining life.

This system was written in Franz Lisp and implemented on a 32 bit engineering workstation. To make the acquisition of expertise of engineers easier, an attempt was made to apply the technique of neural network for the automatic tuning of fuzzy production rules needed in the durability assessment of RC bridge deck (Furuta et. al. 1993). By introducing the fuzzy-set manipulation system into the expert system, it is possible to utilize the knowledge and rules which are expressed in terms of natural language. This enables us to acquire the expertise with ease. Based on the fuzzy reasoning, it is possible to reduce the number of rules necessary for deriving a meaningful conclusion. The reduction is very useful for building a practical expert system.

4.2 Fuzzy Fault Tree Analysis

Up to the present, many bridge failures have been reported, in which various kinds of failure causes were observed. Since bridge structures are large and complicated, it is not easy to identify which failure cause is dominant among the possible causes. Namely, it is difficult to conduct a full-scale experiment or collect satisfactory data to estimate the failure causes or trace the process leading to the failure accident. Difficulty also arises from the fact that the failure accident occurred due to the existence of so-called subjective failure causes such as political, social and economical pressures deriving various kinds of errors.

For the analysis of failure accidents, Fault Tree Analysis (FTA) is a powerful tool and actually has been applied to analyze the accidents of large-scale systems such as high-pressure vessels and plant facilities. In FTA, system success or failure can be evaluated by using a so-called structure function, and the probability of success is used as a measure of system reliability. However, this measure may be inconvenient for special problems, one of which is the analysis of bridge accidents. In this problem, our concern is mainly focused on the inter-relationships between every basic event (i. e., failure cause) and its contribution to the underlying bridge accident. It is difficult to judge even whether a basic event occurred or not through the inquiry, because of the complicated mechanism of bridge accident. Then, it may be convenient to define the occurrence of a basic event by a fuzzy set. Namely, the state of a basic event is characterized by a value belonging to $[0,1]$ (grade of membership) instead of $\{0,1\}$. The state of the top event (failure event) can be estimated by using fuzzy minimum and maximum operations.

In order to deal with such subjective failure causes, an attempt was made (Furuta 1994b) to apply fuzzy logic in the analysis of failure accidents of existing bridges.

Although FTA has succeeded in the various kinds of failure analysis, some special problems require an additional improvement on the description of the state of basic events and top event. For problems assessing bridge accidents, an FTA method was developed

based on fuzzy logic. The method has the following characteristics:
(1) The use of fuzzy minimum and maximum operations enables to simplify the calculation process of structure function.
(2) By using linguistic variables, it is possible to handle the ambiguities involved in the expression of the occurrence of a basic event. In addition, the state of a basic event can be described in a more flexible form, by using the concept of fuzzy logic.
(3) It is possible to discuss the comprehensive safety of bridge by introducing the fuzzy integral into the failure analysis. Using the fuzzy integral, it enables to judge the degree of inevitability of bridge accident in a clear manner. Furthermore, because the fuzzy measure is more flexible than probability measure, the uncertainty and ambiguity involved in the thinking process and way of human beings can be appropriately dealt with.
(4) Considering the comprehensive evaluation value as a weight, it is possible to assess the importance and influence of each failure cause on the bridge accident in a quantitative way.

5. SUMMARY AND DISCUSSIONS

In this paper, the outline of fuzzy logic and its contribution to structural reliability assessment were described and discussed. The probability theory is a useful tool, but it does not cover the whole problems in civil engineering. In the application of the probability theory, an event of interest is usually clearly defined. In other words, the application of the probability theory requires us to deal with events which are collections of outcomes of well-defined actions. It is realized that Bayesian statistics serves the purpose of extending the potential application of probabilistic methods. It is, however, noted that the Bayesian approach is not entirely free from the basic probability axioms, and that it requires subject judgments in the determination of apriori distribution and observation or testing data for updating.

Other statistical methods such as regression analysis, discrimination analysis, and multi-dimensional methods, may also be insufficient to cover all important problems. In any event, it is emphasized here that each method can become more powerful by using it combined with other available methods. The fuzzy logic is one such approach, with which more meaningful solutions may be obtained for complex problems in the state of nature. Real-world problems are usually more complex than their corresponding mathematical models. To compensate the gaps between them, some verbal explanation occasionally adds to the results obtained through the models. The concept of fuzzy sets has been developed to deal with the verbal information which is usually meaningful but not clearly defined.

In this paper, several applications of fuzzy logic in civil engineering are reviewed and discussed from the viewpoint of practical use. Nowadays, the term of "fuzzy" is very popular among Japanese. In Japan, many attempts have been made to apply the fuzzy logic to various engineering problems. In other words, we are in a boom of "fuzzy application". Needles to say, since such boom is easy to run away, we should make clear the essential property and the efficiency of the fuzzy logic.

It may be considered that the advantage of the fuzzy logic lies on the ability of approximation. Introducing the concept of approximation, it is possible to reduce the computation load and time. Furthermore, using the concept of the approximation, it is possible to obtain a practical solution for engineering problems. If the approximate solution is accepted to be satisfactory from practical points of view, it becomes easier to solve complex and large problems of our concern.

One can understand this fact by referring to the damage assessment and reliability evaluation of existing structures (Shiraishi and Furuta 1988). When a structure is damaged, its dominant failure modes may change, because the grade, location and cause of damage can greatly affect the system reliability of the structure. However, it is hardly estimated accurately due to the technical and financial constraints, and it may be better to accept a vague or ambiguous evaluation as a practical and meaningful basis for the reliability assessment of the damaged structure. For this purpose, fuzzy sets are useful, because they can represent the verbal expression of experienced engineers in a clear and informative manner.

It should be emphasized again that civil engineers are pursuing the superiority and efficiency of operation, and therefore they will be satisfied when the fuzzy logic can provide us of shortness and save of load and time, though the logical frame of the fuzzy logic is not clear. In fact, the two examples of tunneling machine controller and cable adjustment system have been applied for practical use and succeeded in making the construction time shorter and the operation simpler. Especially, the latter application, e. g., cable tension adjustment is noteworthy because it is only a practical application of fuzzy logic without using the fuzzy control.

Naturally, it is inevitable to recognize the limit of the fuzzy logic. Since we may not always obtain good results for all problems by the fuzzy logic, we should judge which problem is suitable for the application of fuzzy logic. Moreover, we should keep in improving the theoretical basis and proving its validity through practical investigations. Occasionally, the fuzzy logic has been argued on its arbitrariness, say how to determine the membership function, or how to derive the fuzzy relation. It is evident that there is no general answer for these kinds of question. Since appropriate membership functions or relations should be obtained through sufficient observation and testing. In case that observation and testing data can not be collected, knowledge and engineering judgment are important and practical bases for determining them. One example is the optimum design of earthquake resistant structures in which the validity of the membership functions defined by parabolic functions and the minimum operator in aggregating design constraints were examined and proved through the analysis of the questionnaires distributed to experienced engineers. In the fuzzy control for electric appliances or machine operation, the automatic tuning technique has been utilized so as to get better efficiency, based on such new technologies as neural network, chaos and genetic algorithms. The civil engineering problems require the toughness or robustness of solution, because it includes various uncertainties and ambiguities inherently. Therefore, combining these new technologies, the fuzzy logic is expected to provide us of more flexible way of thinking to seek practical and useful solutions.

ACKNOWLEDGMENTS

The author would like to thank Prof. Naruhito Shiraishi of Kyoto University and Prof. James. T. P. Yao of Texas A&M University for their valuable advice and continuing encouragement to compile this paper.

REFERENCES

Blockley, D.(1975).Predicting the likelihood of structural accidents, Proc. ICE, 59, 659-668.

Blockley, D.(1977).Analysis of structural failures, Proc. ICE, 62, 51-74.

Blockley, D.(1980).The Nature of Structural Design and Safety, Ellis Horwood.

Brown, C.(1979).A fuzzy safety measure, ASCE, J Eng. Mech. Div., 105.

Brown, C. and Leonards, R.(1971).Subjective uncertainty analysis, Preprint No.1388, ASCE National

Structural Meeting, Baltimore.
Brown, C. and Yao, J. eds.(1985).NSF Workshop on Civil Engineering Applications of Fuzzy Sets, W. Lafayette.
Chameau, J. et. al.(1983).Potential applications of fuzzy sets in civil engineering, Int. J. of Man-Machine Stud., 19.
Feng, G. and Liu, X. eds.(1985).Proc. of 1st Int. Sympo. on Fuzzy Mathematics in Earthquake Researches, Beijing.
Furuta, H.(1993).Comprehensive analysis for structural damage based upon fuzzy sets theory, J. of Intelligent and Fuzzy Systems, 1, 55-61.
Furuta, H.(1994a). Application of fuzzy logic in civil engineering in Japan, Proc. of 1st Congress on Computing in Civil Engineering, Washington D.C., 2, 1900-1915.
Furuta, H.(1994b).Analysis of bridge failures using FTA, Proc. of Symposium on Risk Analysis, Ann Arbor.
Furuta, H. et. al.(1983).Optimum aseismic design using fuzzy mathematical programming, IFIP Conference on Computer-Aided Design, Lyon.
Furuta, H. et. al.(1991a).Knowledge-based expert system for damage assessment based on fuzzy reasoning, Computers and Structures, 40, 137-142.
Furuta, H. et. al.(1991b).Seismic reliability analysis of bridge piers using fuzzy probability, Proc. of ICASP-5, Mexico, 2, 896-903.
Furuta, H. et. al.(1993).A fuzzy neural system for repairing bridge decks, Proc. of IABSE Colloquium, Beijing.
Furuta, H. and Shiraishi, N.(1984).Fuzzy importance in fault tree analysis, Fuzzy Sets and Systems, 12, 205-213.
Furuta, H. and Shiraishi, N.(1987).Bridge design planning based on fuzzy multi-attribute analysis, Preprints of Second IFSA Congress, 1, 337-340160-163.
Ishizuka, M., Fu, K. & Yao, J.(1981).Inference procedure with uncertainty for problem reduction method, Tech. Rep. CE-STR-81-24, Purdue Univ.
Kandel, A.(1991).Fuzzy Expert Systems, CRC Press.
Kaneyoshi et. al.(1990).Optimum cable tension adjustment using fuzzy regression analysis, Proc. 3rd IFIP WG7.5 Conference, Berkeley.
Klir, G. and Folger, T.(1988).Fuzzy Sets, Uncertainty, and Information, Prentice-Hall.
Shafer, G.(1976).A Mathematical Theory of Evidence, Princeton University Press.
Shiraishi, N., Furuta, H. & Kawamura, Y(1982).Application of fuzzy set theory to the design of RC beam, Theoretical Applied Mechanics, 31, 173-179.
Shiraishi, N. and Furuta, H.(1983).Reliability analysis based on fuzzy probability, ASCE, J Eng. Mech. Div., 109.
Shiraishi, N., Furuta, H. & Sugimoto, M.(1985). Integrity assessment of bridge structures based on extended multi-criteria analysis, Proc. of ICOSSAR, 1, 505-509.
Shiraishi, N., Furuta, H., Umano, M. & Kawakami, K.(1987). An expert system for damage assessment of reinforced concrete bridge deck, Preprints of Second IFSA Congress, 1, 160-163.
Shiraishi, N. and Furuta, H.(1988).System reliability analysis of damaged structures, in D. Frangopol ed., New Directions in Structural System Reliability.
Shiraishi, N., Furuta, H. & Ozaki, Y.(1988). Application of fuzzy set theory to fatigue analysis of bridge structures, Information Sciences, 45, 175-184.
Sugeno, M.(1972).Fuzzy measure and fuzzy integral, SICE, Japan, 8, 218-226. (in Japanese)
Sugeno, M.(1985).Industrial Applications of Fuzzy Control, North-Holland.
Yao, J.T.P.(1980).Damage assessment of existing structures, ASCE, J Eng. Mech. Div., 106.
Yao, J., Bresler, B. & Hanson, J.(1984).Condition evaluation and interpretation for existing concrete buildings, Annual Convention of ACI, Phoenix.
Yao, J. and Furuta, H.(1986).Probabilistic treatment of fuzzy events in civil engineering, Prob. Eng. Mech., 1, 1, 58-64.
Zadeh, L. A.(1965). Fuzzy sets, Information and Control, 8, 338-353.
Zadeh, L. A.(1975). The concept of linguistic variable and its application to approximate reasoning - part 2, Information Science, 8, 43- 80.

PART TWO

Technical Contributions

5

Experience from the application of Reliability Fatigue Crack Growth Analyses on Real Life Offshore Platform: Parametric Study and Sensitivity Analysis

Alberto ABATE [a] and **Massimiliano ERRIGO** [b]

[a] Systems and Standards Department, Agip S.p.A.
via Emilia 1, 20097 S. Donato Milanese (Milano), Italy
Structural Engineering Department, Politecnico di Milano
Piazza Leonardo da Vinci 32, 20133 Milano, Italy

[b] Offshore Engineering Department, Agip S.p.A.
via Emilia 1, 20097 S. Donato Milanese (Milano), Italy

1. INTRODUCTION

In the present study a requalification scenario of an existing offshore steel jacket platform (fig.1) has been investigated (ref./1/). The requalification process of an offshore structure brings together structural and reliability analyses, inspections data interpretation and rational inspections planning over the operating life; it is a key aspect in the management of a platform for several reasons:

- many platforms have reached their design service life;
- some of them are still temporarily manned;
- structures installed before 1970 have not been designed taking into account fatigue phenomena;
- the employed reliability analyses allow explicit consideration of specific damage conditions observed through inspection;
- it is possible to optimise inspection costs, frequencies and acceptable risk level.

The main goal of this contribution is to show some practical aspects affecting the results of fatigue reliability crack growth analyses on a typical Adriatic sea platform. The focus is on the selection of the suitable characteristics of the basic variables employed in the limit state function and on parametric and sensitivity analysis. In this way it is possible to identify the main sources of uncertainties on which it would be appropriate to concentrate efforts to increase platform reliability. The fracture mechanics approach in a probabilistic formulation offers a basis to develop a rational treatment of such uncertainties.

2. LIMIT STATE FUNCTION MODELLING

The study of the fatigue problem according to the fracture mechanics approach is based on the Paris-Erdogan law (ref./2/). It relates the crack depth increment Δa, during one load cycle, to the stress intensity factor range Δk, in the same load cycle. Since the crack increment for a load cycle is generally very little if compared to the crack dimension itself, the Paris-Erdogan law can be formulated in an incremental way:

$$\frac{da}{dN} = C \cdot (\Delta k)^m \tag{1}$$

where dN is the increment of stress cycles number, whereas C and m are material constants.

The stress intensity factor range Δk is expressed by the following formulation:

$$\Delta k = \Delta\sigma \cdot Y(a) \cdot \sqrt{\pi \cdot a} \tag{2}$$

where Δσ is the far-field stress range, whereas $Y(a)$ is the geometry function and depends on the crack depth a.

The failure criteria is based on the fact that the crack depth a_T, propagated in the time range $(T\text{-}T_0)$, exceeds the critical crack dimension a_C, assumed to be equal to the element thickness. This criteria can be formulated as follows:

$$a_C - a_T \leq 0 \tag{3}$$

Introducing the functions $R(a)$, "structural resistance", and $L(a)$, "load term", and applying the failure criteria of equation (3), the safety margin can be defined as follows (ref./3/):

$$M = R(\alpha) - L(T) \tag{4}$$

$$M = \int_{a_0}^{a_C} \frac{da}{Y(a)^m \cdot \left(\sqrt{\pi \cdot a}\right)^m} - \exp(\ln C) \cdot \frac{T - T_0}{20} \cdot \sum_{i=1}^{4} N_i \cdot A_i{}^m \cdot \Gamma\left(1 + \frac{m}{B_i}\right) = 0 \tag{5}$$

In the above equation, the contribution of four wave direction has been considered in the load term.

3. RELIABILITY UPDATING BASED ON INSPECTION RESULTS

Platforms in service are inspected to detect cracks before they become critical. A crack can be found at the time T_i and its length measured:

$$a(T_i) = a_{m_i} \tag{6}$$

where a_{m_i} is generally random due to measurement uncertainties. Then the safety margin can be defined as:

$$M_i = \int_{a_0}^{a_{m_i}} \frac{da}{Y(a)^m \cdot \left(\sqrt{\pi \cdot a}\right)^m} - \exp(\ln C) \cdot \frac{T - T_0}{20} \cdot \sum_{i=1}^{4} N_i \cdot A_i{}^m \cdot \Gamma\left(1 + \frac{m}{B_i}\right) = 0 \tag{7}$$

A second type of inspection result is that no crack is detected, thus:

$$a(T_i) \leq a_d \tag{8}$$

where a_d is the smallest detectable crack length, depending on the inspection method used.

Similarly:

$$M = \int_{a_0}^{a_d} \frac{da}{Y(a)^m \cdot \left(\sqrt{\pi \cdot a}\right)^m} - \exp(\ln C) \cdot \frac{T - T_0}{20} \cdot \sum_{i=1}^{4} N_i \cdot A_i^m \cdot \Gamma\left(1 + \frac{m}{B_i}\right) \geq 0 \tag{9}$$

4. VARIABLES MODELLING

In the limit state equation (5) the employed variables can be modelled as random or deterministic. The considered model for each of them are reported below and summarised in table 1.

4.1. Initial micro-crack a_0

The variable a_0 has been assumed to be exponentially distributed and its cumulative distribution function has the following expression, ref /4/ and /5/:

$$F_{A_0}(a_0) = 1 - \exp\left[-\lambda \cdot (a_0 - \bar{a})\right] \tag{10}$$

in which a mean value of 0.11 mm has been considered.

4.2. Critical crack depth a_C and thickness t

The Gaussian stochastic model for both the variables a_C and t has been assumed. The inspection results carried out on the platform, have been used to estimate mean value and standard deviation of these variables.

4.3. Initial time T_0 and final time T

The variables T_0 and T are deterministic; in a conservative way it has been considered $T_0=0$, that means to have the crack beginning when the platform was installed (1971). The T value depends on the date on which it is desired to have the reliability results during the parametric study execution.

4.4. Wave numbers N

In equation (5) N_i represents the wave number occurring in the i direction expected in the time range ($T-T_0$); it is considered a deterministic variable.

4.5. Material constants C and m

In order to model the material constants, the results of several studies reported in ref./4/, /6/, /7/ and /8/, have been used. In the mentioned references it seems to be accepted the use of normal distribution for both the variables lnC and m, with the mean values equal to -29.75 and 3.0 respectively, and the corresponding variances 0.5 and 0.09. The numerous experimental data have proved a correlation coefficient ρ equal to -0.9.

4.6. Weibull distribution parameters A and B

To solve equation (5) it is necessary to determine the values of the two Weibull scale and shape parameters A and B that define the cumulative distribution function of the stress range S. It has the following formulation:

$$F_S(S) = 1 - \exp\left[-\left(\frac{S}{A}\right)^B\right] \tag{11}$$

Instead of averaging the effects of the four wave directions considered to obtain only one curve as usually done, four "long term stress range distribution" curves per analysed connection have been considered; each i curve represents the probability that a S value is exceeded for that i wave direction.

In ref./4/, /6/, /7/ and /8/ it is described a suitable stochastic representation of the scale and shape parameters of the Weibull distribution: for the parameters A_i, a lognormal distribution has been assumed and a coefficient of variation equal to 10% has been considered; similarly, for the parameters B_i a lognormal distribution has been used. This choice is justified by the observation that, since B ranges between 0.4 and 0.7 for the Adriatic Sea, using the lognormal distribution it is taken into account that B can not physically assume negative values. A COV of 10% has been used.

4.7. Geometry function Y(*a*)

The geometry function Y(*a*) formulation has been extracted from ref./6/, confirmed also by ref./4/ due to its suitable adhesion to the problem physics reality:

$$Y_{eff,average}(a) = Y_{unw}(a) \cdot M_k(a) \tag{12}$$

where Y_{unw} represents the geometry function that takes into account the assumed semi-elliptical shape of the crack, neglecting the welding effect ("unwelded"), while $M_k(a)$ is a correction factor that considers the influence of the welding in the stress concentration estimate. They have the following expressions, ref./9/ and /10/:

$$Y_{unw}(a) = \left[1.08 - 0.7 \cdot \left(\frac{a}{t}\right)\right] \tag{13}$$

$$M_k(a) = 1.0 + 1.24 \cdot \exp\left[-22.1 \cdot \left(\frac{a}{t}\right)\right] - 3.17 \cdot \exp\left[-357 \cdot \left(\frac{a}{t}\right)\right] \tag{14}$$

To consider the uncertainty related to the above formulation of the $Y_{eff,average}(a)$, a stochastic variable Y_1 has been introduced (ref./8/). Therefore, the final modelling of the geometry function Y(*a*) is the following:

$$Y(a) = Y_1 \cdot Y_{eff,average}(a) \tag{15}$$

Concerning Y_1 a lognormal distribution with unitary mean value and COV equal to 0.1 has been adopted.

4.8 Minimum detectable crack depth a_d

The subsea inspections can not cover all the joints of the platform. Furthermore, due to the uncertainty within the adopted inspection method, a probability exists that, at the time of the inspection T_i, a crack of length 2*c* (or depth *a*) is not detected.

Therefore, the Probability Of Detection curve, POD, assumes a considerable importance and it is obviously a function of the adopted detection method. The inspections are mainly performed using the "Magnetic Particle Inspection" (MPI). Due to the good accuracy of the execution of the inspections, it is possible to use the POD curve proposed by a Brite study mentioned in ref./8/, in which the cumulative distribution function of the crack detection has the following expression:

$$POD(2c) = P_0 \cdot \{1 - \exp[-\beta \cdot (2c - a)]\} \tag{16}$$

where the POD curve is a function of the superficial crack length $2c$, P_0 has a value very close to unity and represents the probability of detection of a long crack, β is the growing speed of the curve, 0.04 mm^{-1} (ref./8/), while α is the minimum detectable crack length, 6.67 mm (ref./11/). A comparison with different formulations from ref. /7/ and /8/ has been made, in which there is the same hypothesis $\frac{a}{2c}$ as constant ratio (semi-elliptical shape crack).

Taking into account some experimental results of offshore tubular joints (ref./8/), it has been possible to assess a more appropriate formulation of the POD curve in which the ratio $\frac{a}{c}$ is deeply conditioned by the relative crack depth $\frac{a}{t}$, see figure 2; in particular, $\frac{a}{c}$ decreases when the ratio $\frac{a}{t}$ grows. Analysing these experimental results, it can be noted that in the initial phase, the crack has nearly a circular shape and only during its development becomes elliptical.

The POD curve proposed in this study is shown in figure 3 together with the other formulations.

5. ANALYSIS OF RESULTS

A summary of the reliability fatigue analysis results and their discussion and interpretation are in the following. It must be pointed out that the reliability index β has been calculated using the second order approximation method (SORM).

In figure 4 the β index behaviour for a K joint is shown. The values have been calculated by the program PROBAN (PROBabilistic ANalysis, Det Norske Veritas Sesam) every six months. Analysing the results, it can be noted that after a few years of service β is below the range of the allowable values. These last ones have been evaluated in accordance with ref./11/, in the case of manned platform (lower line) or not one (upper line).

From these figures it stands out the big increment of the β index after the inspection performed in 1987. This increment, due to the absence of cracks, keeps the values of β above the limit even after the second considered inspection in 1991. It is evident that using the fracture mechanics approach in the reliability fatigue analysis it is possible to get an increment of the safety of the structure due to the possibility to take into consideration added information on the structure state as the time varying.

Figure 5 shows graphically the importance factors of all the variables after 22 years of service. Also the relative importance factors of the B_i parameters has been calculated in order to investigate the relative influence of the single wave direction in the considered joints (fig.6). After this evidence, it is possible to consider that, due to the different location that each joint has in the structure, it is reasonably important to consider separately the single wave direction contributes in the limit state function.

A sensitivity analysis to study the effects of the parameters on the β index, has been developed. A new reliability analysis has been performed every modification of the values of the random variables (considered as the parameter to be changed). Table 2 summarises the different variations in the sensitivity analysis carried out.

Examining figure 7, relevant to a very critical joint, it is evident that the random variables Y_1, lnC (together with m), a_C and t are of very low importance for the original reliability index behaviour vs time. On the opposite, it is more evident the behaviour difference in the case where the parameters A_i and B_i COV have been changed, as shown in figure 8.

Therefore, for a periodic structure monitoring, the determination of these two measures can be very useful for a qualitative assessment of the next inspection survey.

Moreover, due to the possibility, allowed by this analysis methodology, to take into account the information coming from the inspection survey, the structure service life can be extended significantly; furthermore in order to design an inspection planning, the results of this contribution have shown, if no cracks are found, the real possibility to inspect with reasonably increased time ranges and consequently to get inspection costs reduced up to 35 %.

REFERENCES

/1/ A. Abate - L. Losapio (1993), "Aspetti affidabilistici in ambito offshore", Politecnico di Milano, Italy

/2/ P. Paris - F. Erdogan (1963), "A critical analysis of crack propagation laws", J. Basic Engineering Trans. ASME

/3/ A. Almar - Naess, ed. 1985, "Fatigue Handbook for offshore steel structures", Tapir Publishers, Trondheim, Norway

/4/ A. Karamchadani - J.I. Dalane - P. Bjerager (1992), "Systems Reliability Approach To Fatigue of Structures", Journal of Structural Engineering, Vol. 118, No. 3, Paper n. 487, ASCE

/5/ A.H.S. Ang - W.H. Tang (1975 - 1984), "Probability Concepts in Engineering Planning and Design", Voll. 1 e 2, Johhn Wiley & Sons, Toronto

/6/ F. Kirkemo (1988), "Applications of probabilistic fracture mechanics to offshore structures", Appl. Mech. Rev.

/7/ R. Skjong - R. Torhaug (1991), "Rational Methods for Fatigue Design and Inspection Planning of Offshore Structures", Marine Structures Vol. 4, Elsevier Science Publishers, England

/8/ Rapporto D' Appolonia No. 89 - 302 (1990), "Analisi Probabilistica Propagazione Cricca", Vol. 1, Genova

/9/ I.S. Raju - J.C. Newman (1981), "An Empirical stress intensity factor equation for surface crack", Engineering Fracture Mechanics

/10/ I.J. Smith - S.J. Hurworth (1984), "The effect of geometry changes upon the predicted fatigue strength of welded joints", Res. Report 244, Welding Inst.

/11/ Det Norske Veritas (1992), "Structural reliability analysis of Marine Structures", Classification Notes n° 30.6 DNV Classification As, Hovik, Norway

VARIABLE	MODELLING	MEAN VALUE	SCATTER
a_0	EXPONENTIAL	0.11 mm	lower bound 0.0
a_C	NORMAL	10.07 mm	COV = 0.03
T_0	DETERMINISTIC	0 anni	-
T	DETERMINISTIC	1 + 29 YEARS	-
N_1	DETERMINISTIC	38620557	-
N_2	DETERMINISTIC	36381752	-
N_3	DETERMINISTIC	42514976	-
N_4	DETERMINISTIC	36410771	-
lnC	NORMAL	- 29.75	σ = 0.5
m	NORMAL	3.0	σ = 0.09
A_i	LOGNORMAL	DEPENDS ON THE CONNECTION	COV = 0.1
B_i	LOGNORMAL	DEPENDS ON THE CONNECTION	COV = 0.1
Y_1	LOGNORMAL	1.0	COV = 0.1
t	NORMAL	10.07 mm	COV = 0.03
a_d	ASSIGNED	BY	POINTS

TAB. 1

FIG. 1

PARAMETER	MEAN VALUE	COV
Y_1	UNCHANGED	0.0
lnC - m (without correlation)	UNCHANGED	0.0
a_C - t	UNCHANGED	0.0
A_i - B_i	UNCHANGED	0.075
A_i - B_i	UNCHANGED	0.05

TAB. 2

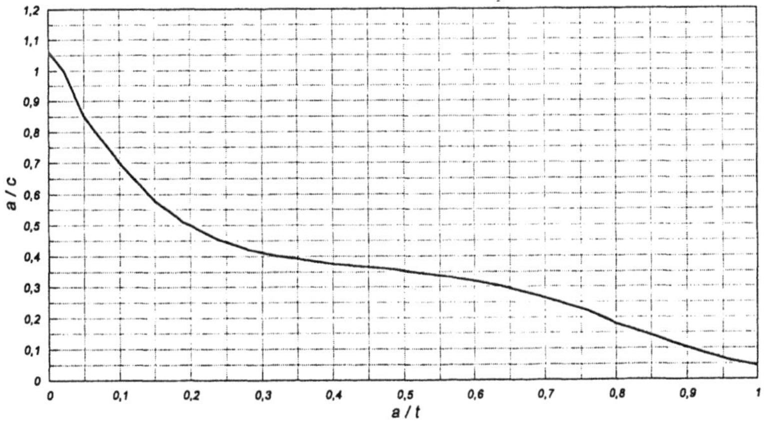

FIG. 2

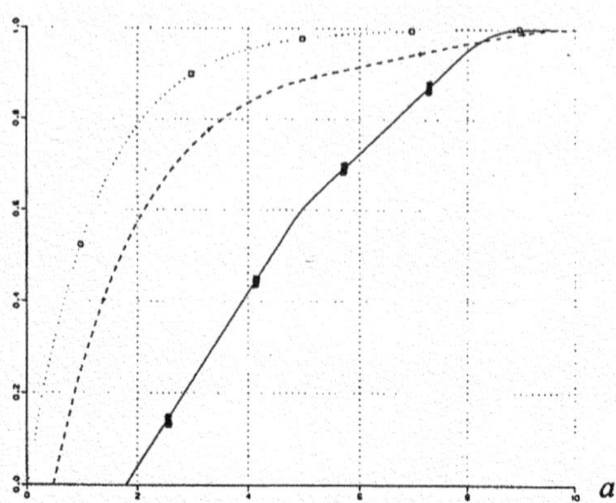

FIG. 3: ▮ POD PROPOSED CURVE

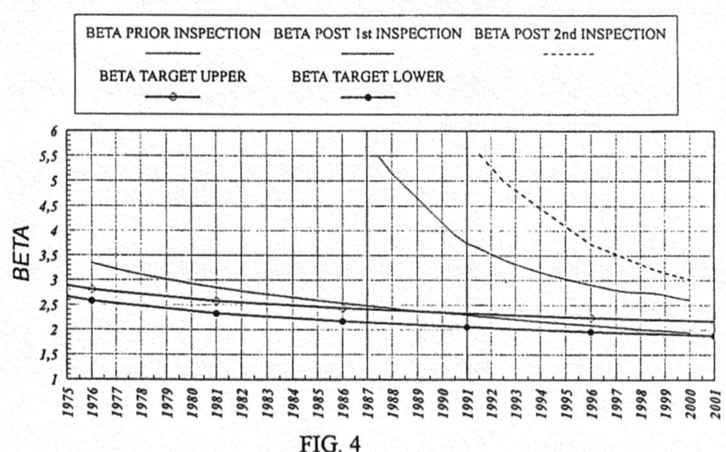

FIG. 4

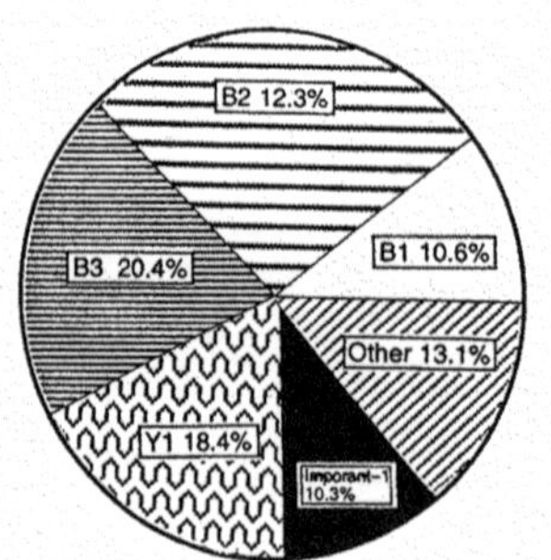

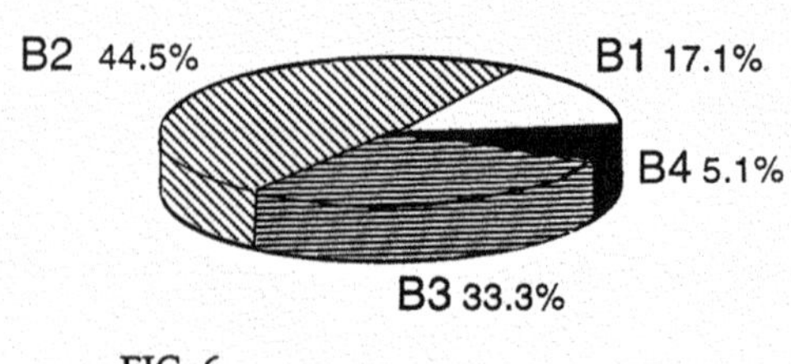

FIG. 6

BETA ORIGINAL
BETA (Y1 not random)
BETA (LnC-m not random)
BETA TARGET UPPER
BETA TARGET LOWER
BETA (ac-th not random)
BETA
3
2,5
2
1,5
1
0,5
0
1975
1976
1977
1978
1979
1980
1981
1982
1983
1984
1985
1986
1987
1988
1989
1990
1991
1992
1993
1994
1995
1996
1997
1998
1999
2000
2001

FIG. 7

BETA ORIGINAL
BETA TARGET UPPER
BETA TARGET LOWER
BETA (COV=0.075 FOR A,B)
BETA (COV=0.05 FOR A,B)
BETA
4
3,5
3
2,5
2
1,5
1
0,5
0
1975
1976
1977
1978
1979
1980
1981
1982
1983
1984
1985
1986
1987
1988
1989
1990
1991
1992
1993
1994
1995
1996
1997
1998
1999
2000
2001

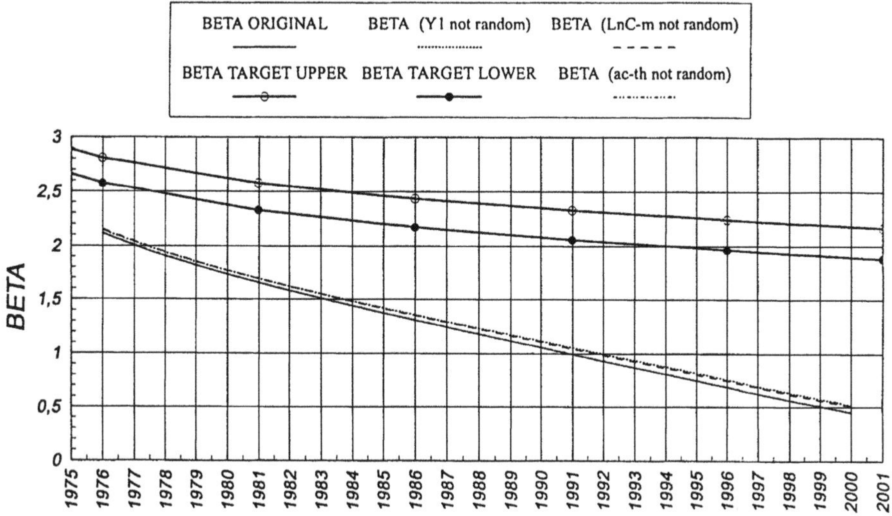

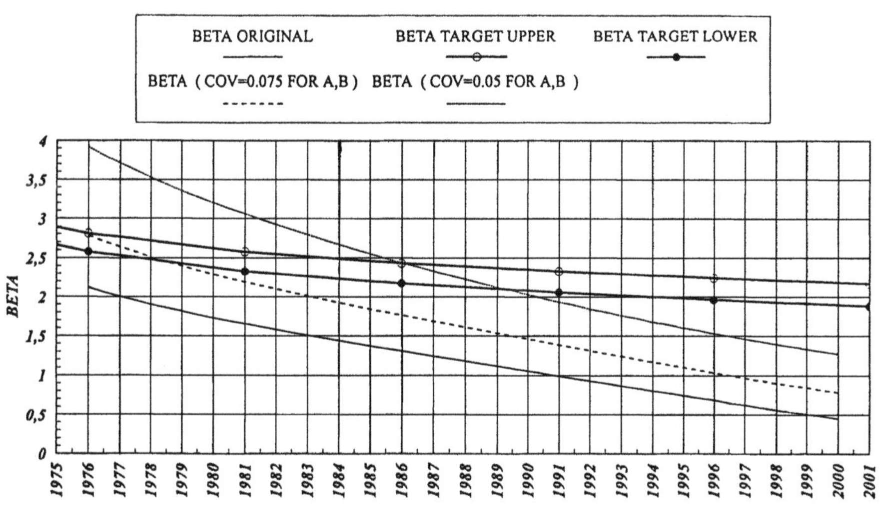

FIG. 8

6

THE USE OF A F. E. CODE FOR MULTICRITERIA RELIABILITY BASED OPTIMIZATION OF COMPOSITE MATERIAL STRUCTURES

A. Borri and E. Speranzini
Energetic Inst. – Structural Section, University of Perugia,
Str. S. Lucia Canetola, 06125 Perugia, Italy.

ABSTRACT

A multicriteria reliability based optimization procedure for laminated composite material structures is presented. In order to solve the optimization problem, a multiobjective optimization criterion was applied which is capable of considering antagonist and non-comparable objective functions. By introducing parameters related to the reliability of the structures, the probability of failure with regard to a predefined limit state was considered among the objectives to optimize. The procedure has been implemented and coupled to a widely diffuse Finite Element code. This code is used both to solve the structural problem and to perform the entire reliability based optimization process by means of the minimization technique contained in the code itself.

As an example, a laminated composite shell in which the objective functions considered represent structural and economic performances is analyzed.

1. RELIABILITY BASED OPTIMAL DESIGN

In the case of laminated composite material structures, it is particularly interesting to perform an analysis that takes into account both structural optimization and structural reliability.

In these stuctures, the process of selecting the optimum solution is highly complex involving qualitative and quantitative factors at the same time. In fact, the structural optimization problem usually presents the difficulty of needing to consider objective functions which are non-comparable and cannot be combined to construct a single objective function. Furthermore, for laminated composite material structures the design variables are numerous so that multicriteria optimization may be an important tool for design.

On the other hand, some aspects of the problem regarding these structures are not deterministic, due also to the peculiar nature of the laminated material, and consequently the optimum design must deal with some uncertainties. In particular, considering random design parameters, it would be interesting to analyse the sensitivity of the optimal solutions with regard to random aspects.

If one introduces structural reliability as a further objective function (for example in terms of the probability of overcoming a predefined limit state), the problem can be posed as a "multicriteria reliability-based optimization" in which the objectives and all performance constraints are expressed as functions of the considered design variables.

In this paper, a procedure that considers the multicriteria optimization problem for composite material structures and the reliability at the same time is presented.

In order to solve the optimization problem, the multi-objective optimization criterion has been applied, which is capable of describing multi-objective function problems where the functions can be not only antagonist but also neither comparable nor commensurable.

As shown in previous papers [1-2], the multi-objective optimization criterion can be based on the concept of Pareto's optimum [3]. In this case, in which the objectives represent structural and economic performances that are non-comparable objective functions, the Pareto optimization problem was considered using the Trade-off method [4]. This method, through changing the vectorial problem into a scalar one, transforms the original problem into a series of simpler minimization problems with only one objective function that conveniently shuts out the others. The result of this optimization process is a set of "Pareto" optimal points, which are optimal in the sense that none of the objectives can be further improved without making at least one of the others worse. Therefore, infinite optimal solutions are possible among which the designer may choose the most applicable according to his necessities, and often on the basis of the indications given by the curve itself.

To introduce in the optimization process some aspects related to structural reliability, one or more objective functions that consider these aspects can be assumed.

In order to perform this process, a procedure has been implemented and added to a widely diffuse Finite Element code (ANSYS [5]) which is normally used in the deterministic field because the code in its standard form doesn't permit the use of random parameters.

In this work the optimization routine has been appropriately employed in order to optimize parameters related to structural reliability. To this aim, the procedure, having transformed the basic random variables into a space of standardized normally distributed variables, determines the minimum distance from the origin to the non-linear failure surface in the set of normalized variables using the minimization technique of the Finite Element code. In this way the code is used both to solve the structural problem and to perform the global optimization process.

2. THE PROCEDURE

The proposed procedure is a point-by-point numerical process able to consider antagonist objective functions and random variables.

To determine the Pareto optimal solutions the methodology described in previous works [1-2] is used. This adopts the Trade-off Method that transforms the vector optimization problem into a scalar substitute one in which only one objective function f_i (preference function or substitute objective function) is minimized while all the others f_j ($j \neq i$) are kept bounded. By solving this minimization problem, a Pareto point is determined; by varying the values of the f_j functions, other Pareto points can be calculated. Each point thus determined is Pareto optimum i.e. there is no other point in which the value of at least one objective function can be reduced without increasing the values of the others.

Since P_{fail} (here defined as the probability of overcoming of a prefixed limit state) has been considered among the objective functions, two different problems have to be tackled:

- evaluating P_{fail} when the value of each design parameter is fixed;
- individuating the values of the design parameters that minimize P_{fail}.

To solve the first problem, the minimum distance from the origin of the space of standardized independent normally distributed variables to a point on the failure surface is determined, having previously defined the failure condition. This minimization problem cannot be solved in closed-form because the analytical expression of the failure surface is not known (only the failure condition is known). However the minimization of the "distance" function between the origin and the points satisfying the failure condition can be numerically performed by the Ansys code. This distance permits to determine $P_{fail} = \Phi(-\beta)$. In any case, the value of this distance can be representative of P_{fail} in the sense that a greater distance corresponds to a smaller P_{fail} and viceversa.

For the second problem, it is sufficient to define the range of the design variables and use the Ansys optimization routine. This computes the point in the space of design variables to which corresponds the minimum value of P_{fail} calculated as above described.

In the following, the algorithm is illustrated in the particular case of two objective functions (where f_1 represents P_{fail} and f_2 is an antagonist function):

1. input parameters are defined;
2. a value for the objective function f_2 is fixed;
3. starting values for the design variables are chosen;
4. the transformation of the basic variables into uncorrelated and standardized normally distributed variables is performed;
5. the probability content in the failure set in standardized normally distributed variable space is evaluated: the optimization routine of the Finite Element code

computes the minimum distance ß from the origin to a point on the failure surface as well as the corresponding $P_{fail} = \Phi(-\beta)$;

6. after having changed the values of the variables, steps 4 and 5 are repeated until the minimum value of P_{fail} is obtained using the optimization routine of the code;
7. the point of P_{fail} and f_2 co-ordinates represents a Pareto optimal point.

The external loop of the algorithm (from step 2 to step 7) is repeated to determine the other Pareto optimal points until the entire range of f_2 is covered; the internal loop (from step 4 to step 6) permits to minimize the function f_1 by varying the variable set.

The Pareto optimal set is therefore determined, the points of which are the optimal solution of the analysed problem.

The procedure is general and can also be applied in the case of basic variables that are not normally distributed and not mutually independent by performing the appropriate transformations.

3. NUMERICAL EXAMPLE

The procedure has been applied to analyse the laminated composite material structure of fig. 1, representing a rectangular shell with a circular hole loaded by uniform pressure.

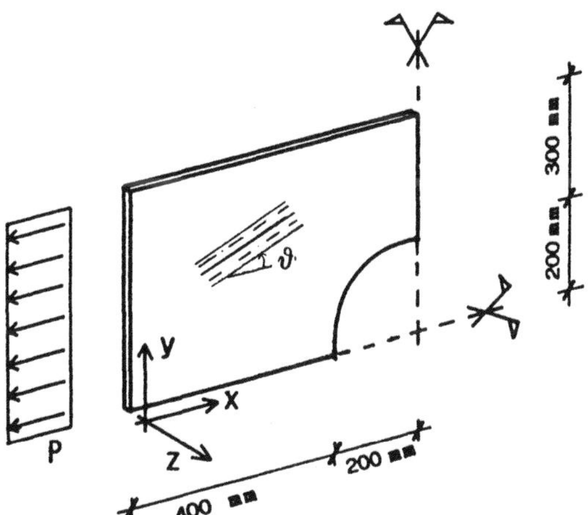

Fig. 1 - Shell composite structure (load : 10^3 N/mm)

The material is a symmetrically cross-ply laminated composite that comprises of five orthotropic stacked layers having the sequence $\theta/90°+\theta/\theta/90°+\theta/\theta$; it consists of E-glass fiber and epoxy resin.

Regarding the composite material, the following basic assumptions have been made [6, 7]:

- the bonds between fibers and matrix are perfect;
- the fibers are regularly spaced and perfectly aligned with each other;
- the material of fiber and matrix is homogeneous, linearly elastic and isotropic;
- the constitutive laws of the fiber and the matrix considered are represented by the stress-strain curves of fig. 2.

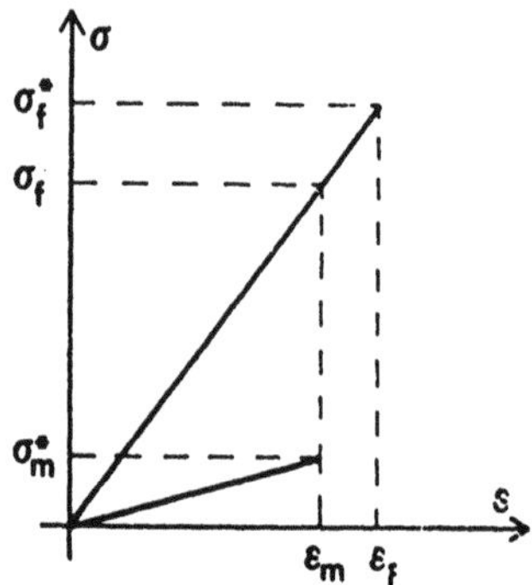

Fig. 2 - Constitutive laws for matrix ad fiber in the case of uniaxial tensile strength

The Pareto optimal solutions are determined by considering two objective functions: the probability of failure with regard to a limit state represented by the overcoming of the critical stress of the composite (using the maximum stress failure criteria), and the volume fraction of the fiber V_F defined as the ratio between the volume of the fibers and the total volume of the composite. The problem is complex due to the characteristics of the material and to the fact that the functions P_{fail} and V_f (which represent the structural performance and the economic aspect respectively) are antagonist and cannot be combined in order to construct a single objective function.

Once V_f is chosen as an objective function, the mechanical characteristics such as the elastic moduli, Poisson's coefficients and the critical stresses have to be determined at each loop of the procedure in terms of the properties of the fiber and the matrix and in terms of the relative fraction of their volume.

Being the considered composite material made up of E-glass fiber and epoxy resin and assuming V_F in the range 0.2-0.6 (appropriate for most commercial laminates) the following failure strength in the composite (failure value with regard to evaluate P_{fail}) have been assumed [7-8]:

$$\sigma_{LT} = \sigma^*_f V_f \qquad \text{longitudinal tensile strength}$$

$$\sigma_{TT} = \sigma^*_m (1-2(V_f/\pi)^{1/2}) \qquad \text{transverse " "} \tag{1}$$

$$\sigma_C = G_m/(1-V_f) \qquad \text{compression strength}$$

where σ^*_f and σ^*_m are the ultimate strength of fiber and matrix respectively and G_m is the shear modulus of matrix. Failure for compressive strength in the fiber direction occurs when the fiber buckling for shear-mode is reached. At high V_f (0.2-0.6) the adjacent fibers buckle in the same wavelength and in phase with one another so that the matrix is subject to shearing deformation; this indicates that σ_C increases with V_f but is controlled by the shear modulus of the matrix. Furthermore for this example it is reasonable to assume a transverse compressive strength comparable with the longitudinal compressive strength [6]. The stresses of eqs. (1) are the failure stresses as regards to which P_{fail} is evaluated.

By taking into account the operative difficulties in the construction of the material, in this example, V_f is considered to vary in a random fashion: consequently the composite strength (eqs. (1)) is also a random variable. Moreover, these aspects directly - and considerably at times - influence not only the critical stresses, but also the elastic moduli and Poisson's coefficients so that in the procedure they have to be adjusted in each step where the V_f varies. V_f is assumed to be normally distributed: the mean value is chosen in the range 0.2-0.6 and the standard deviation is chosen as a fixed percentage of its mean value. So the mean value of the composite strength is the one which corresponds to the assumed value of V_f and the standard deviation is the fixed percentage of this mean value. In order to minimize P_{fail} the mean value of the normal distribution of the fiber direction θ is assumed to vary in the 0°-45° range while the standard deviation is fixed equal to 2°.

Each point of Pareto's curve has been calculated by fixing a value of V_f and then computing the minimum of P_{fail} by means of the procedure presented. It can be noted that, once V_f is fixed , P_{fail} is a function of the random variable θ. For each distribution of θ the procedure computes P_{fail} and than chooses the minimum among the values determined.

Fig. 3 shows the resulting Pareto optimal curve (P_{fail} versus V_f) computed by considering standard deviation of V_f equal to a 20% of its mean value. The points of the curve are all equally optimal, i.e. optimal trade-off points between the structural and economical aspects. The shape of the curve indicates that some regions have greater interest than others, and gives useful quantitative information concerning the choice of a compromise solution. It can be seen, for example, that a choice of V_f in the range 0.3-0.4 can have a great practical interest because considerably small values of P_{fail} can be obtained with a relatively small value of V_f.

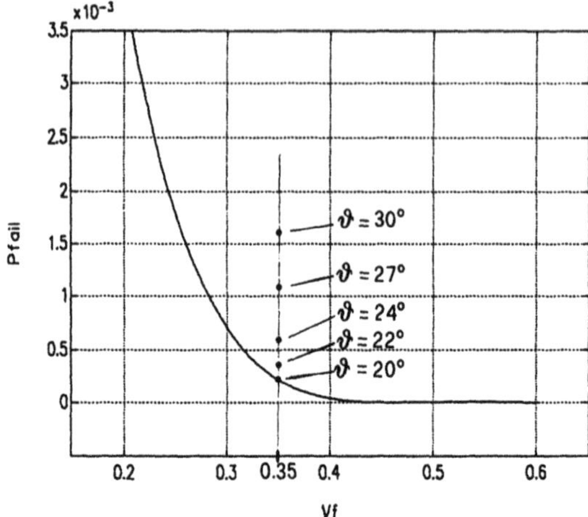

Fig. 3 - Pareto optimal curve

Furthermore, fig. 3 shows that a different value of P_{fail} corresponds to each mean value of θ. In the diagram are shown for example the different values assumed by P_{fail} when θ varies, all corresponding to $V_f = 0.35$. The point on the Pareto curve is that where the minimum values of P_{fail} is obtained, and corresponds to $\theta = 20°$.

In fig. 4 Pareto curves corresponding to different values of the standard deviation of the V_f probability distribution are plotted. It can be seen that the optimal value of the failure probability increases considerably when the standard

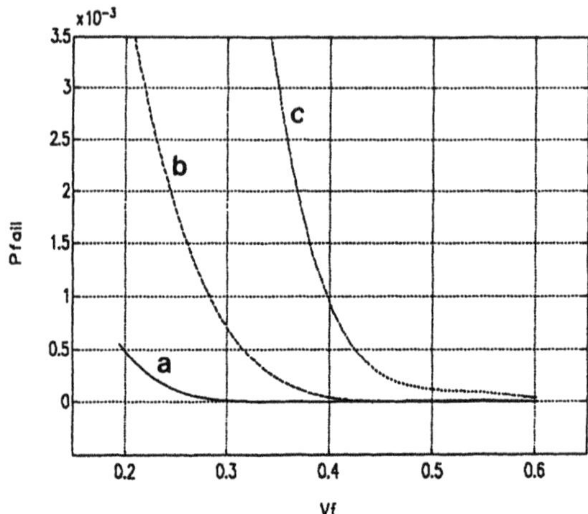

Fig. 4 - Pareto optimal curves (a,b,c curves correspond to V_f standard deviation equal to 15%, 20% and 25 % of its mean value, respectively)

deviation of V_f increases; furthermore, when V_f assumes little values P_{fail} is more affected by the dispersion of the values of the V_f probability distribution.

4. CONCLUSIONS

A procedure that combines a structural optimization problem and a structural reliability one in the case of laminated composite material structures is presented. The procedure is formulated as a multicriteria optimization structural problem in which the probability of failure with respect to a predefined limit state is considered. To this aim, an algorithm that determines the Pareto optimal solutions with the probability of failure as an objective function has been performed. The procedure has been implemented and added to a widely diffuse finite element code considering uncorrelated normally distributed random variables.

Further researches will take into account not normally distributed and not mutually independent random variables and a new procedure for the inverse problem (multicriteria optimization with prefixed reliability) will be performed.

REFERENCES

1. Borri A., Speranzini E., Vetturini R., "Optimum Design of Laminated Composite Material Structures", CADCOMP 92, Delaware, USA, 1992.
2. Borri A., Speranzini E., "Multicriteria Optimization of Laminated Composite Material Structures", Meccanica, No 28, Kluwer Academic Publishers, Netherlands, 1993.
3. Koo D., "Elements of Optimizaton, with Application in Economic and Business", Heidelberg Science Library, 1977.
4. Eschenauer H.A., Kosky J., Osyczka A., "Multicriteria Design Optimization", Springer Verlag, Berlin, 1990.
5. ANSYS, "Seminar Notes: Design Optimization", Swanson Analysis System, Houston, Pa, USA, 1989.
6. Bhagwan D. A., Lawrence J. B., "Analysis and Performance of Fiber Composites", John Wiley & Sons, New York, 1980.
7. Hull D., "An introduction to composite materials", Cambridge University Press, Cambridge, 1981.
8. Engineered Material Handbook, "Composites", Vol 1, ASM International, Metals Park, Hoio.
9. Augusti G., Baratta A., Casciati F., "Probabilistics Methods in Structural Engineering", Chapman and Hall Publishers, New York, 1984.
10. Casciati F., Faravelli L., "Fragility Analisys of Complex Structural Systems", John Wiley and Sons Inc., Tanton, England.

7

Approximation of non-normal responses for drag dominated offshore structures

V. Bouyssy and R. Rackwitz
Technische Universität München, Arcisstr. 21, 80290 München, Germany

Abstract

For deep water offshore structures, the modeling of random waves loads by the Morison equation yields an equation of motion which admits no analytical solution except in few limit cases. This also holds for response moments which have a great influence on the results of reliability analyses. If polynomial approximations of the Morison drag loads are introduced, some procedures are available to obtain the stationary moments of the approximate response. These procedures result in large computer codes and time consuming computations if large structures are considered. Therefore, only first order approximations of the drag nonlinearity usually can be handled. In the paper, the ability of two famous linearization schemes to predict response variance and response fourth moment is numerically investigated. Improvements of these schemes are considered. For quasi-static structures, it is shown that except in very simple cases it is practically impossible to optimize linearization schemes to fit response variance or fourth moment.

1. Introduction

Non-linearities are faced in many physical problems. In the engineering field, they arise for example in the modelling of loads and material properties. If one considers the particular case of deep water offshore structures, the main nonlinearity generally is in the wave loading. Indeed, the force exerted by waves on any tubular member is described by the extended Morison equation which consists of a nonlinear drag term and a linear mass term. Other nonlinearities are often neglected. For instance the sea elevation, water particle velocity and acceleration processes generally are assumed stationary Gaussian with zero-mean during sea states of several hours duration and wave-current usually is neglected. For a discussion of these problems see [1].

In the following we consider a single-degree-of-freedom system. Under the aforementioned assumptions, the equation of motion is

$$m\ddot{y} + c\dot{y} + ky = K_D\,(x - \dot{y})\,|x - \dot{y}| + K_M\dot{x} - K_A\ddot{y} \tag{1}$$

where $|.|$ denotes absolute value; $K_D = \frac{1}{2}\rho D C_D$, $K_M = \frac{1}{4}\pi\rho D^2 C_M$ and $K_A = \frac{1}{4}\pi\rho D^2(C_M - 1)$; C_D and C_M are the drag and the inertia coefficients; ρ is the water density; D is the member diameter; m, c and k are structural parameters; x and y denote the water particle velocity and the structural displacement.

If equation (1) is solved, member forces and stresses can be calculated. Some of these responses may further be used for reliability assessment or in fatigue life estimation. It is well known that results of these analyses are accurate only if a good fit of the extreme value behaviour of these processes is achieved. If the approximation due to Winterstein [2] is used, only the first four order moments of these processes need to be known. For equation (1), however, the estimation of even low order moments of y is not straightforward. The statistical moments of $y(t)$, of course, can be exactly calculated through time domain simulations but this approach is excessively time consuming. Therefore, approximations must be introduced.

If the nonlinear term $(x - \dot{y}) \mid x - \dot{y} \mid$ is approximated by a polynomial, then a Volterra series representation of the response can be obtained [3, 4, 5]. If a spectral description of the response is chosen, then stationary response moments result from the integration of higher order spectra in the frequency domain. This method, however, is very time consuming as it involves laborious multidimensional convolutions. Moreover, it is not clear whether this method could be successfully extended to large problems. So far, only simple linear spectral analysis appears possible in practical cases, i.e. when large structures and numerous load cases (sea states) have to be considered. Then the nonlinear term $(x - \dot{y}) \mid x - \dot{y} \mid$ is simply replaced by a linear term $A\sigma_{x-\dot{y}}(x - \dot{y})$, where A is a constant and $\sigma_{x-\dot{y}}$ denotes the standard deviation of $(x - \dot{y})$ as proposed already by Borgman [3].

It is obvious, however, that any first order approximation is a severe simplification because the response process then is globally assumed to be Gaussian. Also, sub- and superharmonic responses are underpredicted. Indeed, some authors found that linearization procedures may be excessively inefficient in predicting the response variance in some cases (see, e.g., [6]). Before investigating the possibility to extend the aforementioned method to larger systems and to implement such method in appropriate computer codes, it therefore appears necessary to study in detail the error involved in the variance linearization schemes. The emphasis here is on two linearization schemes which set out from random seas and are widely used in offshore engineering. Analysis is performed in the time domain for a standardized form of the equation of motion (1) suggested in [7] so that conclusions are valid in general. Further, following an idea by Naess et al [8] possible extensions of the linearization schemes for estimation of higher order moments are investigated. This idea is quite attractive because if response moments up to the fourth order could be determined, then at most four spectral analyses each one with the appropriate linearization constant would enable to fit the response by non-Gaussian models such as proposed by Winterstein [2].

2. Standardized equation of motion

Introducing the so called hydrodynamic mass $m_{hyd} = K_A$, we rewrite (1) in the form

$$(m + m_{hyd})\,\ddot{y} + c\dot{y} + ky = K_D\,(x - \dot{y})\,|x - \dot{y}| + K_M\dot{x} \tag{2}$$

The left hand side of equation (2) is further standardized in a classical way

$$\ddot{y} + 2\xi_s\omega_s\dot{y} + \omega_s^2 y = \frac{K_D}{m + m_{hyd}}\,(x - \dot{y})\,|x - \dot{y}| + \frac{K_M}{m + m_{hyd}}\dot{x} \tag{3}$$

where ω_s and ξ_s denote the system natural frequency and damping. In the following, it is assumed that a Pierson-Moskowitz or a JONSWAP one-sided spectral density $G_{\eta\eta}(.)$

describes the sea elevation process. Let ω_0 denote some dominant, central frequency of the wave excitation. Then dividing equation (3) by ω_0^2 yields a standardized equation of motion [7]

$$\overline{y}'' + 2\xi_s\overline{\omega}_s\overline{y}' + \overline{\omega}_s^2\overline{y} = (u - \alpha\overline{y}')\,|u - \alpha\overline{y}'| + \beta a \tag{4}$$

where $\overline{y} = (m+m_{hyd})\omega_0^2 y/K_D$ is a standardized displacement; $\overline{y}' = d\overline{y}/d(\omega_0 t)$; $\overline{\omega}_s = \omega_s/\omega_0$; $u = x/p$ and $a = \dot{x}/q$ are the standardized water particle velocity and acceleration with the factors p and q obtained from

$$p^2 = \omega_0^2 G_{\eta\eta}(\omega_0) \quad \text{and} \quad q^2 = \omega_0^5 G_{\eta\eta}(\omega_0) \tag{5}$$

Then it is easily proved that the densities of the standardized water particle velocity $u(t)$ and acceleration $a(t)$ written as functions of ω/ω_0 are independent of the sea state. The two forcing function coefficients α and β in equation (4) are respectively a measure of water-structure interaction and a hydrodynamic ratio. They are given by

$$\alpha = \frac{p}{\omega_0} K_D/(m + m_{hyd}) \quad \text{and} \quad \beta = \frac{q}{p^2} K_M/K_D \tag{6}$$

3. Linearization of the drag term

The choice of the linearization procedure is not straightforward because the first derivative of $x|x|$ vanishes in zero. Numerous linearization procedures exist to approximate this nonlinear term by a linear one but none of them can be shown to be optimal to fit a given low order moment of the solution of equation (4). In practice, two linearization procedures are generally preferred to estimate the response variance. The first approach follows from Caughey's work [9] and consists in replacing the nonlinear forcing function in equation (4) by a linear term which minimizes the mean square error

$$E\left[\left((u - \alpha\overline{y}')\,|u - \alpha\overline{y}'| - A\sigma_{u-\alpha\overline{y}'}\,(u - \alpha\overline{y}')\right)^2\right] \tag{7}$$

Here $E[.]$ denotes mathematical expectation. The second linearization procedure follows from an idea by Bolotin [10] and appears to be less common in applications although it is simpler (see [11]). It involves replacing the original nonlinear drag term by a linear term which has same variance

$$E\left[\left((u - \alpha\overline{y}')\,|u - \alpha\overline{y}'|\right)^2\right] = E\left[\left(A\sigma_{u-\alpha\overline{y}'}\,(u - \alpha\overline{y}')\right)^2\right] \tag{8}$$

Both procedures require iterative estimation of the linearization factor A because the forcing function depends on the structural velocity $\overline{y}'$ solution of equation (4). They always yield different results as can be seen when the time-dependent damping is neglected. In this limit case $\alpha = 0$, (7) and (8) respectively lead to $A = \sqrt{8/\pi}$ and $A = \sqrt{3}$.

These procedures can be generalized by simply changing the value of the exponent in (7) or (8). Hence a possible extension of Caughey's procedure may be to minimize the error with respect to another norm than the euclidian one, i.e. the linearization factor A minimizes

$$E\left[\left|(u - \alpha\overline{y}')\,|u - \alpha\overline{y}'| - A\sigma_{u-\alpha\overline{y}'}\,(u - \alpha\overline{y}')\right|^n\right] \tag{9}$$

whereas an extension of Bolotin's rule (8) may be to require

$$E\left[\left|(u-\alpha\overline{y}')\,|u-\alpha\overline{y}'|\right|^{n}\right]=E\left[\left|A\sigma_{u-\alpha\overline{y}'}\,(u-\alpha\overline{y}')\right|^{n}\right] \tag{10}$$

In both cases n is a positive real. Analytical solutions to both generalized procedures (9) and (10) do not exist in general because of the hydrodynamic interaction. However, if this interaction is negligible ($\alpha = 0$), Bolotin's extension (10), on the one hand, admits a unique analytical solution A for any value of n

$$A=\sqrt{2}\sqrt[n]{\Gamma\left(n+\frac{1}{2}\right)/\Gamma\left(\frac{n}{2}+\frac{1}{2}\right)} \tag{11}$$

where $\Gamma(.)$ denotes the gamma function. On the other hand, an analytical solution to (9) exists for $n = 2$ and 4 only. Extensive computations were performed to estimate A for other values of n. Linearization factors are shown as functions of n on Figure 1.

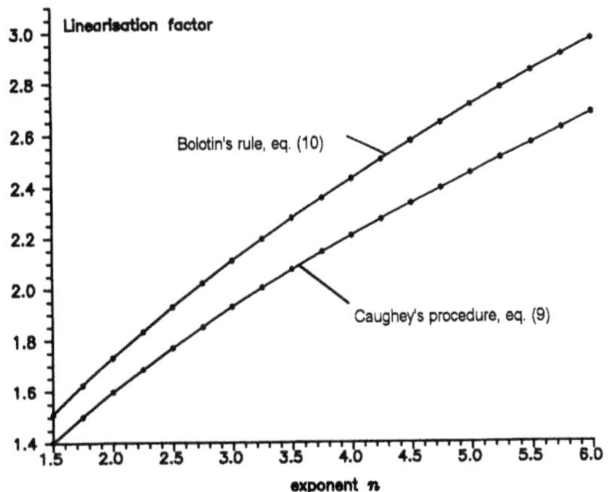

Figure 1. Linearization factor A as function of the exponent n ($\alpha = 0$)

4. Numerical work

In the following we investigate the ability of the classical linearization procedures (7) and (8) to yield approximate solutions with accurate second and fourth order moments. Time histories of the stationary Gaussian standardized water particle velocity u and acceleration a are simulated by using the spectral representation method. A Pierson-Moskowitz spectrum is assumed for the sea-elevation process. Harmonics with random phase and deterministic amplitude are used. Time histories with duration 4096 sec and time step $\Delta t = 0.125$ sec are simulated with N=16384 harmonics so that they are almost perfectly Gaussian according to the central limit theorem. The constant average acceleration step-by-step method proposed by Newmark is used to solve equation (4) numerically. For $\alpha \neq 0$, the forcing function depends on the response process and the "exact" response time history is computed iteratively. Also time histories resulting from the linearized forcing function are computed. Finally, statistics of the response time histories are calculated.

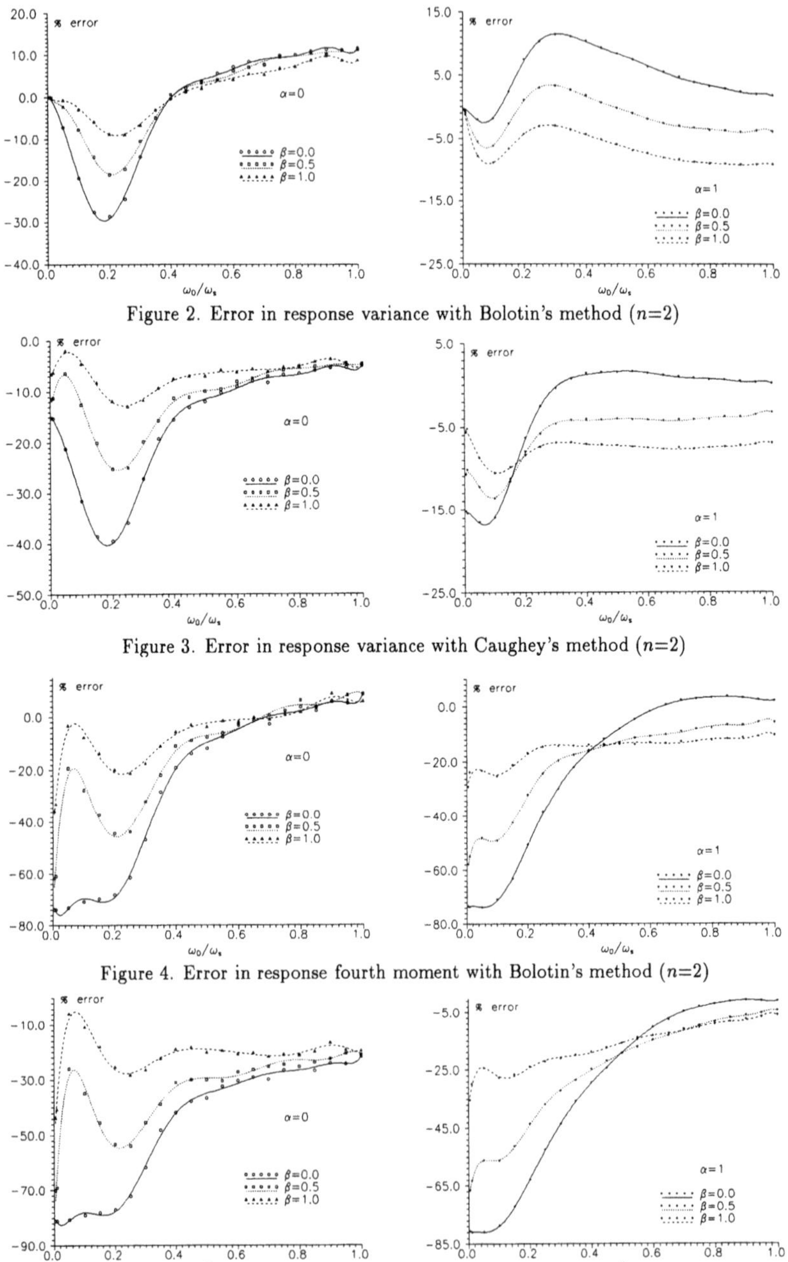

Figure 2. Error in response variance with Bolotin's method (n=2)

Figure 3. Error in response variance with Caughey's method (n=2)

Figure 4. Error in response fourth moment with Bolotin's method (n=2)

Figure 5. Error in response fourth moment with Caughey's method (n=2)

However, it is well known that higher moments of simulated time histories of water velocity and acceleration exhibit some scatter [12]. Therefore, the aforementioned procedure is repeated several times until convergence is reached in the fourth order response moments.

Moments are computed for several values of the frequency ratio ω_0/ω_s and different structural data (α, β, ξ_s). The frequency ratio ω_0/ω_s is taken in (0;1) ranging from static to dynamic excitation. Ratios above 1.0 are not considered because they may only be observed for very flexible structures for which non linearities cannot be handled by simple linearization. The error in response variance, on the one hand, and in response fourth order moment, on the other hand, is reported as function of the frequency ratio ω_0/ω_s in Figure 2-3 and 4-5, for $\alpha = 0$ and 1, $\beta = 0.0$, 0.5 and 1.0, and $\xi_s = 2\%$. Further, the exponent n in (9) and (10) which yields exact second or fourth order response moment was estimated. Results for $\alpha = 0$ are reported on Figures 6 and 7.

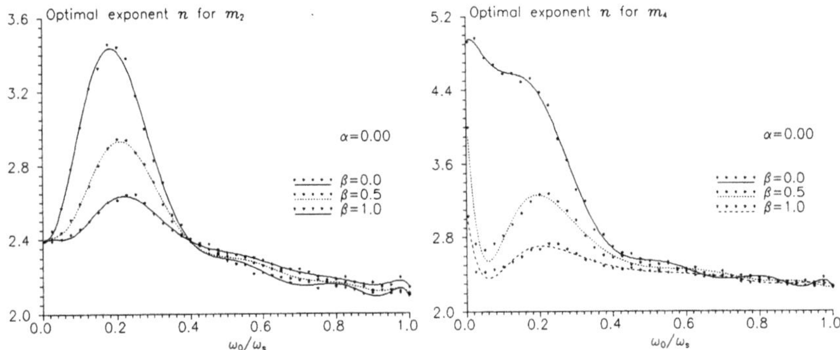

Figure 6. Optimal values of the linearization exponent n in Caughey's extension (9) to fit response second and fourth order moments ($\alpha = 0$)

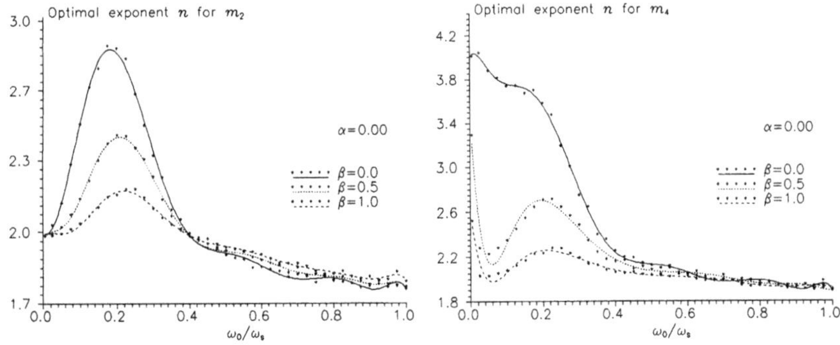

Figure 7. Optimal values of the linearization exponent n in Bolotin's extension (10) to fit response second and fourth order moments ($\alpha = 0$)

5. Discussion and conclusions

It is first observed that classical linearization procedures (7) and (8) provide satisfactory fits of both second and fourth order response moments for high values of the frequency ratio ω_0/ω_s (see Figures 2-5). In fact, a limit theorem by Rosenblatt proves that the response process asymptotically becomes Gaussian as ω_0/ω_s increases to infinity (see [13, 14]). Thus accurate extreme response or fatigue analyses of a dynamic structure can be performed with second order statistics estimated from linear analysis only.

Most structures, however, are designed to behave essentially in a static manner. Then responses can no longer be assumed Gaussian and fourth order moments of approximate responses are excessively inaccurate. Our results further show that linearization procedures may be excessively inefficient to fit the response variance in this case. For example, the classical Caughey procedure (7) strongly underestimates (by 40% for $\beta = 0$, i.e. when drag-dominated) the response variance in the case of quasi-static excitation ($0 < \omega_0/\omega_s < \frac{1}{3}$). In this case Bolotin's method ($n = 2$) also underestimates the variance but it yields exact variance estimates in the static case. This last feature is due to the fact that condition (8) is satisfied by the whole forcing function for $\omega_0/\omega_s = 0$ because the sea elevation process $\eta(t)$ is taken stationary. More generally, as already noticed by Hu et al [7], it is found that linearization procedures strongly underestimate the response variance at the subharmonic $\frac{1}{5}$ and not at the subharmonic $\frac{1}{3}$ as usually expected [6]. As the water-structure interaction α increases, the fit of the response variance improves but predictions of the response fourth moment remain extremely inacurate (see Figures 2-5). It follows that an increase of hydrodynamic interaction yields more non-Gaussian behaviour of the response. This is in agreement with results reported in [15]. As β increases, the nonlinear term however becomes negligible and fairly good results are obtained. This also agrees with our expectations.

If α can be assumed close to zero, optimal linearization schemes can be derived from the reported results. Again, optimal values of the exponent n strongly depend on the forcing coefficient β for quasi-static excitations. It must be recognized that the value of β cannot be estimated accurately because large model uncertainties are attached to the hydrodynamic coefficients. Moreover, it may be difficult to define the frequency ratio ω_0/ω_s for structures having several natural frequencies and experiencing the whole range of possible sea states. If modal decomposition is applied to analyse a multi-degree-of-freedom system then our results can be used under the restricting assumption that each modal equation of motion is of the kind (4), i.e. that modal loads reduce to a single Morison equation. In fact, modal forces are a combination of Morison forces correlated in space. Therefore, it is practically rather difficult to use optimum linearization schemes to fit response statistics of quasi-static structures except in very simple cases. It can be proposed to choose Bolotin's rule ($n = 2$) instead of Caughey's procedure ($n = 2$) to estimate the response variance of such structures as this former proves to be simpler and more conservative. It further yields the exact variance of the quasi-static response.

Acknowledgement

This study was supported by Elf Aquitaine under contract N° EAP-9125 which is highly appreciated.

References

1. Claus G., Lehmann E., Östergaard C. (1992). *Offshore Structures*, I+II, Springer Verlag

2. Winterstein S. (1988). Nonlinear vibration models for extremes and fatigue, *ASCE Journal of Engineering Mechanics*, 114, 1772-1790

3. Borgman L. E. (1967). Spectral analysis of ocean wave forces on piling, *ASCE Journal of the Waterways and Harbors Division*, 93, 129-156

4. Borgman L. E. (1982). Specification of the wave environment, *Ocean Structural Dynamics Symposium*, Corvallis

5. Olagnon M., Prevosto M., Joubert P. (1988). Nonlinear spectral computation of the dynamic response of a single cylinder, *ASME Journal of Offshore Mechanics and Arctic Engineering*, 110, 278-281

6. Eatock Taylor R. and Rajagopalan A. (1982). Dynamics of offshore structures, Part I: Perturbation analysis, *Journal of Sound and Vibration*, 83, 401-416

7. Hu S.-L. J., Tsiatas G., McGrath J. E. (1991). Optimal linearization of Morison-type wave loading, *ASCE Journal of Engineering Mechanics*, 117, 1537-1553

8. Naess A., Galeazzi F., Dogliani M. (1992). Extreme response predictions of nonlinear compliant structures by stochastic linearization, *Applied Ocean Research*, 14, 71-81

9. Caughey T. K. (1963). Equivalent linearization techniques, *Journal of the Acoustical Society of America*, 35, 1706-1711

10. Bolotin V. V. (1969). *Statistical methods in Structural Mechanics*, Holden-Day, San Francisco

11. Karadeniz H. (1989). *Advanced stochastic analysis program for offshore structures*, TU Delft Report

12. Tucker M. J. (1984). Numerical simulation of a random sea: a common error and its effect upon wave group statistics, *Applied Ocean Research*, 6, 118-122

13. Kotulski Z. and Sobczyk K. (1981). Linear systems and normality, *Journal of Statistical Physics*, 24, 359-373

14. Grigoriu M. and Ariaratnam S. T. (1988). Response of linear systems to polynomials of Gaussian processes, *ASME Journal of Applied Mechanics*, 55, 905-910

15. Manuel L. and Cornell C. A. (1993). Sensitivity of the dynamic response of a jack-up rig to support modelling and Morison froce modelling assumptions, *Proc. 12th Int. Conf. on Offshore Mechanics and Arctic Engng.*, 2, 243-250

8

Parameter-dependent integrals: some mathematical tools

K. Breitung [a]

[a] Department of Civil Engineering, University of Calgary
2500 University Drive N.W.
Calgary, Alberta, Canada, T2N 1N4

1. INTRODUCTION

In many reliability problems integrals of the following form

$$F(\boldsymbol{\tau}) = \int_{g(\boldsymbol{x},\boldsymbol{\tau})\le 0} f(\boldsymbol{x},\boldsymbol{\tau})\, d\boldsymbol{x} \tag{1}$$

are of interest. Here $f(\boldsymbol{x},\tau)$ is usually a probability density and $g(\boldsymbol{x},\tau)$ a limit state function. Both functions depend on a parameter vector $\boldsymbol{\tau}$. Then $F(\boldsymbol{\tau})$ denotes the failure probability for the parameter value $\boldsymbol{\tau}$.

In the case that only the integrand depends on the parameter $\boldsymbol{\tau}$ the derivatives are obtained easily by differentiating under the integral sign. But the case that the limit state function depends on a parameter is of importance in reliability problems, especially in optimization. An example of such a problem is given in [7]. The concepts of asymptotic approximation methods for such integrals are outlined in [5].

2. DERIVATIVE OF $F(\tau)$

In [3] is was shown that under some regularity conditions the derivative $F'(\tau)$ is given by

$$F'(\tau) = \int_{D(\tau)} f_\tau(\boldsymbol{x},\tau)\, d\boldsymbol{x} - \int_{G(\tau)} f(\boldsymbol{y},\tau)\frac{g_\tau(\boldsymbol{y},\tau)}{|\nabla_{\boldsymbol{y}} g(\boldsymbol{y},\tau)|}\, ds_\tau(\boldsymbol{y}). \tag{2}$$

Here $D(\tau) = \{\boldsymbol{y}; g(\boldsymbol{y},\tau) < 0\}$, $G(\tau) = \{\boldsymbol{y}; g(\boldsymbol{y},\tau) = 0\}$ and $ds_\tau(\boldsymbol{y})$ denotes surface integration over $G(\tau)$.

A disadvantage of this relation is that the second term is a surface integral, which is in general difficult to compute. Using the divergence theorem, we can transform it into a domain integral. The divergence theorem states that for a continuously differentiable vector function $\boldsymbol{u}(\boldsymbol{x})$ defined on a domain F with boundary G we have for the derivative then the following form

$$\int_F \operatorname{div}(\boldsymbol{u}(\boldsymbol{x}))\ d\boldsymbol{x} = \int_G \langle \boldsymbol{n}(\boldsymbol{y}), \boldsymbol{u}(\boldsymbol{y})\rangle\ ds(\boldsymbol{y}). \tag{3}$$

Here $\operatorname{div}(\boldsymbol{u}(\boldsymbol{x})) = \sum_{i=1}^n \partial u_i(\boldsymbol{x})/\partial x_i$, $\boldsymbol{n}(\boldsymbol{y})$ is the outward pointing normal at $\boldsymbol{y}$ and $\langle \boldsymbol{x}, \boldsymbol{y}\rangle$ is the scalar product of $\boldsymbol{x}$ and $\boldsymbol{y}$.

First we assume that the gradient $\nabla \boldsymbol{y}(\boldsymbol{y}, \tau)$ does not vanish throughout $D(\tau)$. If we define now a vector field $\boldsymbol{u}(\boldsymbol{x})$ by

$$\boldsymbol{u}(\boldsymbol{x}) = \frac{f(\boldsymbol{x},\tau)g_\tau(\boldsymbol{x},\tau)}{|\nabla_{\boldsymbol{x}} g(\boldsymbol{x},\tau)|^2}\nabla_{\boldsymbol{x}} g(\boldsymbol{x},\tau) \tag{4}$$

we get on the surface $G(\tau)$ with outward pointing normal $\boldsymbol{n}(\boldsymbol{y}) = |\nabla_{\boldsymbol{y}} g(\boldsymbol{y},\tau)|^{-1}\nabla_{\boldsymbol{y}} g(\boldsymbol{y},\tau)$ for the scalar product $\langle \boldsymbol{n}(\boldsymbol{y}), \boldsymbol{u}(\boldsymbol{y})\rangle$ the form

$$\langle \boldsymbol{n}(\boldsymbol{y}), \boldsymbol{u}(\boldsymbol{y})\rangle = \frac{f(\boldsymbol{x},\tau)g_\tau(\boldsymbol{x},\tau)}{|\nabla_{\boldsymbol{x}} g(\boldsymbol{x},\tau)|}, \tag{5}$$

and therefore we have that

$$\int_{G(\tau)} f(\boldsymbol{y},\tau)\frac{g_\tau(\boldsymbol{y},\tau)}{|\nabla_{\boldsymbol{y}} g(\boldsymbol{y},\tau)|}\ ds_\tau(\boldsymbol{y}) = \int_F \operatorname{div}\left(\frac{f(\boldsymbol{x},\tau)g_\tau(\boldsymbol{x},\tau)}{|\nabla_{\boldsymbol{x}} g(\boldsymbol{x},\tau)|^2}\nabla_{\boldsymbol{x}} g(\boldsymbol{x},\tau)\right)\ d\boldsymbol{x}. \tag{6}$$

This gives then for the derivative $F'(\tau)$ the form

$$F'(\tau) = \int_{g(\boldsymbol{x},\tau)\leq 0} \left[f_\tau(\boldsymbol{x},\tau) - \operatorname{div}\left(\frac{f(\boldsymbol{x},\tau)g_\tau(\boldsymbol{x},\tau)}{|\nabla_{\boldsymbol{x}} g(\boldsymbol{x},\tau)|^2}\nabla_{\boldsymbol{x}} g(\boldsymbol{x},\tau)\right)\right]\ d\boldsymbol{x}. \tag{7}$$

If now the gradient vanishes at a finite number of points in $D(\tau)$, but the Hessian $\boldsymbol{H}_g(\boldsymbol{x}) = (g^{ij}(\boldsymbol{x},\tau))_{i,j=1,\dots,n}$ is regular at all these points, equation (7) remains valid. We show this for the case of a single point $\boldsymbol{x}^*$ in $D(\tau)$ with vanishing gradient. We consider a domain $D^*(\tau)$ given by $D^*(\tau) = D(\tau) \setminus K(\epsilon)$, where $K(\epsilon)$ is the sphere around $\boldsymbol{x}^*$ with radius ϵ. The value of ϵ is chosen so small that $K(\epsilon) \subset D(\tau)$. Then the divergence theorem is valid for this domain. The boundary of it are the surface $G(\tau)$ and the surface $S(\epsilon)$ of the sphere $K(\epsilon)$. So we obtain, since the outward pointing normal on $S(\epsilon)$ is $|\boldsymbol{x}^* - \boldsymbol{y}|^{-1}(\boldsymbol{x}^* - \boldsymbol{y})$, the following equation

$$\begin{aligned}\int_{G(\tau)} f(\boldsymbol{y},\tau)\frac{g_\tau(\boldsymbol{y},\tau)}{|\nabla_{\boldsymbol{y}} g(\boldsymbol{y},\tau)|}\ ds_\tau(\boldsymbol{y}) &= \int_F \operatorname{div}\left(\frac{f(\boldsymbol{x},\tau)g_\tau(\boldsymbol{x},\tau)}{|\nabla_{\boldsymbol{x}} g(\boldsymbol{x},\tau)|^2}\nabla_{\boldsymbol{x}} g(\boldsymbol{x},\tau)\right)\ d\boldsymbol{x} \\ &\quad \underbrace{- \int_{S(\epsilon)} f(\boldsymbol{y},\tau)\frac{g_\tau(\boldsymbol{y},\tau)}{|\nabla_{\boldsymbol{y}} g(\boldsymbol{y},\tau)|}\langle\frac{\nabla_{\boldsymbol{y}} g(\boldsymbol{y},\tau)}{\nabla_{\boldsymbol{y}} g(\boldsymbol{y},\tau)}, \frac{\boldsymbol{x}^*-\boldsymbol{y}}{|\boldsymbol{x}^*-\boldsymbol{y}|}\rangle\ ds_\epsilon(\boldsymbol{y})}_{=I(\epsilon)}\end{aligned} \tag{8}$$

with $ds_\epsilon(\boldsymbol{y})$ denoting surface integration over $S(\epsilon)$.

For the last surface integral $I(\epsilon)$ we get with $K = \max_{\boldsymbol{y}\in K(\epsilon)} |f(\boldsymbol{y},\tau)g_\tau(\boldsymbol{y},\tau)|$ the upper bound

$$\left| \int\limits_{S(\epsilon)} f(\boldsymbol{y},\tau)\frac{g_\tau(\boldsymbol{y},\tau)}{|\nabla_{\boldsymbol{y}} g(\boldsymbol{y},\tau)|}\langle\frac{\nabla_{\boldsymbol{y}} g(\boldsymbol{y},\tau)}{\nabla_{\boldsymbol{y}} g(\boldsymbol{y},\tau)}, \frac{\boldsymbol{x}^* - \boldsymbol{y}}{|\boldsymbol{x}^* - \boldsymbol{y}|}\rangle \, ds_\epsilon(\boldsymbol{y}) \right| \leq K \int\limits_{S(\epsilon)} |\nabla_{\boldsymbol{y}} g(\boldsymbol{y},\tau)|^{-1} \, ds_\epsilon(\boldsymbol{y}).$$

Making a Taylor expansion of the first derivatives of g at $\boldsymbol{x}^*$ gives

$$\nabla_{\boldsymbol{y}} g(\boldsymbol{y},\tau) = \boldsymbol{H}_g(\boldsymbol{x}^*)(\boldsymbol{y} - \boldsymbol{x}^*) + o(\epsilon). \tag{9}$$

With $\mu_0 = \min |\mu_1|, \ldots, |\mu_n| > 0$, where the μ_i's are the eigenvalues of $\boldsymbol{H}_g(\boldsymbol{x}^*)$ we get for the norm

$$|\nabla_{\boldsymbol{y}} g(\boldsymbol{y},\tau)| \geq \mu_0|\boldsymbol{y} - \boldsymbol{x}^*| + o(\epsilon) = \mu_0\epsilon + o(\epsilon). \tag{10}$$

This gives then for the integral $I(\epsilon)$ that

$$|I(\epsilon)| \leq K\mu_0\epsilon^{-1} \int\limits_{S(\epsilon)} ds_\epsilon(\boldsymbol{y}) = K\mu_0\frac{2\pi^{n/2}}{\Gamma(n/2)}\epsilon^{n-2} + o(\epsilon^{n-2}) = O(\epsilon^{n-2}). \tag{11}$$

Therefore equation (7) remains valid if the Hessian is regular and the dimension of the integration domain is larger than two.

3. QUADRATIC FORMS ON SUBSPACES

In a number of problems the definiteness of a matrix under linear constraints is of interest. Given is an $n \times n$ matrix $\boldsymbol{H}$ and a subspace U spanned by m linearly independent vectors $\boldsymbol{a}_1, \ldots, \boldsymbol{a}_m$. To find if the matrix is positive (or negative) on the subspace orthogonal to U.

We consider first the case that $\boldsymbol{a}_i = \boldsymbol{e}_i$, i.e. the vectors $\boldsymbol{a}_i$ are the first m unit vectors and that $\boldsymbol{H}$ is a diagonal matrix with diagonal elements $\mu_1, \ldots, \mu_n$. Then we have for the quadratic form $\boldsymbol{x}^T\boldsymbol{H}\boldsymbol{x}$ the representation

$$\boldsymbol{x}^T\boldsymbol{H}\boldsymbol{x} = \sum_{i=1}^{n} \mu_i x_i^2 = \sum_{j=1}^{m} \mu_j x_j^2 + \sum_{j=k+1}^{n} \mu_j x_j^2. \tag{12}$$

This quadratic form is positive definite under the constraint $x_1 = \ldots = x_k = 0$ if the last $n - m$ diagonal elements $\mu_{m+1}, \ldots, \mu_n$ are positive. We consider now the quadratic form defined by

$$\boldsymbol{x}^T\boldsymbol{H}^*\boldsymbol{x} = \sum_{j=1}^{k} x_j^2 + \sum_{j=k+1}^{n} \mu_j x_j^2. \tag{13}$$

with $\boldsymbol{H}^*$ a diagonal matrix with diagonal elements $1, \ldots, 1, \mu_{m+1}, \ldots, \mu_n$. This quadratic form is positive definite if the quadratic form $\boldsymbol{x}^T\boldsymbol{H}\boldsymbol{x}$ is positive definite under the constraint $x_1 = \ldots = x_k = 0$. The projection matrix $\boldsymbol{P}$ of projection onto the subspace spanned by $\boldsymbol{e}_1, \ldots, \boldsymbol{e}_m$ is given by the diagonal matrix with the first m elements equal to one and the rest equal to zero and the projection onto the orthogonal subspace by $\boldsymbol{I}_n - \boldsymbol{P}$. Then we can the last equation in the form

$$\boldsymbol{x}^T \boldsymbol{H}^+ \boldsymbol{x} = \boldsymbol{x}^T \left(\boldsymbol{P}\boldsymbol{P} + (\boldsymbol{I}_n - \boldsymbol{P})\boldsymbol{H}(\boldsymbol{I}_n - \boldsymbol{P}) \right) \boldsymbol{x}. \tag{14}$$

In the general case of a subspace spanned by m arbitrary linearly independent vectors $\boldsymbol{a}_1, \ldots, \boldsymbol{a}_m$, we can reduce the problem to the case above by making a rotation such the m-dimensional subspace spanned by these vectors is transformed into the subspace spanned by the first m new coordinate vectors. The projection matrix onto that subspace is given by

$$\boldsymbol{P} = \boldsymbol{A}(\boldsymbol{A}^T \boldsymbol{A})^{-1} \boldsymbol{A}^T \tag{15}$$

with $\boldsymbol{A} = (\boldsymbol{a}_1, \ldots, \boldsymbol{a}_m)$. Then we get for the quadratic form $\boldsymbol{x}^T \boldsymbol{H}^+ \boldsymbol{x}$ the same form as in the last equation. This gives finally: A matrix $\boldsymbol{H}$ is positive definite under the constraint $\boldsymbol{A}^T \boldsymbol{x} = \mathbf{o}_m$ iff the matrix $\boldsymbol{H}^+ = \boldsymbol{P}\boldsymbol{P} + (\boldsymbol{I}_n - \boldsymbol{P})\boldsymbol{H}(\boldsymbol{I}_n - \boldsymbol{P})$ is positive definite. Analogously we have that it is negative definite under these constraints iff $\boldsymbol{H}^- = -\boldsymbol{P}\boldsymbol{P} + (\boldsymbol{I}_n - \boldsymbol{P})\boldsymbol{H}(\boldsymbol{I}_n - \boldsymbol{P})$. If a matrix is positive definite, can be checked easily by making a Cholesky decomposition. If this algorithm does not break down, it is positive definite.

In SORM we need the curvature correction factor for obtaining an asymptotic approximation for the failure probability. It is given in the form

$$\mathrm{P}(F) \sim \Phi(-\beta) \prod_{i=1}^{n-1} (1 - \beta\kappa_i)^{-1/2}. \tag{16}$$

Here the κ_i's are the main curvatures of the limit surface at the beta point $\boldsymbol{x}_0$ (see [2]).

In [6] it is shown that the largest eigenvalue can be obtained during the final stage of a numerical search for the beta point. The last result about quadratic forms on subspacse gives a simple method, which avoids an eigenvalue analysis for computing the main curvatures $\kappa_1, \ldots, \kappa_{n-1}$, for calculating this curvature correction factor $(\prod_{i=1}^{n-1}(1-\kappa_i))^{-1/2}$.

By a rotation of the coordinates it can always be achieved that the x_n axis is in the direction of the normal vector of the surface G at $\boldsymbol{x}_0$ and the tangential space is spanned by the vectors in the directions of the $x_1, \ldots, x_{n-1}$- axes. The square of the curvature factor is then, (see [2] , eq.(25))

$$\prod_{i=1}^{n-1} (1 - \kappa_i) = \det \begin{pmatrix} 1 + \frac{g_{11}(\boldsymbol{x}_0)}{|\nabla g(\boldsymbol{x}_0)|} & \frac{g_{12}(\boldsymbol{x}_0)}{|\nabla g(\boldsymbol{x}_0)|} & \cdots & \frac{g_{1,n-1}(\boldsymbol{x}_0)}{|\nabla g(\boldsymbol{x}_0)|} \\ \frac{g_{21}(\boldsymbol{x}_0)}{|\nabla g(\boldsymbol{x}_0)|} & 1 + \frac{g_{22}(\boldsymbol{x}_0)}{|\nabla g(\boldsymbol{x}_0)|} & \cdots & \frac{g_{2,n-1}(\boldsymbol{x}_0)}{|\nabla g(\boldsymbol{x}_0)|} \\ \vdots & \cdots & \ddots & \vdots \\ \frac{g_{n-1,1}(\boldsymbol{x}_0)}{|\nabla g(\boldsymbol{x}_0)|} & \cdots & \cdots & 1 + \frac{g_{n-1,n-1}(\boldsymbol{x}_0)}{|\nabla g(\boldsymbol{x}_0)|} \end{pmatrix}. \tag{17}$$

The $g_{ij}(\boldsymbol{x}_0)$ are the second derivatives of g at $\boldsymbol{x}_0$ with respect to x_i and x_j. If we add a row and a column with zeros everywhere and only 1 in the main diagonal, the value of the determinant remains the same and we obtain

$$\det \begin{pmatrix} 1 + \frac{g_{11}(\boldsymbol{x}_0)}{|\nabla g(\boldsymbol{x}_0)|} & \frac{g_{12}(\boldsymbol{x}_0)}{|\nabla g(\boldsymbol{x}_0)|} & \cdots & \frac{g_{1,n-1}(\boldsymbol{x}_0)}{|\nabla g(\boldsymbol{x}_0)|} \\ \frac{g_{21}(\boldsymbol{x}_0)}{|\nabla g(\boldsymbol{x}_0)|} & 1 + \frac{g_{22}(\boldsymbol{x}_0)}{|\nabla g(\boldsymbol{x}_0)|} & \cdots & \frac{g_{2,n-1}(\boldsymbol{x}_0)}{|\nabla g(\boldsymbol{x}_0)|} \\ \vdots & \cdots & \ddots & \vdots \\ \frac{g_{n-1,1}(\boldsymbol{x}_0)}{|\nabla g(\boldsymbol{x}_0)|} & \cdots & \cdots & 1 + \frac{g_{n-1,n-1}(\boldsymbol{x}_0)}{|\nabla g(\boldsymbol{x}_0)|} \end{pmatrix}$$

$$= \det \underbrace{\begin{pmatrix} 1+\frac{g_{11}(\boldsymbol{x}_0)}{|\nabla g(\boldsymbol{x}_0)|} & \frac{g_{12}(\boldsymbol{x}_0)}{|\nabla g(\boldsymbol{x}_0)|} & \cdots & \frac{g_{1,n-1}(\boldsymbol{x}_0)}{|\nabla g(\boldsymbol{x}_0)|} & 0 \\ \frac{g_{21}(\boldsymbol{x}_0)}{|\nabla g(\boldsymbol{x}_0)|} & 1+\frac{g_{22}(\boldsymbol{x}_0)}{|\nabla g(\boldsymbol{x}_0)|} & \cdots & \frac{g_{2,n-1}(\boldsymbol{x}_0)}{|\nabla g(\boldsymbol{x}_0)|} & 0 \\ \vdots & \cdots & \ddots & \vdots & \vdots \\ \frac{g_{n-1,1}(\boldsymbol{x}_0)}{|\nabla g(\boldsymbol{x}_0)|} & \cdots & \cdots & 1+\frac{g_{n-1,n-1}(\boldsymbol{x}_0)}{|\nabla g(\boldsymbol{x}_0)|} & 0 \\ 0 & \cdots & \cdots & 0 & 1 \end{pmatrix}}_{=\boldsymbol{D}} . \tag{18}$$

If we set

$$\boldsymbol{H}(\boldsymbol{x}_0) = (\delta_{ij} + |\nabla g(\boldsymbol{x}_0)|^{-1} g_{ij}(\boldsymbol{x}_0))_{i,j=1,\ldots,n}, \tag{19}$$

we get that the matrix $\boldsymbol{D}$ can be written in the form

$$\boldsymbol{D} = \boldsymbol{P}^T \boldsymbol{H}(\boldsymbol{x}_0)\boldsymbol{P} + \boldsymbol{e}_n \boldsymbol{e}_n^T \tag{20}$$

with $\boldsymbol{e}_n = (0, \ldots, 0, 1)^T$ and $\boldsymbol{P} = \boldsymbol{I}_n - \boldsymbol{e}_n \boldsymbol{e}_n^T$. Due to the special choice of the coordinate system, $\boldsymbol{e}_n$ is the normal vector of the surface G at $\boldsymbol{x}_0$ and $\boldsymbol{P}$ is the projection matrix onto the tangential space of G at $\boldsymbol{x}_0$. But this formulation is invariant under linear coordinate changes; we just have to replace $\boldsymbol{e}_n$ by the normal vector $\boldsymbol{n}$ in the new coordinates $\boldsymbol{n} = |\nabla g(\boldsymbol{x}_0)|^{-1}\nabla g(\boldsymbol{x}_0)$. So we have for an arbitrary coordinate system the following expression for the curvature factor

$$\prod_{i=1}^{n-1}(1-\kappa_i) = \det(\boldsymbol{D}) = \det(\boldsymbol{P}^T \boldsymbol{H}(\boldsymbol{x}_0)\boldsymbol{P} + |\nabla g(\boldsymbol{x}_0)|^{-2}\nabla g(\boldsymbol{x}_0)(\nabla g(\boldsymbol{x}_0))^T) \tag{21}$$

with $\boldsymbol{P} = \boldsymbol{I}_n - |\nabla g(\boldsymbol{x}_0)|^{-2}\nabla g(\boldsymbol{x}_0)(\nabla g(\boldsymbol{x}_0))^T$.

Since the function $|\boldsymbol{x}|^2$ has a local minimum at the point $\boldsymbol{x}_0$ under the constraint $g(\boldsymbol{x}) = 0$ the matrix $\boldsymbol{H}(\boldsymbol{x}_0)$ is positive definite under the linear constraint $\boldsymbol{n}^T\boldsymbol{x} = 0$, if the extremum is regular. Therefore in this case the determinant of $\boldsymbol{D}$ can be found from the Cholesky decomposition of $\boldsymbol{P}^T\boldsymbol{H}(\boldsymbol{x}_0)\boldsymbol{P} + |\nabla g(\boldsymbol{x}_0)|^{-2}\nabla g(\boldsymbol{x}_0)(\nabla g(\boldsymbol{x}_0))^T$.

In the same way we can obtain the asymptotic approximation in the case of non-normal random variables with p.d.f. $f(\boldsymbol{x})$. Here the asymptotic approximation is given by

$$\mathrm{P}(F) \sim (2\pi)^{(n-1)/2} \frac{f(\boldsymbol{x}^*)}{|\nabla l(\boldsymbol{x}^*)||\det(\boldsymbol{H}^*(\boldsymbol{x}^*))|^{1/2}} \tag{22}$$

with $l(\boldsymbol{x}) = \ln(f(\boldsymbol{x}))$ the log-likelihood function (see [4]). The matrix $\boldsymbol{H}^*(\boldsymbol{x}^*)$ is defined by

$$\boldsymbol{H}^*(\boldsymbol{x}^*)) = \boldsymbol{A}^T\boldsymbol{H}(\boldsymbol{x}^*)\boldsymbol{A}. \tag{23}$$

Here $\boldsymbol{H} = (l^{ij}(\boldsymbol{x}^*) - |\nabla l(\boldsymbol{x}^*)||\nabla g(\boldsymbol{x}^*)|^{-1}g^{ij}(\boldsymbol{x}^*))_{i,j=1,\ldots,n}$ and $\boldsymbol{A} = (\boldsymbol{a}_1, \ldots, \boldsymbol{a}_{n-1})$, where the $\boldsymbol{a}_i$'s form an orthonormal basis of the tangential space of the limit state surface at $\boldsymbol{x}^*$.

If the log-likelihood function has a regular maximum with respect to the failure domain at $\boldsymbol{x}^*$, then we can compute $\det(\boldsymbol{H}^*(\boldsymbol{x}^*))$ again as

$$\det(\boldsymbol{H}^*(\boldsymbol{x}^*)) = \det\left(\boldsymbol{P}\boldsymbol{H}(\boldsymbol{x}^*)\boldsymbol{P} - \boldsymbol{n}(\boldsymbol{x}^*)\boldsymbol{n}(\boldsymbol{x}^*)^T\right). \tag{24}$$

Here $\boldsymbol{P} = \boldsymbol{I}_n - \boldsymbol{n}(\boldsymbol{x}^*)\boldsymbol{n}(\boldsymbol{x}^*)^T$.

3. PARAMETER DEPENDENCE OF THE PML-POINT

We consider the problem that the reliability problem is a function of a parameter τ. Then the point of maximal likelihood (PML) depends on this parameter value. The change of the beta value in FORM theory under parameter changes was treated in [1]. We consider an integral as in equation (1), but with only a scalar parameter τ.

Following from the results of the last paragraph the failure probability is approximated making a Taylor expansion of the log-likelihood function around the PML and using the Laplace method. We asssume that for all feasible values of τ there is exactly one PML $\boldsymbol{x}^* = \boldsymbol{x}^*(\tau)$ on the limit state surface $G(\tau)$.

The gradients of these functions with respect to the first n variables $x_1, \ldots, x_n$ are denoted by $\nabla_{\boldsymbol{x}} f$ (resp. $\nabla_{\boldsymbol{x}} g$) and the partial derivative with respect to τ by f_τ (resp. g_τ). The Hessian of a function $f(\boldsymbol{x}, \tau)$ with respect to the first n variables is written as $\boldsymbol{H}_f$.

We assume that for a fixed value of τ there is a unique PML $\boldsymbol{x}^*(\tau)$ and we write in a shorthand notation $\boldsymbol{x}^*$. The vector of the first derivatives of $\boldsymbol{x}^*$ with respect to τ is written as $\boldsymbol{x}^*_\tau$. This point is a stationary point of the Lagrangian function $L(\boldsymbol{x}, \lambda, \tau)$ defined by

$$L = f - \lambda g. \tag{25}$$

Therefore for the point $\boldsymbol{x}^*(\tau)$ the following equation system

$$\begin{aligned} \nabla_{\boldsymbol{x}} f - \lambda \nabla_{\boldsymbol{x}} g &= \mathbf{o}_n, \\ g &= 0. \end{aligned} \tag{26}$$

must be fulfilled.

To find the derivatives of the coordinates of the PML with respect to parameter changes, we differentiate this system with respect to τ and set all derivatives equal to zero. This gives then

$$\begin{aligned} \boldsymbol{H}_f \boldsymbol{x}^*_\tau + \nabla_{\boldsymbol{x}} f_\tau - \lambda_\tau \nabla_{\boldsymbol{x}} g - \lambda(\boldsymbol{H}_g \boldsymbol{x}^*_\tau + \nabla_{\boldsymbol{x}} g_\tau) &= \mathbf{o}_n, \\ \langle \nabla_{\boldsymbol{x}} g, \boldsymbol{x}^*_\tau \rangle + g_\tau &= 0. \end{aligned} \tag{27}$$

Rearranging the terms gives

$$\begin{aligned} (\boldsymbol{H}_f - \lambda \boldsymbol{H}_g)\boldsymbol{x}^*_\tau + \frac{\partial}{\partial \tau}(\nabla_{\boldsymbol{x}} f - \lambda \nabla_{\boldsymbol{x}} g) &= \lambda_\tau \nabla_{\boldsymbol{x}} g \\ \langle \nabla_{\boldsymbol{x}} g, \boldsymbol{x}^*_\tau \rangle &= -g_\tau. \end{aligned} \tag{28}$$

Since always $\nabla_{\boldsymbol{x}} f - \lambda \nabla_{\boldsymbol{x}} g = \mathbf{o}_n$, we get deleting this term

$$\begin{aligned} (\boldsymbol{H}_f - \lambda \boldsymbol{H}_g)\boldsymbol{x}^*_\tau &= \lambda_\tau \nabla_{\boldsymbol{x}} g \\ \langle \nabla_{\boldsymbol{x}} g), \boldsymbol{x}^*_\tau \rangle &= -g_\tau. \end{aligned} \tag{29}$$

This gives then for the vector $\boldsymbol{x}^*_\tau$ the form

$$\boldsymbol{x}^*_\tau = \lambda_\tau (\boldsymbol{H}_f - \lambda \boldsymbol{H}_g)^{-1} \nabla_{\boldsymbol{x}} g. \tag{30}$$

To determine the value of λ_τ we compute the scalar product of $\boldsymbol{x}^*_\tau$ and $\nabla_{\boldsymbol{x}} g$, yielding

$$\langle \nabla_{\boldsymbol{x}} g, \boldsymbol{x}^*_\tau \rangle = \lambda_\tau (\nabla_{\boldsymbol{x}} g)^T (\boldsymbol{H}_f - \lambda \boldsymbol{H}_g)^{-1} \nabla_{\boldsymbol{x}} g. \tag{31}$$

From equation (29) follows then that the lefthand side is equal to $-g_\tau$ and so we get

$$g_\tau = -\lambda_\tau (\nabla_{\boldsymbol{x}} g)^T (\boldsymbol{H}_f - \lambda \boldsymbol{H}_g)^{-1} \nabla_{\boldsymbol{x}} g. \tag{32}$$

This gives then for the derivative λ_τ then

$$\lambda_\tau = -\left[(\nabla_{\boldsymbol{x}} g)^T (\boldsymbol{H}_f - \lambda \boldsymbol{H}_g)^{-1} \nabla_{\boldsymbol{x}} g \right]^{-1} g_\tau \tag{33}$$

Inserted into equation (30) we obtain

$$\boldsymbol{x}_\tau^* = -\frac{g_\tau}{(\nabla_{\boldsymbol{x}} g)^T (\boldsymbol{H}_f - \lambda \boldsymbol{H}_g)^{-1} \nabla_{\boldsymbol{x}} g} (\boldsymbol{H}_f + \lambda \boldsymbol{H}_g)^{-1} \nabla_{\boldsymbol{x}} g. \tag{34}$$

The change of the value $f^* = f(\boldsymbol{x}^*(\tau), \tau)$ is given by

$$f_\tau^* = \langle \nabla_{\boldsymbol{x}} f, \boldsymbol{x}_\tau^* \rangle + f_\tau. \tag{35}$$

Due to the Lagrange multiplier theorem, we can replace the gradient of f by the gradient of g, giving

$$f_\tau^* = \lambda \langle \nabla_{\boldsymbol{x}} g, \boldsymbol{x}_\tau^* \rangle + f_\tau. \tag{36}$$

From equation (29) we get then

$$f_\tau^* = -\lambda g_\tau + f_\tau. \tag{37}$$

If $f_\tau = 0$, the Lagrange multiplier λ gives the change of the value of f^* relative to the negative change of g with respect to τ. If the constraint is given in the form $g(\boldsymbol{x}) - \tau = 0$ and the p.d.f. depends not on τ, we obtain the simple form

$$f_\tau^* = \lambda. \tag{38}$$

If we consider the approximation for the failure probability given in equation (22) and we neglect the change of the quantities in the denominator we get approximately

$$\frac{\partial \mathrm{P}(F)}{\partial \tau} \approx \frac{(2\pi)^{-(n-1)/2}}{|\nabla l(\boldsymbol{x}^*)| |\det(\tilde{\boldsymbol{H}}(\boldsymbol{x}^*))|^{1/2}} (-\lambda g_\tau + f_\tau). \tag{39}$$

References

[1] P. Bjerager and S. Krenk. Parametric sensitivity in first order reliability theory. *Journal of the Engineering Mechanics Division ASCE*, 115:1577–1582, 1989.

[2] K. Breitung. Asymptotic approximations for multinormal integrals. *Journal of the Engineering Mechanics Division ASCE*, 110(3):357–366, 1984.

[3] K. Breitung. Parameter sensitivity of failure probabilities. In A. Der-Kiureghian and P. Thoft-Christensen, editors, *Reliability and Optimization of Structural Systems '90, Proceedings of the 3rd IFIP WG 7.5 Conference, Berkeley, California*, pages 43–51, New York, 1991. Springer. Lecture Notes in Engineering 61.

[4] K. Breitung. Probability approximations by log likelihood maximization. *Journal of the Engineering Mechanics Division ASCE*, 117(3):457–477, 1991.

[5] K. Breitung. *Asymptotic Approximations for Probability Integrals*. Springer, Berlin, 1994. Lecture Notes in Mathematics, Nr. 1592.

[6] A. Der Kiureghian and M. De Stefano. Efficient algorithm for second-order reliability analysis. *Journal of the Engineering Mechanics Division ASCE*, 117(12):2904–2923, 1991.

[7] A. Der Kiureghian, Y. Zhung, and Ch.-Ch. Li. Inverse reliability problem. *Journal of the Engineering Mechanics Division ASCE*, 120(5):1154–1159, 1994.

9

Optimization of Foundation of Bridge on Soft Ground

Y.Demura* and **M.Matsuo****

* Department of Civil Engineering, Ishikawa National College of Technology, Japan

** Department of Geotechnical and Environmental Engineering, Nagoya University, Japan

Abstract

Presented is a procedure of optimizing the design of bridge–pier foundations constructed on soft ground which is likely to experience the long–time deformation due to the weight of the bridge. The whole structure of a bridge consisting of the superstructure and the foundations should be designed as a whole in such a way that the total expected cost of the whole structure become minimum. This procedure is totally different from the current design method in which the superstructure and the foundations are treated as a set of separate systems rather than as a total system consisting of subsystems, i.e., the superstructure and the foundations. In evaluating the total expected cost of a bridge, the construction cost both of the superstructure and the foundations as well as the damage occurrence probability should be taken into account.

Keywords: Foundation of Bridge, Soft Ground, Optimum Design, System Reliability, Bayes' Theorem

1. INTRODUCTION

Figure 1 shows a sketch of a bridge placed on pile–supported piers resting on the bearing stratum overlain by the soft clay layer. The pier will settle by amount of s due to the consolidation of the ground. The settlement is induced by the penetration of the pile–tip into the bearing stratum. The piles are drawn down by the negative friction caused by the consolidation of the clay layer loaded by the weight of the embankment. The purpose of this study is to propose the methodology of optimizing the foundation of structure on soft ground.

Suppose we have two bridges A and B, one of which, bridge A is designed with relatively low safety factor of foundation against the settlement, while the other, bridge B is designed with relatively high safety factor of foundation. The foundation of bridge

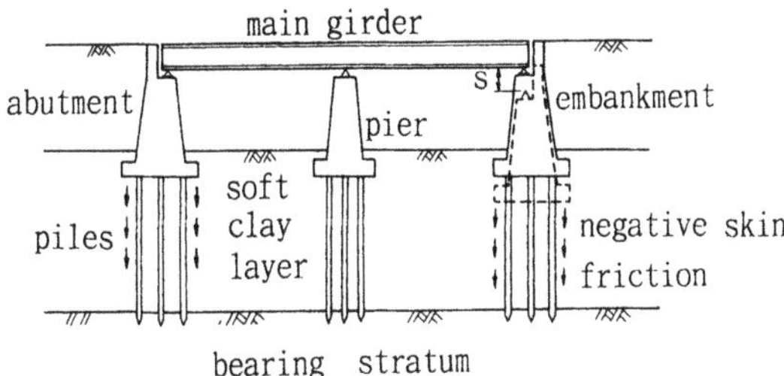

Figure 1 Bridge Constructed on Soft Ground

A is inexpensive, but likely to suffer from the unfavorable settlement with high probability. The settlement of foundation results in the additional stresses in the main girder, i.e., the main girder will have high probability of failure. Hence, the maintenance cost of main girder is expensive. The maintenance cost includes the repair works to be needed due to the future settlement. In the case of the other bridge B, the construction cost of foundation is expensive, but the maintenance cost of main girder is inexpensive. The comparison of the bridges A and B indicates the existence of the safety factor against the settlement which corresponds to the minimum summation of the construction cost and the maintenance cost. Figure 2 shows the relationship between the safety factor of foundation G_{sub} and the costs of the main girder and foundation. It should be recognized that the maintenance cost of main girder varies as a function of the safety factor of main girder.

In the procedure described in this paper, (i) we consider the whole structure as a system consisting of two subsystems, i.e., the superstructure (main girder) and substructure (foundation), and (ii) we choose the optimum design so as to realize the minimum of the total expected cost, i.e., the summation of the construction cost and the expected loss of the whole system.

An accurate prediction of the settlement of the piers is unavoidably needed in such a

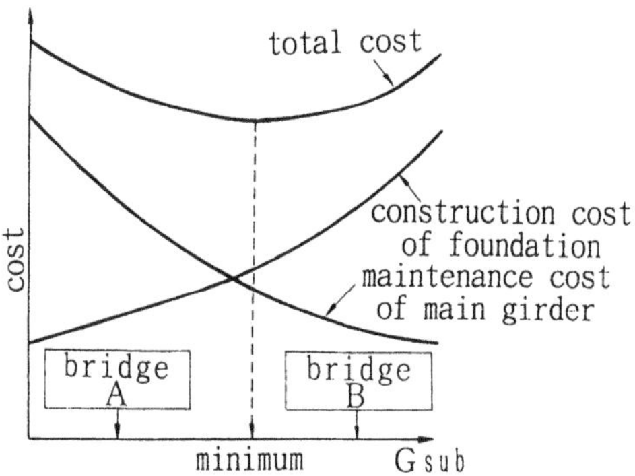

Figure 2 Relationship between Safety Factor and Cost

procedure. The model proposed in this paper includes the probabilistic settlement prediction method developed by collecting a number of case records of the settlement of bridge piers.

2. OPTIMIZATION PROCEDURE

The objective function of the system to be optimized is in principle given as

$$E[C_T]=C_{C,sup}(A_{sup})+C_{C,sub}(A_{sub})+\sum C_F(D_K)P(D_K;A_{sup},A_{sub}) \quad (1)$$

in which $E[C_T]$ denotes the total expected cost, A_{sub} the design variable of the substructure, A_{sup} the design variable of superstructure, $C_{C,sub}$ the construction cost of substructure, $C_{C,sup}$ the construction cost of the superstructure, and D_K denotes the combination of the damages done to the superstructure and to the substructure. The settlement–caused damages to the superstructure are assumed to be dependent from the settlement–caused damages to the substructure. D_K should be evaluated by taking the mechanical and functional interactions between the superstructure and substructure into account. An example will be presented later. $P(D_K)$ is the occurrence probability of D_K, and $\sum C_F(D_K)P(D_K;A_{sup},A_{sub})$ is the expected loss produced by D_K.

The optimum design choice is given by

$$(A^*_{sup},A^*_{sub})=\min E[C_T(A_{sup},A_{sub})] \quad (2)$$

in which A^*_{sup} and A^*_{sub} are the optimum design variables of the superstructure and the substructure selected out of many design alternatives, A_{sup} and A_{sub}.

3. OCCURRENCE PROBABILITY OF SETTLEMENT

Suppose a bridge shown in Figure 3. The differential (uneven) settlement δ is

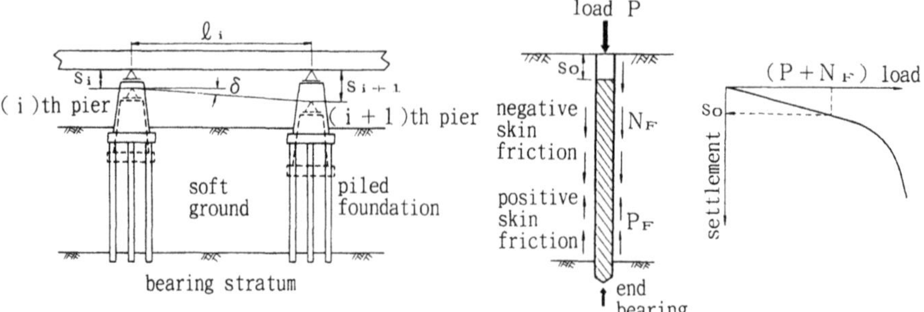

Figure 3 Differential Settlement

Figure 4 Estimation of Settlement of Pier

defined by

$$\delta = \frac{s_{i+1} - s_i}{l_i} \tag{3}$$

in which s_i is the settlement of i-th pier and l_i is the span length between i-th and ($i+1$)-th piers.

Let us estimate the occurrence probability of differential settlement of the piers.

(i) Calculate the differential settlement δ_0 by

$$\delta_0 = \frac{s_{0,\,i+1} - s_{0,\,i}}{l_i} \tag{4}$$

in which the settlement $s_{0,\,i}$ is to be estimated from the results of vertical loading test of a pile by means of the conventional method (see Figure 4).

(ii) The differential settlement estimated by Eq.(4) is known to be somewhat different from the average of the actually observed differential settlement δ through a number of past case histories. The comparison of δ_0 and δ leads to a correction factor H , see the example shown later. Then, using thus obtained empirical factor H , calculate the differential settlement δ which is most likely to take place in the field,

$$\delta = H \cdot \delta_0 \tag{5}$$

(iii) Estimation of δ_0 is accompanied by many sources of uncertainty, such as the error in estimating the friction between the piles and the subsoil (needed in estimation of the negative skin friction), the error in estimating the settlement of each pile etc. Then let us assume that δ_0 and H are random variables, and hence, δ should also be a random variable.

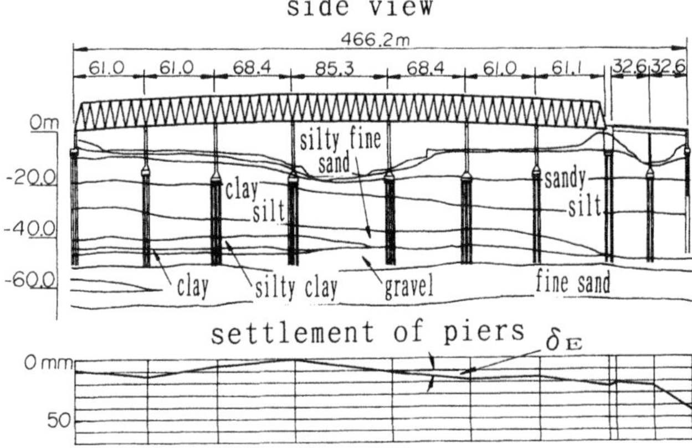

Figure 5 Example of Measured Settlement of Pier

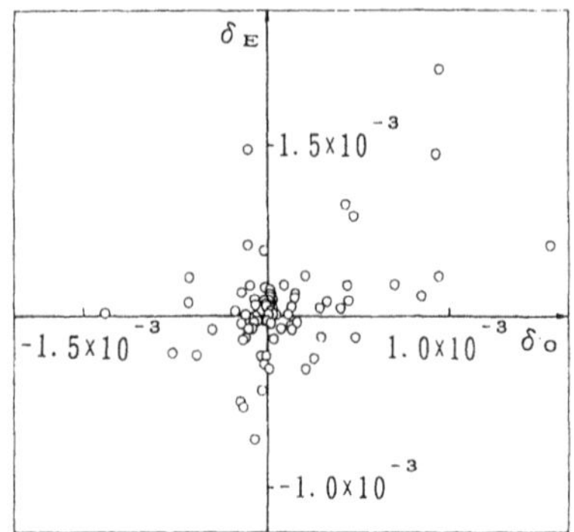

Figure 6 Comparison between Observed Values and Calculated Values

(iv) Calculate the probabilistic distribution of δ_0 from the probabilistic distributions of errors arising from the above mentioned sources of error. The probabilistic parameters of the correction factor H is obtained, as seen later, based on a number of settlement records processed through Bayes' theorem. The settlement records were collected from many bridge piers which suffered from the differential settlement in the past. Considering those observed differential settlements δ_E as the samples of δ , we get the distribution of parameter θ_H of the probabilistic distribution of correction factor H by using the equation (6).

$$\zeta^1(\theta_H) = \frac{\prod_{nS}^{i=1} f_{\delta i}(\delta_i = \delta_{Ei} \mid \theta_H)\zeta^0\theta_H}{\int \prod_{nS}^{i=1} f_{\delta i}(\delta_i = \delta_{Ei} \mid \theta_H)\zeta^0\theta_H d\theta_H} \tag{6}$$

In Eq.(6), $\zeta^0(\theta_H)$ denotes the prior distribution of θ_H while $\zeta^1(\theta_H)$ the posterior distribution of θ_H . And $f_{\delta i}(\delta_i = \delta_{Ei} \mid \theta_H)$ is the conditional distribution of δ for a given value of θ_H derived from the probabilistic distribution of H and δ . "i " is the

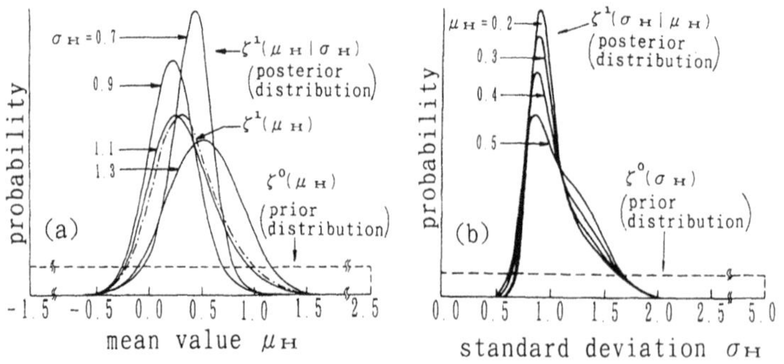

Figure 7 Posterior Distribution of Probabilistic Parameters of Correction Factor

numerical order of the observed case record. n_S is the number of observed case records.

Figure 5 shows an example of a bridge which suffered from the differential settlement. δ_E is the observed differential settlement. In Figure 6, δ_E is plotted against the differential settlement δ_0 calculated by the conventional method. Each data point indicates the mean value of random variables δ_0 and the sample of δ. There exists a correlation between the observed values and the calculated values. Those results indicate that Eq. (5) is acceptable and the correction factor H should be random variable.

Figures 7 (a), (b) show the posterior distributions $\zeta^1(\theta_H)$ obtained by assuming that the prior distributions of the mean value of H and the standard deviation of H are uniform. μ_H is the mean value of H, and σ_H is the standard deviation. Those posterior distributions will be used in the estimation of occurrence probability of differential settlement needed in the optimization procedure.

4. EXAMPLE

The bridge to be used here as an example is shown in Figure 8. It is a plate girder bridge constructed on a soft ground. As the design alternative of the super structure, the flange thickness of girder is taken. As the design alternative of the substructure, the area of the thin film covering the pile is used. This film is fixed to the pile surface to reduce the friction between the pile and soil aiming at the reduction of the negative skin friction. The input data for the numerical calculation are listed in tables 1 and 2.

The damage possibly done to the foundation by the negative skin friction is classified as $D_{sub,1}$ and $D_{sub,2}$. $D_{sub,1}$ denotes the situation in which the excessive settlement takes place resulting in the necessity of reconstruction of the foundation. $D_{sub,2}$ denotes the complement event of $D_{sub,1}$ in which the addition of the stress in the girder flange is more than negligible. $D_{sup,1}$ denotes the situation in which the main girder have yielding or buckling in the main girder. $D_{sup,2}$ denotes the complement event of $D_{sup,1}$.

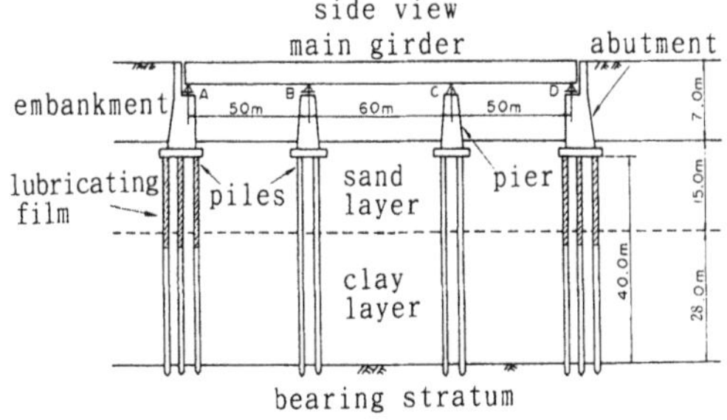

Figure 8 Numerical Example

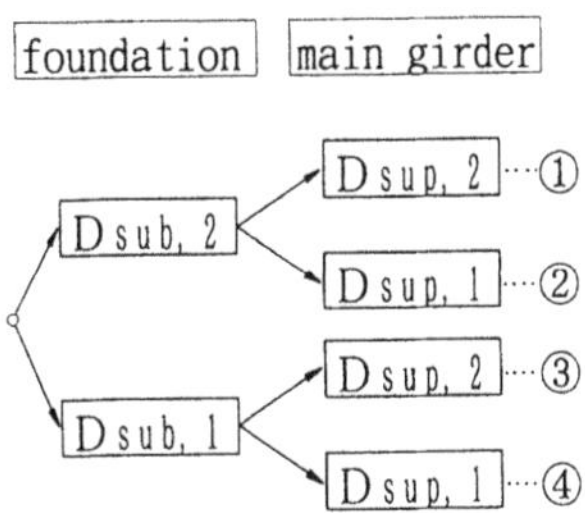

Figure 9 Combination of Damages

Table 1 Probabilistic Input

		actual value / nominal value	coefficient of variation
foundation	negative friction	0.87	0.33
	settlement of pile	0.38	0.087
main girder	stress	0.48	0.36
	strength	1.15	0.10

Table 2 Cost (yen)

foundation	construction	2.8×10^{3} /m³
	maintenance	5.0×10^{7}
main girder	construction	4.0×10^{4} /kN
	maintenance	62.5×10^{9}

The expected loss is estimated for the cases depending on the combination (see Figure 9) of degree of the differential settlement and the degree of damage to the main girder. For example, see the case ① shown in Figure 9 . In this case, the differential settlement was of class $D_{sup,2}$. The maintenance of the foundation is needed. The expected loss for this case is estimated by Eq.(7).

$$E[C_F]_{①}=P(D_{sub,2})[C_{MF}+C_{MG}P(D_{sup,2})] \tag{7}$$

$E[C_F]_{①}$ denotes the expected loss for the case ①. C_{MF} and C_{MG} are the costs for the repair of foundation and main girder. $P(D_{sub,2})$ denotes the occurrence probability of the differential settlement $D_{sub,2}$. $P(D_{sup,2})$ denotes the probability of failure of the main girder. In the case ③, the differential settlement was of class $D_{sub,1}$ resulting in the appreciable addition of stresses in the flanges which yielded or buckled (damage $D_{sup,2}$). The expected loss for this case is estimated by

$$E[C_F]_{③}=P(D_{sub,1})[C_{MG}P(D_{sup,1}|D_{sub,1})] \tag{8}$$

in which $E[C_F]_{③}$ denotes the expected loss for the case ③, $P(D_{sub,1})$ denotes the occurrence probability of the differential settlement $D_{sub,1}$, $P(D_{sup,2}|D_{sub,1})$ denotes

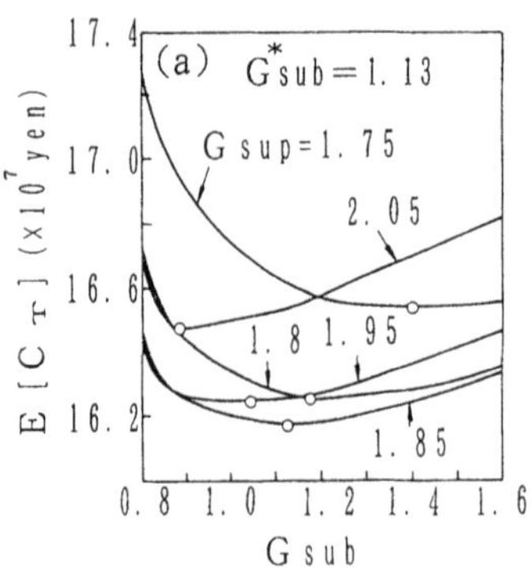

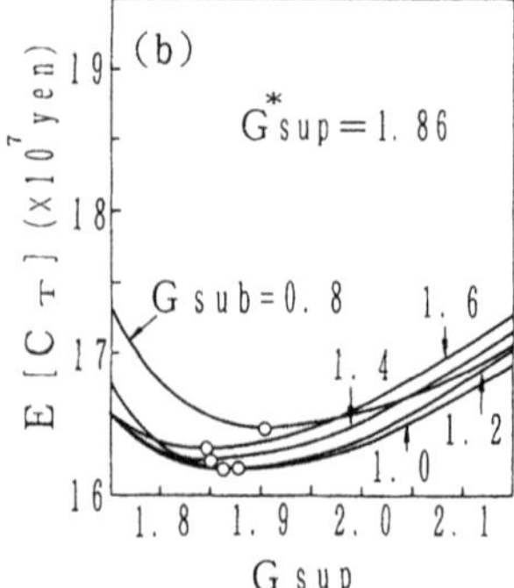

Figure 10 Optimum Solutions

the probability of failure of the main girder subjected to the additional stresses. Each case shown in Fig. 9 is handled in the same fashion. The summation of the expected losses for all the cases plus construction cost is the objective function which we try to minimize by properly choosing the design alternatives, A_{sup} and A_{sub}.

Figure 10 show the final results of the above mentioned optimization procedure. The abscissa of Figure 10(a) is the safety factor G_{sub} against the differential settlement of the foundation, while the ordinate is the total expected cost $E[C_T]$ plotted against G_{sub} with the safety factor G_{sup} of the main girder as a parameter. The abscissa of Figure 10(b) is the safety factor G_{sup} of the main girder and the parameter is the safety factor of foundation. The safety factors at which the total expected cost becomes minimized are $G_{sup}=1.86$ and $G_{sub}=1.13$. These two values are the optimum combination of two safety factors for the super and sub structure. It may be interesting to compare to these two values with the safety factors required by the current conventional design codes, i.e., $G_{sup}=1.7$ and $G_{sub}=1.4$. The safety factor of foundation in the optimum design is smaller than the safety factor in the current design code. The safety factor of main girder in the optimum design is larger than the safety factor in the current design code. These results are due to the settlement of foundation at the optimum design which is larger than the settlement allowable in the current design code.

5. CONCLUSIONS

The optimization procedure for the bridge design is briefly outlined and an example of the application of the optimization procedure is presented. As the conclusions, followings should be noted.

(1) The use of the objective function derived for the total system including both the superstructure and the substructure leads to the optimum design more rational than the design optimized separately for the superstructure and substructure.

(2) The example presented in this paper resulted the safety factors for the superstructure and substructure which happened to be fairly close to the safety factors required by the conventional design codes.

(3) The proposed method seems to be useful in seeking the bridge designs with much harmony in the whole system of the superstructure and substructure.

REFERENCES

1. M.Matsuo and Y.Demura, Proc. of Japan Society of Civil Engrg. Vol.340/ III -4, pp.129-138, 1984.12 (in Japanese).
2. M.Matsuo and Y.Demura, Proc. of Japan Society of Civil Engrg. Vol.364/ III -4, pp.215-224, 1985.12 (in Japanese).

10

RELIABILITY OF ANCHORED MONOLITH STRUCTURES UNDER CORROSION EFFECTS

Dan M. Frangopol[a], Milan Chakravorty[a], Reed L. Mosher[b] and Jan E. Pytte[a]

[a] Department of Civil, Environmental, and Architectural Engineering, University of Colorado, Boulder, Colorado 80309-0428, USA
[b] USAE Waterways Experiment Station, 3909 Halls Ferry Road, Vicksburg, MS 39180-6199, USA

A research program at the University of Colorado at Boulder investigated time-variant reliability assessment of concrete gravity structures founded on rock for different loading conditions. In the course of this investigation, a methodology was developed for reliability analysis of anchored monolith structures under corrosion effects. This paper presents a brief overview of the methodology for time-variant reliability assessment of existing concrete gravity structures under corrosion effects.

1. INTRODUCTION

As stated recently in a U.S. Army Corps of Engineers technical letter [1] "Inland navigation structures are a vital link in the national infrastructure. Over 40% of these facilities are more than 50 years old and the demands for rehabilitation are increasing". In the United States, safety of existing navigation structures has historically been evaluated deterministically, using design criteria for new structures [2]. This often resulted in inconsistent estimates.

Over time, changes in loading conditions as well as material degradation introduce uncertainties which deterministic analysis does not effectively take into consideration. Creep of grout, relaxation of prestressed bar/tendon and possible underground corrosion contribute to the uncertainties in evaluation of existing concrete gravity structures. Therefore, reliability assessment for these structures has to be based on probabilistic methods.

An overview of a methodology for time-variant reliability assessment of existing rock-anchored concrete gravity structures developed at the University of Colorado at Boulder is presented in this paper.

2. ANCHORED MONOLITH STRUCTURES

2.1. Loading Conditions

In the United States, there are many anchored monolith structures (also called gravity structures). The elevation of such a structure is schematically given in Fig.1. Some of them are about 100 years old. One of these structures was considered for reliability assessment. The detailed description of this structure is given in [3]. It was built at the beginning of this century, and after about seventy years in service the structure was strengthened with prestressed rock anchors. The details of the anchors are shown in Fig.2.

The loading conditions used in reliability analyses of existing gravity structures are generally the same as those used in the design of these structures [2]. They consist of four loading situations including normal operating condition (most critical combination of pools that routinely occur due to fluctuations in river flow), maintenance condition (dewatering), extreme operating conditions (higher pools), and catastrophic loading conditions (loss of lower pool or earthquake).

2.2. Corrosion

Over the years, monolith structures strengthened with prestressed rock anchors may be subject of various types of corrosion activities. This investigation deals exclusively with corrosion of anchor bars, assuming that the structure itself does not have appreciable deterioration. A survey of failure of rock anchors by Xanthakos [4] indicates that the life of a rock anchor is mainly controlled by corrosion which largely affects the unbonded or free length of anchor. The rate of corrosion penetration in anchor bar/tendon is governed by the aggressivity of the environment and the protection provided to the anchor in question.

The corrosion initiation is related to the presence of aggressive chemicals, such as chloride ions, beyond a threshold value (generally believed to be between 0.2% to 0.4% by weight of cement of chloride concentration) at the surface of steel. Immediately after reaching the threshold value (i.e., corrosion initiation time t_0) the propagation of corrosion results in material loss.

The material loss function is adapted from a power function fitted to experimental data of atmospheric corrosion penetration in steel coupons by Townsend and Zoccola[5]:

$$C = a\, t_1^{\mathrm{b}} \tag{1}$$

where, C is corrosion penetration (μm) per side, and a and b are regression coefficients.

Assuming analogy between atmospheric corrosion and underground corrosion penetration, eqn.(1) is used with a modifier factor α, $C = \alpha\, a\, t_1^{\mathrm{b}}$. The modifier factor α takes into consideration the difference between the material loss expected in an underground environment and that in an atmospheric environment [3,6,7]. A close corroboration of observed data of material loss in sheet piles in ground by Buslov [8] supports this assumption.

2.3. Prestress

Longterm observation of dams shows that total prestress loss in rock anchors reach a maximum of 10% [9]. This value was also suggested by Littlejohn and Bruce [10] and Eberhardt and Veltrop [11] to cover the worst possible case of prestress loss, including losses due to creep in cement grout. With 10% loss due to relaxation and creep, the prestress loss model may be expressed as follows:

$$P_1(t) = P_o \times [0.9 + 0.1e^{(-\psi t)}] \tag{2}$$

where, P_o is the initial prestress force, $P_1(t)$ is the prestress force at time t (years), and ψ is a decay constant. With relaxation loss of 3% over 7 days, the resulting value of ψ is

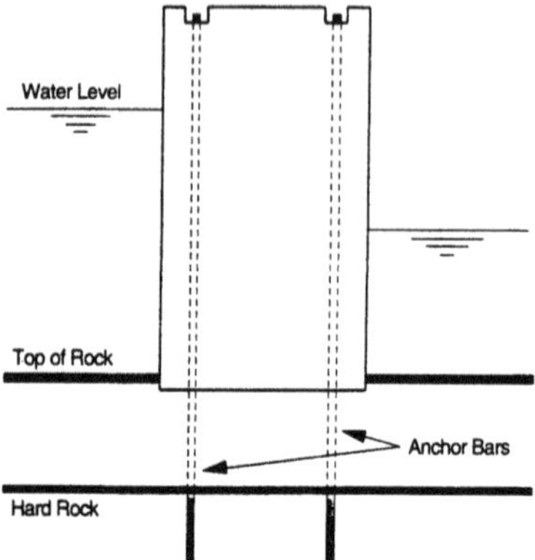

Figure 1. Schematic Elevation of a Monolith Structure

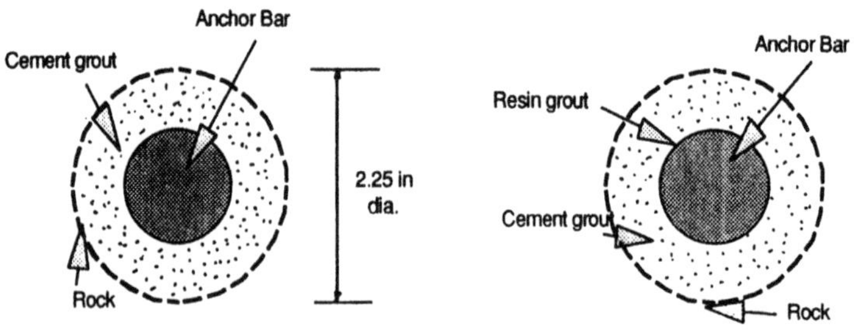

Figure 2. Details of Anchors

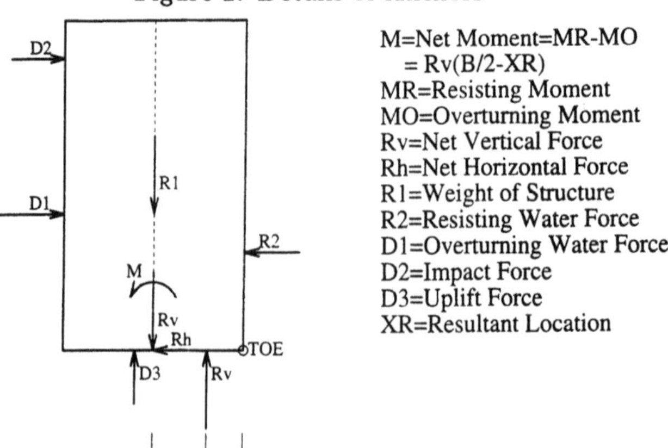

Figure 3. Resultant Location

18.585.

Assuming that the reduction of prestress force due to relaxation and creep occurs before corrosion loss takes effect, the relationship between the prestress force after loss due to relaxation, creep and corrosion at any point in time, $P_2(t)$, and the prestress force after loss only due to relaxation and creep, $P_1(t)$, may be established, from strain compatibility, as [3]:

$$P_2(t) = P_1(t)\frac{l_1 + l_2}{l_2 \frac{A(o)}{A(t)} + l_1} \tag{3}$$

where $A(o)$ = original cross-sectional area of the anchor bar, $A(t)$ = average cross-sectional area of the corroded anchor bar at time $t = t_0 + t_1$, l_1 = part of the free length of the anchor bar not affected by corrosion, l_2 = part of the free length of the anchor bar affected by corrosion, and $P_1(t)$ = prestressing force after loss due to relaxation and creep. If the total free length is affected by corrosion, then $P_2(t) = P_1(t)\, A(t)/A(o)$.

2.4. Variables

The dimensions of anchors, underground strata conditions (characteristics and physical properties), corrosion parameters, physical properties of concrete and prestressing strands, and prestressing forces can be all characterized as random variables with known mean values and standard deviations. A list of such variables used for reliability analysis with respect to resultant location of an anchored monolith structure is included in Table 1.

Table 1
List of variables

Variable	Mean	Standard Deviation
Density of fractured base rock, γ_f (kcf)	0.160	0.016
Density of monolith concrete, γ_{con} (kcf)	0.140	0.007
Uplift factor, E	-0.25	0.20
Angle of internal friction of base rock, ϕ_{rock} (degree)	30	6
Diameter of anchor bar, d (in)	1.25	0.0375
Regression coefficient, a	37.8	7.56
Regression coefficient, b	0.749	0.1498
Initial prestress force, P_o, in anchor type I (kips)	112	8.96
Initial prestress force, P_o, in anchor type II (kips)	93	7.44
Yield stress of anchor bar, f_y (ksi)	129	3.225
Part of free length not affected by corrosion, l_1 (ft)	34.3	0.34
Part of free length affected by corrosion, l_2 (ft)	12.0	0.60
Ultimate bond stress factor, k	0.15	0.075
Time of exposure, t (years)	0 - 100	0
Model coefficient, α	1; 2; 3	0

It may be noted that while many random variables have been considered in the reliability analysis, the level of water for various loading cases is considered as deterministic.

In fact, four loading cases have been considered in the analysis. These cases are associated with the difference between the level of water in the upper and lower pools of 6.4ft, 8.2ft, 25.4ft, and 31ft. Both reliability with respect to resultant location including overturning (rotational stability) and sliding reliability have been considered. In this paper, however, owing to space limitations, only the reliability with respect to resultant location will be briefly presented.

3. ROTATIONAL STABILITY

The probability that the resultant of all vertical forces, R_V, is located at the point of rotation (i.e., at the toe of the structure) is equivalent to the probability that the resisting moment equals the overturning moment. This situation of impending overturning is a true limit state. However, under these circumstances, overstressing of the structure or the foundation may result in tensile stresses at the base causing separation of the base from the foundation. As such, it is desirable to investigate the effect of varying resultant location X_R measured from the toe (see Fig.3). If the resultant is located more than B/3 from the toe (B is the width of the structure/foundation at base), then the structure at the base will have 100% compression in the foundation contact area. Similarly, resultant locations, X_R, more than B/6 and B/4 from the toe correspond to different percentages of base compression. The limit states associated with these resultant locations can best be categorized as several transition states prior to reaching a true limit state, when the structure fails by overturning (i.e., $X_R = 0$). General formulation of these transition state functions for a structure having $m \times n$ anchors subjected to overturning about the **Y**-axis of rotation (see Fig.4) considers the following assumptions: (a) the monolith structure is rigid; (b) linear strain relationship exists among anchors in one row parallel to **X**-axis; (c) all anchors have the same free length and same fixed length; and (d) all anchors get corroded at the same time and with the same corrosion rate.

The general limit state function against rotational instability is:

$$g = M_R + \sum_{i=1}^{m} \sum_{j=1}^{n} [x_m\, f_y\, A_{m,j}(t) + x_i\, P_{2\,ij}(t)] - M_O - R_V\, X_R = 0 \tag{4}$$

where, M_R is the resistance moment of monolith structure without anchors, M_O is the overturning moment, R_V is the resultant vertical force, $P_{2\,ij}(t)$ is obtained from eqn.(3), f_y is defined in Table 1, and the other terms are defined in Fig.3.

In what follows, the limit state of resultant location is used as a two-dimentional stability analysis to replace overturning analysis [2]. In fact, ref.[2] states "overturning is unlikely as a pure mode of failure, as foundation bearing and/or sliding failure would occur before the resultant reaches the toe".

4. RELIABILITY ANALYSIS

The general purpose structural reliability analysis program RELTRAN, based on first order reliability method, developed at the University of Colorado at Boulder [12], has been used to compute reliability indices of anchored monolith structures. The general

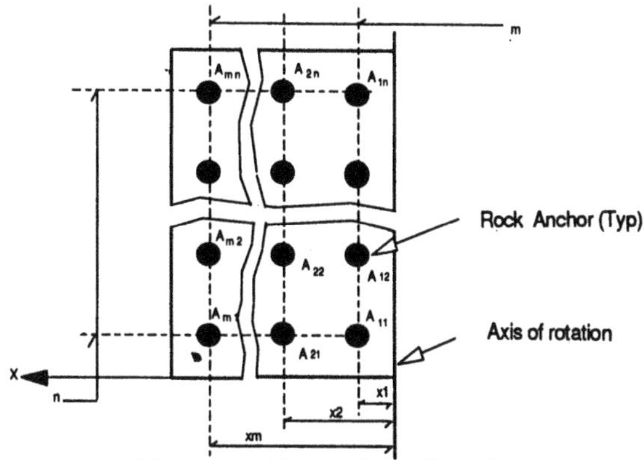

Figure 4. General Anchor Layout

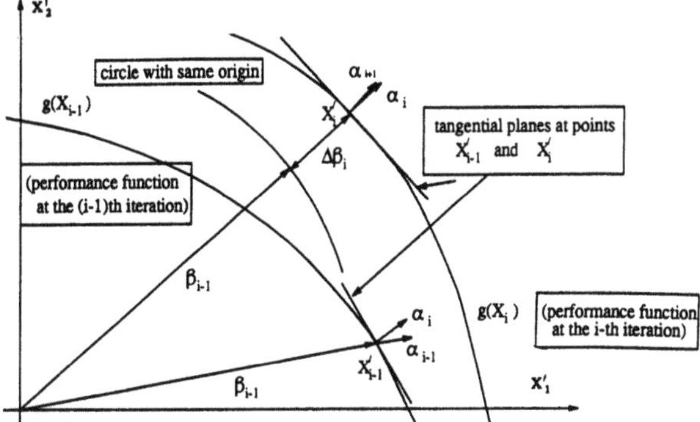

Figure 5. Geometrical Representation of RELTRAN Algorithm [12]

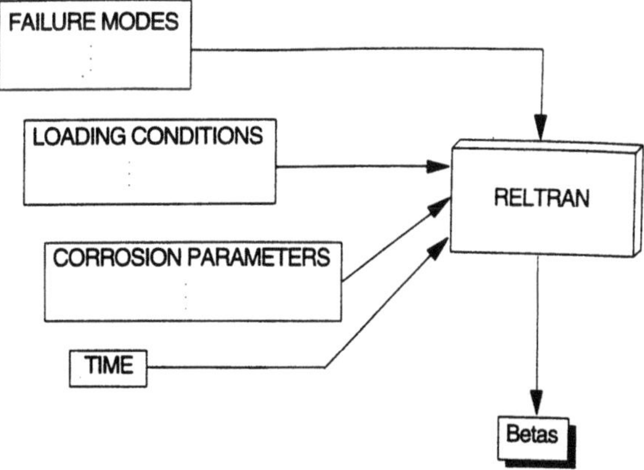

Figure 6. Use of RELTRAN in Time-Dependent Reliability Analysis

illustration of RELTRAN's algorithm is shown in Fig.5 [12]. Time-variant reliability assessment is accomplished through repetitive use of RELTRAN for each time increment (see Fig.6).

5. RESULTS

For three loading conditions, Table 2 shows the results of time-variant reliability analyses of the anchored monolith structure whose characteristics are defined in Table 1, for four cases of resultant location (measured from the toe, see Fig.3): $X_R = 0$ (overturning), $X_R = B/6$ (resultant being in the middle two-thirds of the base), $X_R = B/4$ (resultant being in the middle one-half of the base), and $X_R = B/3$ (resultant being within the kern, 100% compression at the foundation-base area).

Table 2
Reliability Index for Resultant Location

Loading Conditions	Resultant Location (X_R)	Reliability Index β at Time t					
		0 yr	10 yr	20 yr	30 yr	40 yr	50 yr
	0	10.80	10.41	10.37	8.49	8.28	8.05
Normal	B/6	8.49	8.03	7.95	5.30	4.76	4.41
Operating	B/4	6.22	5.68	5.57	4.05	3.62	2.99
	B/3	1.80	1.08	0.97	0.85	0.74	0.63
	0	7.39	6.97	6.93	6.69	5.77	4.85
High	B/6	4.29	3.78	3.72	3.64	3.22	2.84
Water	B/4	1.24	0.64	0.56	0.49	0.42	0.35
	B/3	< 0	< 0	< 0	< 0	< 0	< 0
	0	7.48	7.17	7.14	7.11	6.44	5.58
Maintenance	B/6	4.61	4.26	4.21	4.17	4.11	3.37
	B/4	2.06	1.68	1.62	1.58	1.53	1.48
	B/3	< 0	< 0	< 0	< 0	< 0	< 0

The three loading conditions considered included normal operating, high water, and maintenance. These conditions were associated with different water levels of the upper and the lower pools [3]. The corrosion model coefficient assumed in computations was $\alpha = 2$, and the parameters a and b were considered random variables with means $\bar{a} = 37.8$, $\bar{b} = 0.749$, coefficients of variations $V(a) = V(b) = 20\%$, and correlation coefficient ρ(a,b)=0.5.

6. CONCLUSIONS

In conclusion, it is emphasized that although the proposed time-dependent reliability-based assessment methodology for anchored monolith structures appears to be adequate, much efforts are needed in connection with : (a) developing resistance models for damaged monolith structures, (b) developing practical risk-based criteria for reliability assessment of existing monolith structures under time-dependent effects, and (c) developing a suitable methodology (possibly based on multi-objective reliability-based optimization) for safety management of navigation structures by considering risk-benefit tradeoffs.

In order to realize these ambitious goals, interdisciplinary research interfacing structural analysis, reliability, and optimization techniques is necessary.

ACKNOWLEDGEMENT

The support of this research by the U.S. Army Corps of Engineers under Contract No. DACW 39-92-K-0032 is gratefully acknowledged. The results, opinions, and conclusions expressed in this paper are solely those of the authors and do not necessarily represent those of the sponsor.

REFERENCES

1. ETL-1110-2-532, U.S. Army Corps of Engineers, Washington, D.C. (1992).
2. ETL-1110-2-321, U.S. Army Corps of Engineers, Washington, D.C. (1993).
3. D. M. Frangopol, M. Chakravorty, and J. E. Pytte, Time-Variant Reliability of Rock Anchors for Navigation Structures, Final Research Report for the U.S. Army Corps of Engineers, University of Colorado, Boulder, Colorado (1993).
4. P. P. Xanthakos, Ground Anchors and Anchored Structures, John Wiley and Sons, New York (1991).
5. H. E. Townsend and J. C. Zoccola, Eight-year Atmospheric Corrosion Performance of Weathering Steel in Industrial, Rural and Marine Environments, ASTM, STP No.767, Penn.(1982).
6. M. Chakravorty, D. M. Frangopol, R. L. Mosher, and J. E. Pytte,Time-Dependent Reliability of Rock-Anchored Structures, Reliability Engineering & System Safety (1995) accepted.
7. D. M. Frangopol, M. Chakravorty, R. L. Mosher, and J. E. Pytte, Reliability Assessment of Rock-Anchored Gravity Structures, Proceedings ICASP 6, Paris, France, (1995) accepted.
8. V. Buslov, Corrosion of Steel Sheet Piles in Port Structures, Journal of Waterway Port Coastal and Ocean Engineering, ASCE, Vol.109, No.1 (1983).
9. M. P. Collins and D. Mitchell D, Prestressed Concrete Structures, Prentice Hall, New Jersey, (1991).
10. G. S. Littlejohn and D. A. Bruce, State-of-the Art - Rock Anchors, Foundation Publications Ltd., Sussex, England (1977).
11. A. Eberhardt and J. A. Veltrop, Prestressed Anchorage for Large Trainter Gate, Journal of Structural Div., ASCE, 90(ST6), (1964).
12. Y-H. Lee, S. Hendawi, and D. M. Frangopol,"RELTRAN: A Structural Reliability Analysis Program: Version 2.0", Rep. No. CU SR-9316, University of Colorado, Boulder, Colorado (1993).

11

Reliability of Flexible Structures Exposed to Non-Gaussian Wind Pressure

V. Gusella and A.L. Materazzi

Istituto di Energetica, Università di Perugia, via S. Lucia - Canetola, 06125 Perugia, Italy

The influence of the wind pressure model on the structural response is investigated. The wind turbulence is assumed to be a homogeneous Gaussian field. For the pressure process three models of increasing complexity are used: the classical linearized Gaussian model, the Gaussian quadratic model and the 'exact' non-Gaussian model. The estimate of the peak distribution allows to compare the classical and the proposed procedure on the basis of numerical examples.

1. INTRODUCTION

The assessment of safety of structural components exposed to turbulent wind asks for the use of a suitable model of the pressure fluctuations. Field observations agree to state the Gaussian character of the wind velocity. For the pressure fluctuations, which are roughly proportional to the square of the air velocity, this assumption is no longer true, as it is confirmed by the experimental evidence (Holmes [1]).

Nevertheless in the wind engineering applications it is often assumed, as a first approximation, the wind pressure to be a Normally distributed random process.

Vaicaitis et al. [2] suggested that the errors inherent in this assumption may, in some cases, be significant. Soize [3] extended the procedure proposed for the linear case by Davenport [4], including the effect of non-linear pressure terms, thus defining an analytical expression for the gust response factor of buildings.

The non-normal stochastic response of linear systems was firstly analysed by Lutes et al. [5]. In order to investigate the non-normality of structural response, due to a non-normal excitation, he proposed to extend the conventional analysis by using the third and the fourth moment of the response. The same problem was faced by Grigoriu et al. [6] which analysed the moments and mean crossing rate of the response of linear system subject to polynomials forms of the Gauss-Markov process and by Bucher and Schueller who used spectral analysis to compute the structural response [7].

In the present paper the theory of the structural response to a non-Gaussian excitation in the frequency domain is firstly reviewed. Then comparisons are made between the response computed utilising the classical linearized Gaussian model, the Gaussian model including quadratic terms and the 'exact' non-Gaussian model, in order to verify the consequences of the choice of the model on the evaluation of the structural reliability.

2. MODEL OF THE WIND PRESSURE

The velocity of the moving air particles is well represented by a three dimensional weakly-homogeneous random field following the Normal distribution:

$$\vec{V}(t) = [\overline{V}+v_x(t)]\vec{i}+v_y(t)\vec{j}+v_w(t)\vec{k} \tag{1}$$

where $\overline{V}$ is the mean wind speed and $v_x(t)$, $v_y(t)$, $v_w(t)$ are the random components of the turbulence in the directions x, y and z respectively, which are zero mean normal processes. Then the velocity random field is completely described by the matrix of its power spectral density, which naturally includes all the cross components.

Limiting the attention on the along-wind component of the velocity, the wind induced pressure over a fixed obstacle may be expressed as:

$$p(t) = \frac{1}{2}\rho C_D\, V^2 = \frac{1}{2}\rho C_D\left[\overline{V}+v(t)\right]^2 = \frac{1}{2}\rho C_D\left[\overline{V}^2+2\overline{V}\,v(t)+v^2(t)\right] \tag{2}$$

In most wind engineering applications the validity of the assumption $v_x(t) << \overline{V}$ is silently accepted as a rule, although it is probably hardly verified for mild winds. This hypothesis allows to neglect the second order term in Eq. (2), thus considering the pressure process as being Gaussian. When the squared component is taken into account the resulting process is no longer Normally distributed and its probabilistic structure may not be described using only the first two moments.

3. PROBABILISTIC STRUCTURE OF THE PRESSURE PROCESS

As the pressure process is non-Gaussian, its probabilistic structure may be described specifying the complete set of the statistic moments of ascending order.

The first moment is easy found to be constant:

$$\overline{p} = E\{p(t)\} = E\{cV^2\} = c[\overline{V}^2+\sigma_v^2]\text{, where } c=\frac{1}{2}\rho C_D \tag{3}$$

The second order autocovariance $R_{pp}(\tau)$ is:

$$\begin{aligned} R_{pp}(\tau) &= E\{[p(t)-\overline{p}][p(t+\tau)-\overline{p}]\} = E\{[cv^2(t)-\overline{p}][cv^2(t+\tau)-\overline{p}]\} \\ &= c^2\left[4\overline{V}^2 R_{vv}(\tau)+2R_{vv}^2(\tau)\right] \end{aligned} \tag{4}$$

By setting $\tau = 0$, the Eq. 4 yields the variance of the pressure process:

$$\text{var}\{p^2(t)\} = c^2[\overline{V}^4+2\overline{V}^2\sigma_v^2] \tag{5}$$

The third order autocovariance $R_{ppp}(\tau_1,\tau_2)$ may be expressed in terms of the autocovariance of v(t), $R_{vv}(\tau)$, taking advantage of the Gaussian properties of v(t), which allows to obtain the joint moments of v(t) by differentiating the joint characteristic function:

$$\phi_v(t)=\exp\left[-\frac{1}{2}t^T\Gamma t\right] \tag{7}$$

where $\mathbf{t}^T = [t_1,t_2,\ldots,t_n]$, and setting $\mathbf{t} = 0$.

The generic joint moment of v(t) is given by:

$$E\{v_1^{m_1}\, v_2^{m_2} \ldots\ldots v_n^{m_n}\}=i^{-(m_1+m_2+\ldots+m_n)}\left[\frac{\partial^{m_1+m_2+\ldots+m_n}}{\partial t_1^{m_1}\partial t_1^{m_2}\ldots\ldots\partial t_1^{m_n}}\phi_v(t)\right]_{t=0} \tag{8}$$

The task may be performed by using symbolic computer algebra, thus reducing the expression of $R_{ppp}(\tau_1,\tau_2)$ to the form:

$$\begin{aligned}R_{ppp}(\tau_1,\tau_2)=8c^3\,\overline{V}^2\left[R_{vv}(\tau_1)R_{vv}(\tau_2)+R_{vv}(\tau_1)R_{vv}(\tau_1-\tau_2)+R_{vv}(\tau_2)R_{vv}(\tau_2-\tau_1)\right]+\\+8c^3\,R_{vv}(\tau_1)R_{vv}(\tau_2)R_{vv}(\tau_1-\tau_2)\end{aligned} \tag{9}$$

The frequency-domain counterpart of this formulation includes the power spectral density of the pressure process $S_{pp}(f)$ and the third order power spectral density $S_{ppp}(f_1, f_2)$ which may be computed in closed form if $\sigma_v^2 \ll \overline{V}^2$. In that case the second term in the right hand side of Eq. 9 gives a negligible contribution to the spectral density, which becomes (fig. 1):

$$S_{ppp}(f_1,f_2)=8c^3\,\overline{V}^2\left[S_{vv}(f_1)S_{vv}(f_2)+S_{vv}(f_1)S_{vv}(f_1+f_2)+S_{vv}(f_2)S_{vv}(f_1+f_2)\right] \tag{10}$$

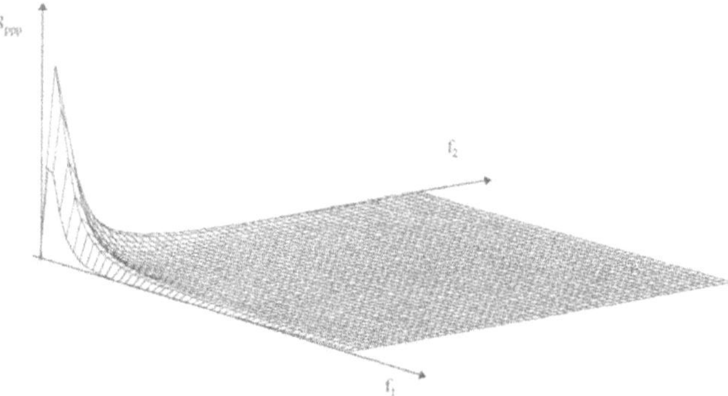

Figure 1. Typical shape of the bi-dimensional power spectral density of pressure $S_{ppp}(f_1,f_2)$.

As a well known property of the Gaussian processes is to have the third moment equal to zero, the function $S_{ppp}(f_1, f_2)$ gives valuable information on the non-Gaussian characteristics of the pressure process. For this reason in the first part of the research, presented in this paper, the analysis is truncated at the third order moments.

4. RESPONSE OF LINEAR SYSTEMS TO THE WIND PRESSURE

4.1 Statistical characteristics of the response process

Pointing the attention over a single-degree-of-freedom linear time-invariant system, the dynamic response under wind excitation may be computed solving the dynamic equilibrium equation in the frequency domain, thus evaluating the power spectral densities of the response.

So it is easy to find:

$$S_{xx}(f)=S_{pp}(f)H(f)\overline{H}(f) \tag{12}$$

$$S_{xxx}(f_1,f_2)=S_{ppp}(f_1,f_2)H(f_1)H(f_2)\overline{H}(f_1+f_2) \tag{13}$$

where H(f) is the frequency response function of the system, and the overbar denotes its complex conjugate (Fig. 2). The corresponding statistical moments may be found by integration:

$$\mu_{2,x}=\sigma_x^2=\int_{-\infty}^{+\infty} S_{xx}(f)df \; ; \; \mu_{3,x}=\int_{-\infty}^{+\infty}\int_{-\infty}^{+\infty} S_{xxx}(f_1,f_2)df_1\,df_2 \tag{14}$$

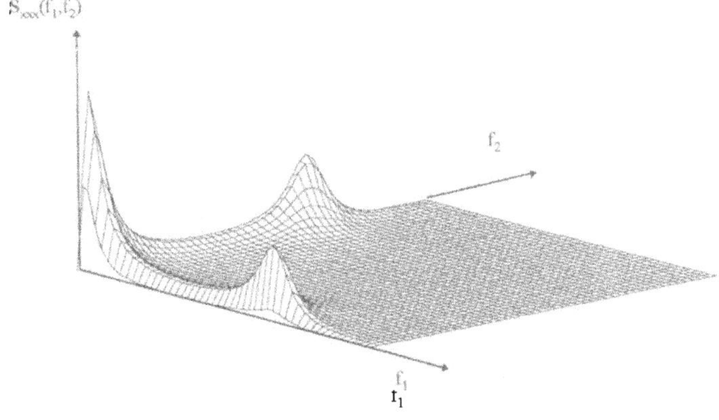

Figure 2. Typical shape of the bi-dimensional power spectral density of the response $S_{xxx}(f_1,f_2)$.

4.2 Distribution of the response x(t)

The probability density function of the structural response under wind usually only slightly departs from the Gaussian shape. Among the methods available for its characterisation on the basis of the knowledge of the response statistical moments we remember the Series Expansion Method and the Maximum Entropy Method. In this paper the first approach is adopted and the function $f_x(x)$ is approximated with the Gram-Charlier series expansion:

$$f_X(x) = N_{m,\sigma}(x)\left[1 + \sum_{j=3}^{n} \frac{C_j}{j!} H_j(x)\right] \tag{15}$$

where $N_{m,\sigma}(x)$ is the Gaussian distribution with given mean and variance, $H_j(x)$ are the Hermite polynomials of order j, and C_j are suited coefficients, computed to fulfil the condition that the distribution $f_X(x)$ has the specified statistical moments.
Considering the first three moments, the density of the response becomes:

$$f_X(x) = \frac{1}{\sigma\sqrt{2\pi}} \exp\left[-\frac{1}{2}\left(\frac{x}{\sigma}\right)^2\right] \cdot \left[1 + \frac{\mu_{3,x}}{6\sigma^3} \cdot \left(\frac{x^3}{\sigma^3} - \frac{3x}{\sigma}\right)\right] \tag{16}$$

5. RELIABILITY ANALYSIS

The first-excursion problem is still an open problem when the response process is non-Gaussian. Only approximate solutions have been proposed.

Let x(t) be a non stationary stochastic process, F (x;0) the first-order distribution of x(t) at t=0 and ν(x,t) the mean rate at which x(t) crosses from below (upcrosses) the threshold x at the time t. In order to estimate the largest value distribution of the response x(t), in the time interval τ, the following expression may be used:

$$F_\tau(x) = F(x;0)\exp\left(-\int_0^\tau \nu(x;t)\,dt\right) \tag{17}$$

where τ is the length of the time interval in which the mean wind speed is being computed.

When the expectation of the positive values of $\dot{x}(t)$ conditional to $x(t) = x$ is denoted by:

$$e(x;t) = E\,[\dot{X}(t) + |X(t) = x] \tag{18}$$

then the mean upcrossing rate at time t is

$$\nu(x;t) = e(x;t)\,f(x;t) \tag{19}$$

where f(x,t) is the density of x(t). The previous functions depend on the time t but for homogeneous processes they are time invariant so that:

$$F_\tau(x) = F(x;0)\exp(-\nu(x)\cdot\tau) = F(x;0)\exp(-e(x)\cdot f(x)\cdot\tau) \tag{20}$$

In order to define e(x), the processes x(t) and $\dot{x}(t)$ and the characteristic function of the vector $[x(t), \dot{x}(t)$ should be specified in closed form. Practical analytical difficulties often arise, so that this expression cannot generally be obtained.

Among the several approximated approaches that have been proposed, in this paper the hypothesis that x(t) and $\dot{x}(t)$ are independent random variables and $\dot{x}(t)$ follows the Gaussian distribution is assumed. As a consequence e(x) can be approximated by:

$$e(x) \cong \frac{\dot{\sigma}}{\sqrt{2\pi}} \tag{21}$$

where $\dot{\sigma}$ may be expressed in terms of the second moment of the response:

$$\dot{\sigma} = \int_{-\infty}^{+\infty} \omega^2 S_{XX}(\omega)\, d\omega \tag{22}$$

More suitable tools have been proposed by Grigoriu [6], that used the translation process:

$$X_T(t) = F^{-1}[\Phi(Z(t))] = g[Z(t)] \tag{23}$$

in which $g(.) = (F^{-1} o \Phi)(.)$, F(.) is the first order distribution of X(t) and Z(t) is a zero-mean, unit variance differentiable Gaussian process. Denoting $g' = \frac{dg(x)}{dz}$ and supposing that x(t) and $X_T(t)$ have similar crossing characteristics, e(x) can be approximated by $e_T(x)$ so that:

$$e(x) \cong e_T(x) = E[\dot{X}(t) + |X(t) = x] \cong \frac{\dot{\sigma} g'(g^{-1}(x))}{\sqrt{2\pi}} \tag{24}$$

6. NUMERICAL EXAMPLE

The proposed methodology is applied to a water tower which is a typical point-like structure, for which the longitudinal velocity fluctuations may be considered to be totally correlated over all the surface exposed to the wind.

The relevant structural data are: height of the pier h = 80. m, surface exposed to the wind S = 20. m^2, drag coefficient C_D = 1.00, shaft stiffness k = 85300. N/m, concentrated mass m = 1300 Kg, natural frequency f_0 = 0.4 Hz. The reference mean wind speed is assumed to be 30 m/s, with a standard deviation of 4.6 m/s. For the analytical expression of the power spectral density of the wind speed, the one proposed by Davenport is used.

A numerical investigation is developed, considering the following parameters:

- *five values of the relative structural damping*: ξ = 0.01, 0.03, 0.05, 0.07, and 0.09.
- *three types of roughness*, the type A (exponent α = 0.15 in the power expression of the boundary layer), the type B (α = 0.28). and the type C (α = 0.40).

The first three moments of the response are considered. The peak distributions are estimated for a duration of 3600 s and for a non-exceedance probability of 90%. In addition to the non Gaussian method, two other methods are applied. Firstly the one which neglects the second-order term in Eq. 2 and assumes the pressure and the response processes to be Gaussian (Gaussian linear method). Then a third procedure is used, which uses the full expression of the pressure, but considers it and the response process as Gaussian (Gaussian non-linear method).

A synthesis of the numerical results are presented in the following Figures 5 to 9, where the maximum dynamic displacements, along with the total displacements (static + dynamic) are reported. The Figures 5, 6 and 7 show the maximum dynamic displacement of the top of the tower for the wind climates A, B and C respectively. The Figures 8 and 9 show the relative difference between the responses computed with the non-Gaussian method and with the Gaussian linear one.

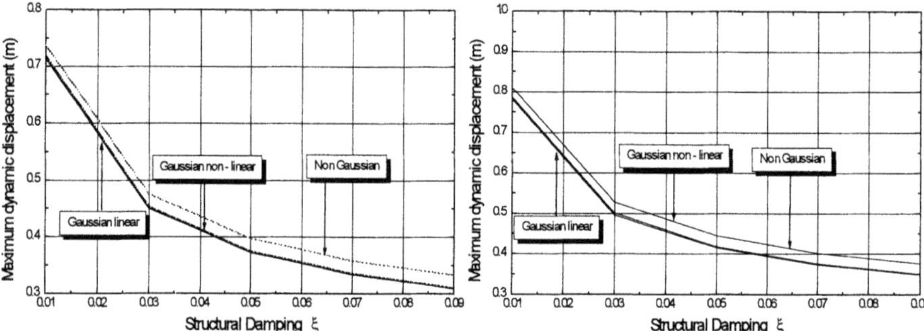

Figure 5. Wind climate A.
Maximum dynamic displacement.

Figure 6. Wind climate B.
Maximum dynamic displacement.

The first figure refers to the maximum dynamic displacement, while the second one refers to the total displacement (static + dynamic component).

While the difference between the results obtained with the Gaussian linear method and the Gaussian non-linear one is in most cases negligible, the effects of the application of the non-Gaussian procedure becomes important for increasing values of the structural damping and of the intensity of the turbulence.

For the roughness type A, the increase of the dynamic part of the response ranges from 3 % ($\xi = 0.01$) to over 7 % ($\xi = 0.09$), while for the type B it ranges from 4 to 8 % and for the type C it ranges from 5 to 9%.

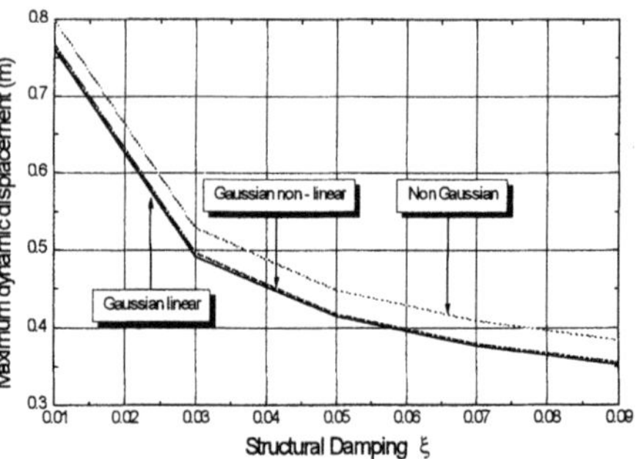

Figure 7. Wind climate C.
Maximum dynamic displacement

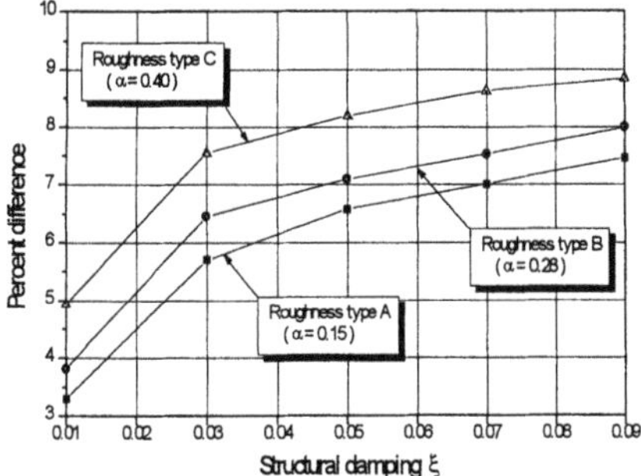

Figure 8. Relative difference between the non-Gaussian method and the Gaussian linear one. Max. dynamic displacement.

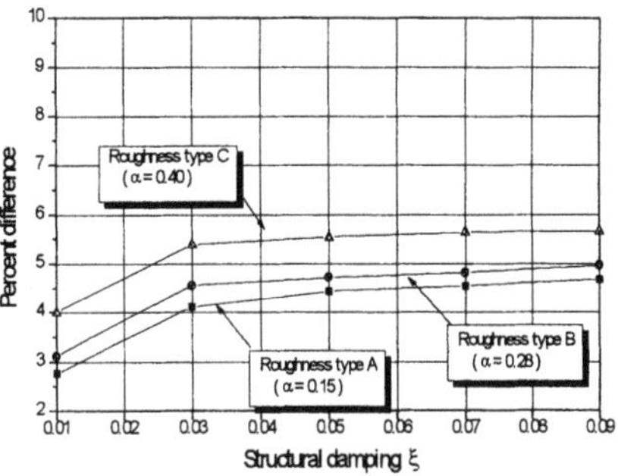

Figure 9. Relative difference between the non-Gaussian method and the Gaussian linear method. Total top displacement.

7. CONCLUDING REMARKS

An investigation has been presented into the effects of the non-Gaussian nature of the wind pressure over the assessment of the safety of wind-exposed structures, effects which are usually neglected by the classical methods used for calculating the along-wind response.

The proposed method, which operates in the frequency domain, has been applied to some case studies modelled as single degree of freedom systems, but is suited to be easily extended to the case of multi-degree-of-freedom systems.

The results of the numerical examples pointed out that the classical procedure, which assumes the wind pressure to be a Gaussian process, can, in conditions of high intensity of the turbulence and high value of the structural damping, underestimate the peak values of the dynamic part of the response by over 8 %.

REFERENCES

1. J.D. Holmes, "Non-Gaussian Characteristics of Wind Pressure Fluctuations", Jour. of Wind Engineering and Industrial Aerodynamics, Vol. 7, (1981), pp. 103-108.
2. R. Vaicaitis and E. Simiu, "Nonlinear Pressure Terms and Alongwind Response", J. Struct. Div., ASCE, 103, No.ST4, (1977), pp. 903-906.
3. C. Soize, "Gust Loading Factors with Nonlinear Pressure Terms", J. Struct. Div., ASCE, 104, No.ST6, (1978), pp. 991-1007.
4. A.G. Davenport, "Note on the distribution of the largest value of a random function with application to gust loading", (1964) , Proc. Instn. Civ. Engrs., 24.
5. L.D. Lutes and S.-L. J. Hu, "Non-Normal Stochastic Response of Linear Systems", J. Engg. Mech., ASCE, 112, (1986), pp. 127-141.
6. M. Grigoriu, "Crossings of Non-Gaussian Translation Processes", J. Engg. Mech. Div., ASCE, 110, No.6, (1986), pp. 610-20.
7. C.G. Bucher and G.I. Schueller, "Non-Gaussian Response of Linear Systems", in Recent Advances in Structural Dynamics (G.I. Schueller Ed.), Springer Verlag, 1991.

12

A Stochastic Crack Growth Model with Propagation Resistance as a Random Field*

Hiroshi Ishikawa[a], **Hiroaki Tanaka**[b] and **Satoshi Wakasa**[c]

[a] Dept. of Information Science, Kagawa University, Takamatsu, Kagawa 760 Japan
[b] Dept. of Applied Mathematics & Physics, Kyoto University, Kyoto 606-01 Japan
[c] Miura Institute of Research & Development, Matsuyama, Ehime 799-26 Japan

A new stochastic crack growth model is proposed, in which the propagation resistance appearing in the empirically obtained crack growth law is mathematically modeled as a random field. First, a mathematical methodology to obtain the solution of the randomized crack growth equation is discussed. Next, an approximate method to obtain a probability distribution of the residual life is proposed. Finally, the method is applied to the random propagation problem of an edge crack under uniform tensile stressing, where numerical evaluation of the residual life distribution in this case is performed.

1. Introduction

In order to investigate the probability distribution of the fatigue crack propagation life due to the randomness caused by microscopic inhomogeneity of the material, it has been often the case to introduce a mathematical model based upon the well-known Paris-Erdogan crack growth law such that

$$\frac{da}{dn} = C_0(\Delta K)^m, \tag{1}$$

with C_0, the crack propagation resistance, being a variable of stochastic nature, and m being a material constant. In the model developed by Tsurui and Ishikawa [1][2] (termed as TI-model) and other models proposed to date [3], the propagation resistance C_0 has been treated as a random process with temporal variation. That is, the growth equation (1) has been extended into a random differential equation, which is a differential equation with a temporally random input.

However, since the randomness induced in the actual materials will cause spatial random fluctuations in C_0, it is more appropriate, in reference to the engineering reality, to

*This research was supported in part by a Scientific Grant-In-Aid in 1993 (No. 05750066) from the Ministry of Education, Science and Culture, Japan.

model C_0 as a random field than as a random process. Hence, Eq.(1) needs to be extended into the following random differential equation:

$$\frac{da}{dn} = \frac{C_0}{C(a)}(\Delta K)^m \quad (C_0 : \text{const.}), \tag{2}$$

where $C(a)$ is a random field showing a spatially random variation. The reciprocal form is selected so as to express the "resistance"clearly [4][5].

The random differential equation (2) is a differential equation defined on a random field, and it has a special feature that the variable to indicate the location of the random field is combined concurrently with a dependent variable of the equation. In this respect, a mathematical methodology to obtain the solution to the random differential equation in the form of Eq.(2) has been newly developed [6].

In this paper, according to a mathematical methodology developed by Tanaka and Tsurui [6] to define a solution for such a random differential equation of special type, we construct a new probabilistic model to obtain the probability distribution of the random growth of fatigue cracks. The result will be compared with TI-model through numerical computation of the residual life distribution, which plays a crucial role in the reliability-based design.

2. Mathematical analysis of randomized crack growth equation

In this section, we give a survey of the mathematical methodology first developed by Tanaka and Tsurui [6] to analyze the randomized crack growth equation (2).

2.1.Definition of the solution

For the sake of convenience in analysis, we use the dimensionless crack length $X = a/a_r$, where a_r represents the reference length. Suppose that the stress intensity factor range ΔK can be expressed as

$$\Delta K = \alpha \Delta S_0 \sqrt{X} f(X), \tag{3}$$

where ΔS_0 is a stress amplitude (it is assumed to be constant), the function $f(\,\cdot\,)$ is a correcting function expressing the size effect of the component and α is a constant determined according to the geometry of the crack. Substituting Eq.(3) into Eq.(1), we can obtain

$$\frac{dX}{dt} = \varepsilon g(X) \qquad (g(X) \equiv \{\sqrt{X} f(X)\}^m), \tag{4}$$

in which we use a new time variable t such that $\varepsilon a_r t = C_0 \alpha^m (\Delta S_0)^m n$.

As mentioned in the preceding section, the microscopic inhomogeneity of the material gives rise to a spatially distributed random variation of the coefficient ε. Hence, we need to extend the differential equation (4) to the following random differential equation:

$$\frac{dX}{dt} = \frac{\varepsilon}{C(X)} g(X), \tag{5}$$

where $C(x)$ is a random field. The solution of Eq.(5) will show a temporally random variation according to the spatially random variation of the random field $C(x)$, even if the initial value $X(0) = x_0$ is deterministic.

If the material is macroscopically homogeneous, the random field $C(x)$ will be a homogeneous random field. That is, its first and second moments are given as

$$\mathrm{E}[C(x)] = M_C(\text{const.}) \tag{6}$$

$$\mathrm{E}\Big[\{C(x) - \mathrm{E}[C(x)]\}\{C(x+x') - \mathrm{E}[C(x+x')]\}\Big] = \sigma_C^2 \rho_C(x') \quad (\text{independent of } x) \tag{7}$$

It should be noted that we can assume, without loss of generality, that the mean value M_C takes on unity.

Integrating Eq.(5) formally under the initial condition $X(0) = x_0$, we have

$$\int_{x_0}^{X(t)} \frac{C(x)}{g(x)} dx = \varepsilon t. \tag{8}$$

It is natural to define the solution $X(t)$ of Eq.(5) as a function $X(t)$ satisfying the integral equation (8) provided that the left-hand side integral exists. However, if there is a non-zero probability that $C(x)$ takes a negative value, we can not uniquely define $X(t)$, since the left-hand side of Eq.(8) is not a monotonically increasing function. Such a situation should be taken into account even if the distribution range of $C(x)$ is restricted to the positive area, since we need to introduce the Markov approximation method [1], which is a kind of diffusion-approximation, in our practical analysis. Hence, introducing a new random field such that

$$Z(x; x_0) \equiv \int_{x_0}^{x} \frac{C(x')}{g(x')} dx', \tag{9}$$

we define the solution of Eq.(5), by making a partial revision on Eq.(8), as

$$X(t) \equiv \inf_{x > x_0} \{Z(x; x_0) = \varepsilon t\}. \tag{10}$$

If the transformed random field $Z(x; x_0)$ is almost certainly continuous, $X(t)$ can be uniquely defined with probability one.

Figure 1 gives a conceptual illustration of the definition of the solution $X(t)$, which is defined as a first passage location [7] of the random field $Z(x; x_0)$ to a level εt.

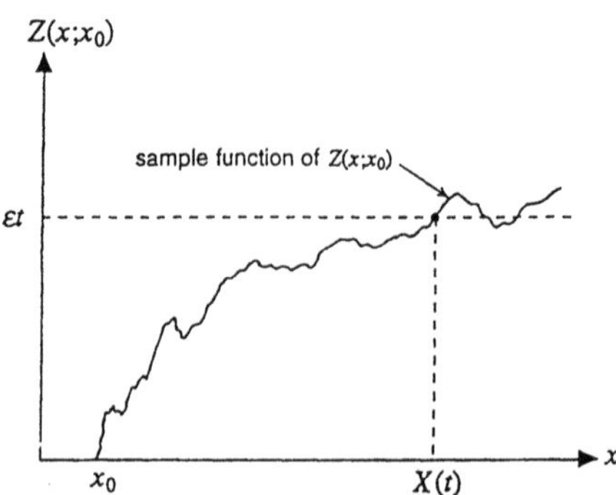

Fig.1 Relation between the solution $X(t)$ and the sample field of $Z(x; x_0)$.

2.2. Death point

If the function $g(x)$ rapidly diverges as $x \to \infty$, the transformed random field $Z(x; x_0)$ may not diverge as $x \to \infty$ under the assumption that the random field $C(x)$ is homogeneous. In fact, if the exponent m in the Paris law is greater than 2, we have to take into account such a possibility [1].

At the time t when εt is greater than the supremum of $Z(x; x_0)$, it is natural to consider that the solution $X(t)$ has already diverged to infinity. Hence, for a given t, we define that the solution $X(t)$ lies in the state **D** if

$$\sup_{x > x_0} Z(x; x_0) < \varepsilon t, \tag{11}$$

is satisfied. According to TI-model [1], we call the state **D** the death point.

3. Probability distribution of the solution process and residual life distribution

According to the definition of the solution $X(t)$ discussed in the preceding section, we need to solve a first passage problem for the transformed field $Z(x; x_0)$ to obtain the probability distribution of the solution process $X(t)$. Unfortunately, however, since we do not have a complete methodology to solve a first passage problem of general type, it is very difficult to derive the exact probability distribution of the solution process $X(t)$. Hence, in this section, we give an approximate method [6] to derive the transition probability distribution of the solution $X(t)$ and the residual life distribution.

3.1. Transition probability distribution of the solution process

With the aid of Eqs.(6) and (7), we can obtain the mean and variance of the transformed field $Z(x; x_0)$ as follows:

$$\mu_Z(x; x_0) \equiv \mathrm{E}[Z(x; x_0)] = \int_{x_0}^{x} \frac{dx'}{g(x')}, \tag{12}$$

$$\sigma_Z^2(x; x_0) \equiv \mathrm{E}\Big[\{Z(x; x_0) - \mathrm{E}[Z(x; x_0)]\}^2\Big] = \sigma_C^2 \int_{x_0}^{x} \frac{dx'}{g(x')} \int_{x_0}^{x} \frac{\rho_C(x' - x'')}{g(x'')} dx''. \tag{13}$$

If the random field $C(x)$ is a Gaussian field with independent increment, the transformed field $Z(x; x_0)$ is also a Gaussian field, whose probabilistic properties can be completely specified by Eq.(12) and (13). Even under the situation in which $C(x)$ does not have such a property, we can treat $Z(x; x_0)$ approximately as a Gaussian field by applying the Markov approximation method [1]. Hence, we can obtain an approximate form of the probability distribution function of $Z(x; x_0)$ in the following form:

$$P(z, x \mid x_0) = \Pr[Z(x; x_0) \le z] = \Phi\left[\frac{z - \mu_Z(x; x_0)}{\sigma_Z(x; x_0)}\right], \tag{14}$$

where $\Phi(\,\cdot\,)$ is the standardized normal distribution function.

It should be noted that the result gives a non-zero probability that $Z(x; x_0)$ decreases as a function of x. However, we can almost neglect it if the spatial interval $|x - x_0|$ is sufficiently large enough in comparison to the spatial correlation distance of the random field $C(x)$. In this case, according to the definition given by Eq.(10), we can obtain an approximate relationship as

$$W(x,t \mid x_0,) \equiv \Pr[X(t) \leq x \mid X(0) = x_0] \simeq 1 - \Pr[Z(x;x_0) \leq \varepsilon t]. \tag{15}$$

Hence, the transition probability distribution function of the solution process $X(t)$ is given as follows:

$$W(x,t \mid x_0) = \Phi\left[\frac{\mu_Z(x;x_0) - \varepsilon t}{\sigma_Z(x;x_0)}\right], \tag{16}$$

which is a probability distribution of special type, whose domain is restricted to $x_0 < x$. If the function $g(x)$ rapidly diverges as $x \to \infty$, $W(x,t \mid x_0)$ does not tend to unity as $x \to \infty$, since there is a non-zero probability that the solution $X(t)$ lies in the death point **D**.

3.2.Residual life distribution

Let $T(x_0, x_c)$ be a time when the sample function with the initial condition $X(0) = x_0$ arrives at a prescribed critical level $x_c(> x_0)$ for the first time. We call it the residual life of $X(t)$, and denote its probability distribution function as

$$H(t \mid x_0, x_c) \equiv \Pr[T(x_0, x_c) \leq t]. \tag{17}$$

According to Eq.(16), the probability that $X(t)$ decreases is exactly zero. Hence, we obtain

$$H(t \mid x_0, x_c) = 1 - \Pr[X(t) \leq x_c \mid X(0) = x_0] = \bar{\Phi}\left[\frac{\mu_Z(x_c;x_0) - \varepsilon t}{\sigma_Z(x_c;x_0)}\right], \tag{18}$$

where $\bar{\Phi}(\,\cdot\,)$ is the complementary function of the standardized normal distribution. If the cracked component fails when the crack grows to the critical length x_c, $H(t \mid x_0, x_c)$ gives a probability of failure at time t. Thus, it must be kept at a very small value so as to make the component highly reliable in actual situations.

3.3.Residual life distribution in TI model

In TI-model, the random field $C(x)$ has been transformed into a random process reflecting a certain typical value of the solution as a function of time, that is, Eq.(5) has been transformed into

$$\frac{dX}{dt} = \varepsilon \check{C}(t) g(X), \qquad \check{C}(t) \equiv \frac{1}{C(\hat{\mathrm{E}}[X(t)])}, \tag{19}$$

where $\hat{\mathrm{E}}[X(t)]$ expresses a certain typical value of the solution $X(t)$. Equation (19) is a differential equation driven by a temporally random noise $\check{C}(t)$, which does not have such a special feature as Eq.(5).

By use of the Markov approximation method [1], the transition probability density function of the solution $X(t)$ of Eq.(19) can be obtained as a solution of the generalized Fokker-Planck equation. Moreover, by neglecting a probability that the crack length decreases, the residual life distribution function has been obtained in an analytical form [2] as

$$H(t \mid x_0, x_c) = \bar{\Phi}\left[\frac{\mu_Z(x_c; x_0) - \varepsilon \tilde{M}_c t}{\sqrt{2G(t)}}\right], \tag{20}$$

$$G(t) = \frac{\varepsilon \tilde{\sigma}_C^2 \xi_0}{\tilde{M}_C} \int_0^t \frac{dt'}{g(\hat{X}(t'))}, \tag{21}$$

$$\tilde{M}_C \equiv \mathrm{E}[\tilde{C}(t)], \qquad \tilde{\sigma}_C^2 \equiv \mathrm{Var}[\tilde{C}(t)], \tag{22}$$

where $\hat{X}(t)$ is the solution of the following ordinary differential equation:

$$\frac{d\hat{X}}{dt} = \varepsilon \hat{M}_C g(\hat{X}), \qquad \hat{X}(0) = x_0. \tag{23}$$

The constant ξ_0 corresponds to a spatial correlation distance of the random field $C(x)$.

4. Numerical example

In this section, we apply the result obtained in the preceding section to the random growth of an edge crack in an infinite plate under uniform tensile stressing, schematically illustrated in Fig.2. The result will be compared with TI-model through numerical calculations for the residual life distribution.

The geometrical factor α and the correcting function $f(x)$ appearing in Eq.(3) are given as $\alpha = 1.12$ and $f(x) = 1$, respectively, for the edge crack. The spatial correlation function $\rho_C(x)$ is assumed to be exponential, that is,

$$\rho_C(x) = \exp(-\frac{|x|}{L_C}), \tag{24}$$

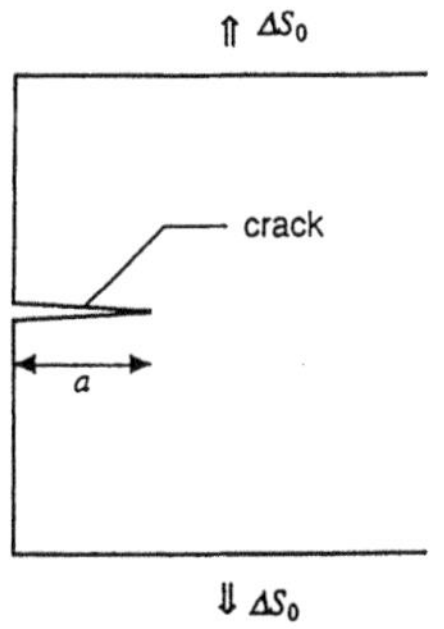

Fig.2 An edge crack in an infinite plate.

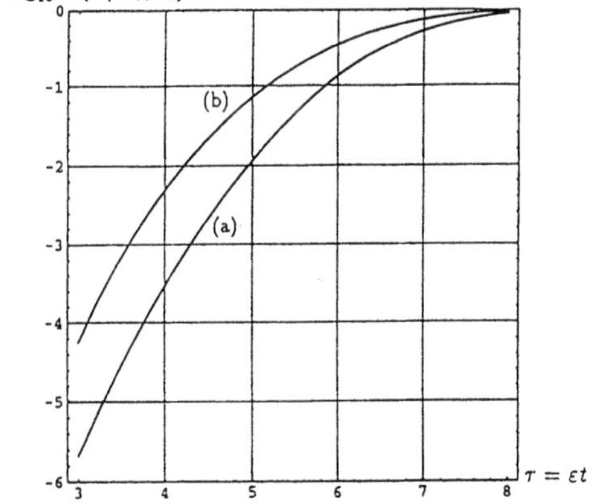

Fig.3 Residual life distributions by (a) the present method and (b) TI-model.

($x_0 = 0.05$, $x_c = 1.0$, $m = 3.0$, $\sigma_C = 0.3$, $L_C = \xi_0 = 0.03$)

in which the constant L_C represents the spatial correlation distance of $C(x)$, which is assumed here to be equal to the constant ξ_0 in TI-model.

Although $1/C(x)$ does not generally follow the same distribution as $C(x)$, if we can assume that the probability distribution of $C(x)$ is approximated by a log-normal distribution, $1/C(x)$ obeys the log-normal distribution with slightly different mean and variance as

$$\mathrm{E}[1/C(x)] = 1 + \sigma_C^2, \quad \mathrm{Var}[1/C(x)] = \sigma_C^2(1+\sigma_C^2)^2, \tag{25}$$

under the condition that the mean and variance of $C(x)$ are given by Eqs.(6) and (7),respectively. In what follows, we make a brief comparison between the present method and TI-model provided that Eq.(25) holds.

Figures 3 ~ 5 show the residual life distributions as a function of $\tau \equiv \varepsilon t$ calculated by (a) the present method (given by Eq.(16)) and (b) TI-model (given by Eq.(20)), in which the vertical axes are plotted on logarithmic scale to show the precise behaviors of the so-called high-reliability region. In each figure, the parameter values of x_0, x_c and m are fixed to be $x_0 = 0.05$, $x_c = 1.0$ and $m = 3.0$, respectively, and those of σ_C and L_C are chosen as in the figures. The parameter values for Fig.5 are selected in such a way that the quantity $\tilde{\sigma}_C^2 = {\sigma_C}^2(1+{\sigma_C}^2)^2\xi_0$ takes on the same value as that for Fig.3. In other word, with parameter values thus chosen, the uncertainty factor in TI-model becomes nearly the same in both cases.

From these figures we can observe that the residual life distribution for the present method shifts to a longer side in comparison to that for TI-model. This might be caused by the approximations introduced in the present model such that (1) the distribution of $Z(x;x_0)$ is approximated to be Gaussian, and (2) the probability that $Z(x;x_0)$ decreases is neglected in the first passage problem of $Z(x;x_0)$.

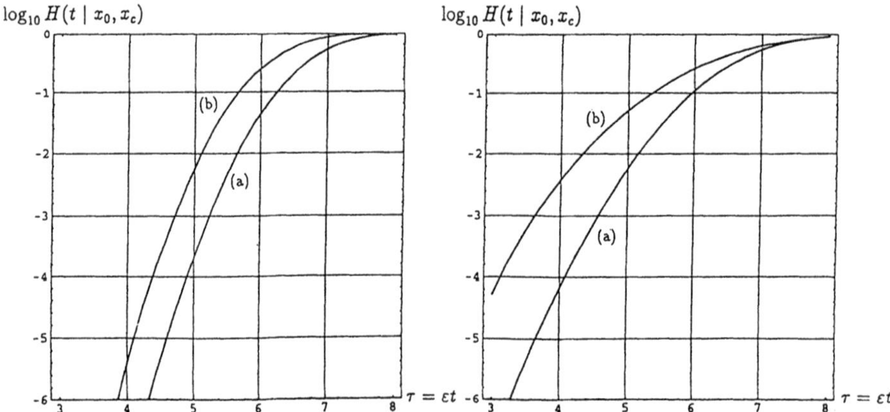

Fig.4 Residual life distributions by (a) the present method and (b) TI-model. ($x_0 = 0.05$, $x_c = 1.0$, $m = 3.0$, $\sigma_C = 0.3$, $L_C = \xi_0 = 0.01$)

Fig.5 Residual life distributions by (a) the present method and (b) TI-model. ($x_0 = 0.05$, $x_c = 1.0$, $m = 3.0$, $\sigma_C = 0.2$, $L_C = \xi_0 = 0.074$)

However, in TI-model the approximations are also adopted such that (1) the introduced shape of the correlation function of $\tilde{C}(t)$ is not theoretical but is assumed, and (2) the decrease probability is also neglected in the first passage problem of $X(t)$. In this respect, it is impossible to determine which model gives a closer value to the true value. We have no other way but to utilize simulation techniques to prove this issue.

We can also observe that the difference in the result between these two models becomes a little larger when the values of σ_C and ξ_0 become larger. In fact, when σ_C assumes a value of nearly 0.1, the difference becomes very small, which is unfortunately not shown in the figure for simplicity. Further, judging from the comparison between Figs.3 and 5, TI-model reduces to neary the same result as long as the quantity $\tilde{\sigma}_C^2\xi_0$ takes on the same value. On the contrary, the present model produces a little difference.

5. Concluding remarks

In this paper, we have constructed a new mathematical model to describe the random fatigue crack growth by modeling the propagation resistance as a random field.

Although our present method can reflect the more precise engineering reality compared with TI-model, the more difficulties arise in solving the mathematical equations. However, these difficulties can be alleviated by use of the Markov approximation method, and we can derive the residual life distribution as well as the probability distribution of the crack length. In future studies, the accuracy of the approximation should be verified through computer simulation procedure.

References

1. A. Tsurui and H. Ishikawa, Application of Fokker-Planck Equation to a Stochastic Fatigue Crack Growth Model, *Structural Safety*, Vol. **4**, pp. 15-29 (1986).
2. H. Ishikawa and A. Tsurui, A Stochastic Model of Fatigue Crack Growth in Consideration of Random Propagation Resistance, *Trans. of JSME, Ser. A*, Vol. **50**, pp. 1309-1315 (1984) (in Japanese).
3. K. Sobczyk, Modeling of Random Fatigue Crack Growth, *Engineering Fracture Mechanics*, Vol. **26**, pp. 609-623 (1986).
4. H. Itagaki, T. Ishizuka, T. Seki and M. Fujinami, Effect of Load Profile on Fatigue Crack Propagation under Narrow Band Random Loading, *Proc. of JCOSSAR'91*, pp. 411-418 (1991) (in Japanese).
5. T. Sasaki, S. Sakai and H. Okamura, Stochastic Modeling of Fatigue Crack Growth and its Numerical Analysis, *Proc. of JCOSSAR'91*, pp. 575-578 (1991) (in Japanese).
6. H. Tanaka and A. Tsurui, A Random Differential Equation of Special Type Defined on a Random Field and Its Application to Stochastic Crack Growth, *Proc. of 12th Symposium on Reliability of Materials and Structures (JSMS)*, pp. 113-118 (1993) (in Japanese).
7. O. Ditlevsen, Random Fatigue Crack Growth – A First Passage Problem –, *Engineering Fracture Mechanics*, Vol. **23**, No. 2, pp. 467-477 (1986).

13

Algorithms for Reliability-Based Optimal Design[†]

C. Kirjner-Neto[*], E. Polak[*], and A. Der Kiureghian[+]

[*] Department of Electrical Engineering and Computer Sciences, University of California, Berkeley, CA 94720, USA.

[+] Department of Civil Engineering, University of California, Berkeley, CA 94720, USA.

We present a new formulation for a class of structural optimization problems with reliability constraints, which makes it possible to use a new outer approximations algorithm for their solution. Our numerical results support the viability of our approach.

1. INTRODUCTION

This paper deals with the problem of minimizing the initial cost of a structure subject to a minimum reliability requirement. As a measure of reliability we make use of the first-order reliability index, i.e., the minimum distance from the origin to the limit-state surface in the standard normal space [8]. This measure is readily applicable to component reliability problems governed by a single, nearly flat limit-state surface, and this case is the main focus of this paper. However, the proposed formulation is also applicable to series system problems that are governed by a multitude of nearly flat limit-state surfaces, provided the reliability constraint is expressed in terms of separate constraints on each failure mode.

Structural optimization problems with a reliability constraint of the type described above have been studied in numerous previous papers (see [10] for a comprehensive bibliography). However, in virtually all of these previous studies a straightforward approach is used with "off-the-shelf" algorithms.

In most cases a two-level optimization problem is solved (see, e.g., [1]). For a given value of the design parameters, the reliability index is obtained by minimizing the distance from the origin to the limit-state surface, and its gradient with respect to the design parameters is computed. These are used in an outer algorithm to guide the search for a new value of the design parameters that attempts to minimize the cost function while satisfying the reliability constraint. To our knowledge, no proof of convergence of such algorithms has been presented. An alternative approach has been to replace the minimization in the computation of the reliability index with first order optimality conditions, as a constraint. Unfortunately, this approach requires second derivatives of the limit-state functions. Furthermore, as far as we know, no serious attempt has been made to take advantage of the special form of the structural reliability problem.

In this paper we present a new mathematical formulation of the optimization problem with a reliability constraint and propose an ''outer approximations'' algorithm for solving it. The algorithm does not require repeated computation of the reliability index or its sensitivities, nor does it require the second derivatives of the cost or limit-state functions. Most importantly, the algorithm has proven convergence properties, which include an accounting

[†] The research reported herein was sponsored by the CUREe-Kajima grant M1975 and the NSF grant ECS 9302926.

for approximations in the various steps of the algorithm.

Initially, the formulation is presented for a component reliability problem. The formulation is then extended to a series system problem with a separate reliability constraint on each of its modes. Numerical results highlight the efficiency and robustness of the proposed algorithm.

Given $x \in \mathbb{R}^n$, $n > 1$, we denote its components by $x^1, x^2, \cdots, x^n$. Subscripts are used to denote elements of a sequence, as in $\{x_i\}_{i=0}^{\infty}$.

2. STATEMENT OF THE PROBLEM

We begin by developing two alternative mathematical formulations of the simplest structural optimization problems with reliability constraints. Let $f^0 : \mathbb{R}^d \to \mathbb{R}$ be a cost function, and $g : \mathbb{R}^n \times \mathbb{R}^d \to \mathbb{R}$ be a limit-state function defined on the n-dimensional standard normal space, obtained by transforming the basic random variables of the problem, and parametrized by a vector $p \in \mathbb{R}^d$ of design parameters. We assume that $f^0(\cdot)$ and $g(\cdot,\cdot)$ are twice continuously differentiable, and that the design variables must be chosen so that they satisfy m constraints of the form

$$f^j(p) \le 0, j \in \mathbf{m}, \tag{2.1}$$

where the $f^j : \mathbb{R}^d \to \mathbb{R}$ are twice continuously differentiable functions, and for any positive integer m, $\mathbf{m} \triangleq \{1, \ldots, m\}$.

For each design vector $p \in \mathbb{R}^d$, let $\beta(p)$ denote the corresponding first-order reliability index, i.e.,

$$\beta(p) \triangleq \min_{u \in \mathbb{R}^n} \{ \|u\| \mid g(u, p) \le 0 \}. \tag{2.2}$$

Let $r > 0$ be a required lower bound on the first order reliability index. Then the simplest formulation of an optimal design problem, subject to a reliability constraint has the form

$$\mathbf{P}_1 \quad \min \{ f^0(p) \mid \beta(p) \ge r, \; f^j(p) \le 0, \; j \in \mathbf{m} \}. \tag{2.3}$$

For each design vector $p \in \mathbb{R}^d$ let $\mathbf{F}_p$ denote the corresponding failure domain, i.e.,

$$\mathbf{F}_p \triangleq \{ u \in \mathbb{R}^n \mid g(u, p) \le 0 \}. \tag{2.4}$$

In view of (2.2), we have that

$$\beta(p) \ge r \text{ if and only if } \mathbf{F}_p \cap \mathbf{B}_r = \emptyset, \tag{2.5a}$$

where $\mathbf{B}_r$ denotes the open ball of radius r and center at the origin. To obtain a mathematical reformulation of $\mathbf{P}_1$ using (2.5a) we introduce the following technical assumption which rules out a most unlikely situation:

Assumption 2.1 For all $p \in \mathbb{R}^d$ and $u \in \mathbb{R}^n$ such that $g(u, p) = 0$, $\nabla_u g(u, p) \ne 0$ (i.e., for all points on the limit-state surface, the gradient of the limit-state function is nonzero).□

When Assumption 2.1 is satisfied, $\mathbf{F}_p \cap \mathbf{B}_r = \emptyset$ if and only if

$$\min_{u \in \overline{\mathbf{B}}_r} g(u, p) \ge 0, \tag{2.5b}$$

where $\overline{\mathbf{B}}_r$ denotes the closed ball of radius r centered at the origin. Hence problem $\mathbf{P}_1$ is equivalent to the following mathematical alternative

$$\mathbf{P}_{1,\infty} \quad \min \{ f^0(p) \mid \max_{u \in \overline{\mathbf{B}}_r} \tilde{g}(u,p) \le 0, \; f^j(p) \le 0, \; j \in \mathbf{m} \}, \tag{2.6}$$

where $\tilde{g}(\cdot,\cdot) \triangleq -g(\cdot,\cdot)$.

3. OUTER APPROXIMATIONS ALGORITHMS

Problem $\mathbf{P}_{1,\infty}$ is a nonlinear programming problem with an infinite number of constraints, since the inequality

$$\max_{u \in \overline{\mathbf{B}}_r} \tilde{g}(u,p) \le 0 \tag{3.1a}$$

is equivalent to the infinite system of inequalities

$$\tilde{g}(u,p) \le 0, \; \forall u \in \overline{\mathbf{B}}_r . \tag{3.1b}$$

Optimization problems with such constraints are known as semi-infinite optimization problems, because the design vector is finite-dimensional, but the number of constraints is infinite. There is a rather large literature dealing with the numerical solution of such problems (see, e.g., the review papers [3,9]). A major source of difficulty in solving $\mathbf{P}_{1,\infty}$ is the fact that in structural design the ball $\overline{\mathbf{B}}_r$ is high dimensional, i.e., $\overline{\mathbf{B}}_r \subset \mathbb{R}^n$ with n large. Hence $\mathbf{P}_{1,\infty}$ cannot be solved using discretization techniques as in [9].

An alternative approach is provided by the method of outer approximations, see, e.g., [2]. First observe that given any set $\mathbf{U}_k \triangleq \{u_1, u_2, \cdots, u_k\} \subset \overline{\mathbf{B}}_r$, the problem

$$\mathbf{P}_{1,k} \quad \min \{ f^0(p) \mid \max_{u \in \mathbf{U}_k} \tilde{g}(u,p) \le 0, f^j(p) \le 0, j \in \mathbf{m} \}, \tag{3.2a}$$

has finitely many constraints. Instead of solving the original problem $\mathbf{P}_{1,\infty}$, we will solve a sequence of problems of the form $\mathbf{P}_{1,k}$. The method is called outer-approximations because the feasible set for each $\mathbf{P}_{1,k}$ contains the feasible set for $\mathbf{P}_{1,\infty}$. Next, assume that $\hat{p}_k$ solves $\mathbf{P}_{1,k}$. Then $\hat{p}_k$ also solves $\mathbf{P}_{1,\infty}$ if

$$\max_{u \in \mathbf{U}_k} \tilde{g}(u,\hat{p}_k) = \max_{u \in \overline{\mathbf{B}}_r} \tilde{g}(u,\hat{p}_k). \tag{3.2b}$$

At the start, we do not know how to choose a set $\mathbf{U}_k$ such that (3.2b) is satisfied. Hence methods of outer approximations accumulate a satisfactory approximation to such a set by using points that satisfy approximately the relationship

$$u_j \in arg \max_{u \in \overline{\mathbf{B}}_r} \tilde{g}(u,p_j), \tag{3.2c}$$

with p_j an approximate solution to $\mathbf{P}_{1,\infty}$, as we will explain more precisely below by stating an algorithm. In practice, methods of outer approximations have been found to be very efficient.

It is easier to grasp the nature of outer approximations algorithms by first viewing them in a simple, "conceptual" form, such as the following algorithm for solving $\mathbf{P}_{1,\infty}$. We call this algorithm conceptual because it contains steps (Step 1 and Step 2) that can be performed only if global optimization techniques are used, which is completely out of the question for most real life structural optimization problems of the form $\mathbf{P}_{1,\infty}$.

Conceptual Algorithm 3.1.

Data. $p_1 \in \mathbb{R}^d$, $\mathbf{U}_0 = \emptyset$.

Step 0. Set $i = 1$.

Step 1. Compute a maximizer u_i for the problem

$$\max_{u \in \bar{\mathbf{B}}_r} \tilde{g}(u, p_i). \tag{3.3a}$$

Step 2. Set $\mathbf{U}_i = \mathbf{U}_{i-1} \cup \{ u_i \}$, and compute a minimizer p_{i+1} for the problem

$$\min_{p \in \mathbb{R}^d} \{ f^0(p) \mid \tilde{g}(u, p) \leq 0,\ u \in \mathbf{U}_i,\ f^j(p) \leq 0,\ j \in \mathbf{m} \}. \tag{3.3b}$$

Step 3. Replace i by $i + 1$ and go to Step 1. □

Conceptual Algorithm 3.1 has the following convergence property:

Theorem 3.2 Suppose that Conceptual Algorithm 3.1 has constructed an infinite sequence $\{ p_i \}_{i=1}^{\infty}$. Then every accumulation point of the sequence $\{ p_i \}_{i=1}^{\infty}$ is a solution for $\mathbf{P}_{1,\infty}$. □

The next step is to replace the exact maximization (3.3a) and minimization (3.3b) by appropriate approximations, so as to obtain an implementable algorithm that retains the convergence properties of the Conceptual Algorithm, to the extent that it can be shown to compute stationary points for the problem $\mathbf{P}_{1,\infty}$. We will carry out these modifications in two steps.

First, instead of computing a global minimizer in Step 2 of Conceptual Algorithm 3.1, we will compute an approximate feasible stationary point p_{i+1} for (3.3b). Second, instead of computing a global maximizer in Step 1, we will compute an approximate feasible stationary point u_i for (3.3a). In view of the flatness of the limit-state surface, prevalent in many problems of interest, we introduce the following assumption:

Assumption 3.3. For all $p \in \mathbb{R}^d$ such that $\beta(p) \geq r$, any feasible stationary point of the problem $\min_{u \in \bar{\mathbf{B}}_r} g(u, p)$ is a global minimizer of that problem. □

It is well-known that in the absence of convexity assumptions, nonlinear programming algorithms can only be shown to compute stationary points. Traditionally stationary points are characterized by systems of equations (such as the Kuhn-Tucker or Fritz John conditions). However, for our purposes it is convenient to characterize stationary points as zeros of a continuous, negative-valued optimality function. We will define two families of optimality functions. One for the maximization problems (3.3a), and one for the minimization problems (3.3b). Hence, given $p \in \mathbb{R}^d$, the optimality function $\theta_p : \mathbb{R}^n \to \mathbb{R}$, for (3.3a), is defined by

$$\theta_p(u) \triangleq -\min_{\mu \in \Sigma_2} \{ -\mu^2(\| u \|^2 - r^2) + \tfrac{1}{2} \| \mu^1 \nabla_u g(u, p) + 2\mu^2 u \|^2 \} - \max \{ 0, \| u \|^2 - r^2 \} \tag{3.4a}$$

where for any positive integer q, Σ_q is the q-simplex, i. e.,

$$\Sigma_q \triangleq \{ \mu = (\mu^1, \cdots, \mu^q) \in \mathbb{R}^q \mid \mu^j \geq 0, j \in \mathbf{q}, \text{ and } \sum_{j=1}^{q} \mu^j = 1 \}. \tag{3.4b}$$

Note that at any $u \in \overline{\mathbf{B}}_r$, the minimand in (3.4a) is the sum of two non-negative terms. Hence it is zero if and only if both of these terms are zero, which, in turn, implies that the Fritz John optimality conditions for (3.3a) are satisfied. Next, given a set $\mathbf{U}_i = \{ u_1, \cdots, u_i \} \subset \overline{\mathbf{B}}_r$, let the functions $f^j: \mathbb{R}^d \to \mathbb{R}$, $j = m+1, \cdots, m+i$ be defined by

$$f^{m+j}(p) \triangleq \tilde{g}(u_j, p), \quad j \in \mathbf{i}. \tag{3.5}$$

Then (3.3b) can be rewritten as $\min \{ f^0(p) \mid f^j(p) \leq 0,\ j = 1, \ldots, m+i \}$. We define the optimality function $\Theta_{\mathbf{U}_i}: \mathbb{R}^d \to \mathbb{R}$, for (3.3b), as follows:

$$\Theta_{\mathbf{U}_i}(p) \triangleq - \min_{(\mu^0, \mu) \in \Sigma_{m+i+1}} \{ - \sum_{j=1}^{m+i} \mu^j f^j(p) + \tfrac{1}{2} \| \sum_{j=0}^{m+i} \mu^j \nabla f^j(p) \|^2 \} - \psi(p)_+, \tag{3.6}$$

where

$$\psi(p)_+ \triangleq \max_{1 \leq j \leq m+i} \{ 0, f^j(p) \}. \tag{3.7}$$

Again, note that at any feasible point p, the minimand in (3.6) is the sum of two non-negative terms. Hence it is zero if and only if both of these terms are zero, which, in turn, implies that the Fritz John optimality conditions for (3.3b) are satisfied.

The key properties of $\theta_p(\cdot)$ and $\Theta_{\mathbf{U}_i}(\cdot)$ are given by the following result:

Theorem 3.4 Given any $p \in \mathbb{R}^d$ and $\mathbf{U}_i = \{ u_1, u_2, \ldots, u_i \} \subset \overline{\mathbf{B}}_r$, we have

(a) The functions $\theta_p: \mathbb{R}^n \to \mathbb{R}$ and $\Theta_{\mathbf{U}_i}: \mathbb{R}^d \to \mathbb{R}$ are well-defined, continuous and take values in $(-\infty, 0]$.

(b) $\theta_p(u) = 0$ if and only if $u \in \mathbb{R}^n$ satisfies the Fritz John optimality conditions for problem (3.3a), that is, $\theta_p(u) = 0$ if and only if u is a stationary point for the maximization problem in Step 1 of Conceptual Algorithm 3.1.

(c) $\Theta_{\mathbf{U}_i}(p) = 0$ if and only if $p \in \mathbb{R}^d$ satisfies the Fritz John optimality conditions for problem (3.3b), that is, $\Theta_{\mathbf{U}_i}(p) = 0$ if and only if p is a stationary point for the minimization problem in Step 2 of Conceptual Algorithm 3.1. □

It is clear from Theorem 3.4 that if the value of the optimality function at a point is close to zero, then that point is an approximate stationary point. We use this fact to construct the following implementable algorithm:

Algorithm 3.5.

Parameters. $C_1, C_2 \in (0, \infty)$.

Data. $p_1 \in \mathbb{R}^d$, $\mathbf{U}_0 = \emptyset$.

Step 0. Set $i = 1$.

Step 1. Compute $u_i \in \overline{\mathbf{B}}_r$ such that $\theta_{p_i}(u_i) \geq -C_1/i$.

Step 2. Set $\mathbf{U}_i = \mathbf{U}_{i-1} \cup \{ u_i \}$, and compute $p_{i+1} \in \mathbb{R}^d$ such that $\Theta_{\mathbf{U}_i}(p_{i+1}) \geq -C_2/i$.

Step 3. Replace i by $i + 1$ and go to Step 1. □

The convergence properties of Algorithm 3.5 are given by the following result:

Theorem 3.6. Suppose that Assumption 3.3 is satisfied and that Algorithm 3.5 has constructed an infinite sequence $\{p_i\}_{i=1}^{\infty}$. Then any accumulation point of $\{p_i\}_{i=1}^{\infty}$ is a stationary point for $\mathbf{P}_{1,\infty}$. □

Algorithm 3.5 is an implementable algorithm since there exist mathematical programming algorithms capable of performing Steps 1 and 2 in a finite number of iterations. In Section 5, where we present some computational results, we use CFSQP ([4]) to perform Steps 1 and 2 of Algorithm 3.5.

4. EXTENSION TO MULTIPLE LIMIT-STATE FUNCTIONS

It is possible to extend our implementable algorithm and the corresponding theory to more general problems. In this section we show how to apply the ideas described above to the optimization of a series system with a separate reliability constraint on each of its modes.

Let $g^k:\mathbb{R}^n \times \mathbb{R}^d \to \mathbb{R}$, $k \in \mathbf{q}$, be twice continuously differentiable limit-state functions defined in the n-dimensional standard normal space and parametrized by $p \in \mathbb{R}^d$. Given $r \in \mathbb{R}^q$, with $r^k > 0$, $k \in \mathbf{q}$, we will consider optimal design problems of the following form:

$$\mathbf{P}_2 \quad \min\{f^0(p) \mid \beta^k(p) \geq r^k\,,\ k \in \mathbf{q},\ f^j(p) \leq 0,\ j \in \mathbf{m}\}\,, \tag{4.1}$$

where, for each $k \in \mathbf{q}$, $\beta^k(p)$ is the first-order reliability index for the limit-state function $g^k(\cdot,p)$, defined as in (2.2). Each of the q reliability constraints in (4.1) can be dealt with using the same ideas applied to deal with the unique reliability constraint in (2.3).

For each $p \in \mathbb{R}^d$, let $\mathbf{F}_p^k$ denote the failure domain of the k-th component, i.e.,

$$\mathbf{F}_p^k \triangleq \{u \in \mathbb{R}^n \mid g^k(u,p) \leq 0\}\,,\quad k \in \mathbf{q}. \tag{4.2}$$

Then we have

$$\min_{k \in \mathbf{q}} \beta^k(p) \geq r^k \text{ if and only if } \mathbf{F}_p^k \cap \mathbf{B}_{r^k} = \emptyset\,,\ \text{for all } k \in \mathbf{q}. \tag{4.3a}$$

If all limit-state functions $g^k(\cdot,\cdot)$, $k \in \mathbf{q}$, satisfy Assumption 2.1, then $\mathbf{F}_p^k \cap \mathbf{B}_{r^k} = \emptyset$, for all $k \in \mathbf{q}$, if and only if

$$\min_{k \in \mathbf{q}}\ \min_{u \in \overline{\mathbf{B}}_{r^k}} g^k(u,p) \geq 0 \tag{4.3b}$$

from which follows that $\mathbf{P}_2$ is equivalent to the following problem:

$$\mathbf{P}_{2,\infty} \quad \min\{f^0(p) \mid \max_{k \in \mathbf{q}}\ \max_{u \in \overline{\mathbf{B}}_{r^k}} \tilde{g}^k(u,p) \leq 0,\ \max_{j \in \mathbf{m}} f^j(p) \leq 0\}\,, \tag{4.4}$$

where $\tilde{g}^k(\cdot,\cdot) \triangleq -g^k(\cdot,\cdot)$, $k \in \mathbf{q}$. □

The following conceptual algorithm solves $\mathbf{P}_{2,\infty}$:

Conceptual Algorithm 4.1.

Data. $p_1 \in \mathbb{R}^d$, $\mathbf{U}_0^k = \emptyset$, $k \in \mathbf{q}$.

Step 0. Set $i = 1$.

Step 1. Compute maximizers u_i^k, $k \in \mathbf{q}$, for the problems

$$\max_{u \in \bar{\mathbf{B}}_{r^k}} \tilde{g}^k(u, p_i), \quad k \in \mathbf{q}. \tag{4.5a}$$

Step 2. Set $\mathbf{U}_i^k = \mathbf{U}_{i-1}^k \cup \{u_i^k\}$, $\tilde{\mathbf{U}}_i = \bigcup_{k=1}^{q} \mathbf{U}_i^k$ and compute a minimizer p_{i+1} for the problem

$$\min_{p \in \mathbb{R}^d} \{ f^0(p) \mid \tilde{g}^k(u, p) \le 0,\ u \in \tilde{\mathbf{U}}_i,\ f^j(p) \le 0,\ j \in \mathbf{m} \}. \tag{4.5b}$$

Step 3. Replace i by $i + 1$ and go to Step 1. □

It can be shown that any accumulation point of a sequence $\{p_i\}_{i=1}^{\infty}$ constructed by Conceptual Algorithm 4.1 is a stationary point for $\mathbf{P}_{2,\infty}$.

To obtain an implementable algorithm, we follow the pattern set in Section 3 and replace the exact maximizations in Step 1 and minimization in Step 2 of Conceptual Algorithm 4.1 by appropriate implementable approximations.

Recall that in Section 3 for a given $p \in \mathbb{R}^d$ we defined an optimality function $\theta_p(\cdot)$ for problem (3.3a). In (4.5a) we have q problems of the same form as (3.3a), and hence we define q optimality functions $\theta_{k,p}: \mathbb{R}^n \to \mathbb{R}$, $k \in \mathbf{q}$, similarly to (3.4a).

Next, given a finite set $\tilde{\mathbf{U}}_i \subset \bigcup_{k=1}^{q} \bar{\mathbf{B}}_{r^k}$, we define an optimality function $\bar{\Theta}_{\tilde{\mathbf{U}}_i}: \mathbb{R}^d \to \mathbb{R}$ following the pattern set in (3.6), (3.7). It can be shown that the optimality functions $\theta_{k,p}(\cdot)$ and $\bar{\Theta}_{\tilde{\mathbf{U}}_i}(\cdot)$ have properties similar to those stated in Theorem 3.4 for $\theta_p(\cdot)$ and $\Theta_{\mathbf{U}_i}(\cdot)$ respectively. These optimality functions can be used to construct an implementable algorithm as shown below.

Algorithm 4.3.

Parameters. $C_1, C_2 \in (0, \infty)$.

Data. $p_1 \in \mathbb{R}^d$, $\mathbf{U}_0^k = \emptyset$, $k \in \mathbf{q}$.

Step 0. Set $i = 1$.

Step 1. Compute $u_i^k \in \bar{\mathbf{B}}_{r^k}$, $k \in \mathbf{q}$, such that $\theta_{k,p_i}(u_i^k) \ge -C_1/i$, $k \in \mathbf{q}$,

Step 2. Set $\mathbf{U}_i^k = \mathbf{U}_{i-1}^k \cup \{u_i^k\}$, $\tilde{\mathbf{U}}_i = \bigcup_{k=1}^{q} \mathbf{U}_i^k$ and compute $p_{i+1} \in \mathbb{R}^d$ such that $\Theta_{\tilde{\mathbf{U}}_i}(p_{i+1}) \ge -C_2/i$,

Step 3. Replace i by $i + 1$ and go to Step 1. □

The convergence properties of Algorithm 4.3 are given by the following result:

Theorem 4.4. Suppose that all $g^k(\cdot,\cdot)$, $k \in \mathbf{q}$, satisfy Assumption 3.3 and that Algorithm 4.3 has constructed an infinite sequence $\{p_i\}_{i=1}^{\infty}$. Then any accumulation point of $\{p_i\}_{i=1}^{\infty}$ is a stationary point for $\mathbf{P}_{2,\infty}$. □

5. IMPLEMENTATION AND NUMERICAL RESULTS

We have used Algorithm 3.5 to solve the problem of determining the depth h, and the width b of a short column with rectangular cross section, so that the total mass is minimized, while the first-order reliability index for the fully plastic mode of failure is kept greater than or equal to 2.5. We assumed that the column is made of elastic-perfectly-plastic material with yield stress Y in both directions. Let M and P denote the bending moment and axial force applied to the column. The limit-state function in terms of the variables (P, M, Y) is given by

$$G(x, p) = 1 - \frac{4M}{bh^2Y} - \frac{P^2}{(bhY)^2}. \tag{5.1}$$

where $x \triangleq (P, M, Y)^T$. We assumed that (P, M, Y) has a Nataf-type (see [5]) joint distribution with marginals and correlation coefficients as shown in the table below:

variable	statistics	correlation coefficients		
		P	M	Y
P	N(500,100)	1.0	0.5	0.0
M	N(2000,400)	0.5	1.0	0.0
Y	LN(5,0.5)	0.0	0.0	1.0

We defined the limit-state function on 3-dimensional standard normal space by introducing an invertible transformation $T: \mathbb{R}^3 \to \mathbb{R}^3$ (see [5] for details), mapping $x = (P, M, Y)$ into the 3-dimensional standard normal space. The limit-state function $g: \mathbb{R}^3 \times \mathbb{R}^2 \to \mathbb{R}$ we obtained is of the form

$$g(u, p) \triangleq G(T^{-1}(u), p). \tag{5.2}$$

Finally, we assumed that the width b and the depth h had to satisfy $5 \le b \le 15$ and $15 \le h \le 25$. The mathematical formulation of our example optimal design problem is as follows:

$$\min_{p=(b,h)} \{ bh \mid \beta(p) \ge 2.5,\ 5 \le b \le 15,\ 15 \le h \le 25 \}. \tag{5.3}$$

We used the program CalRel [7] to implement the transformations T and T^{-1} and to evaluate the limit-state function $g(\cdot,\cdot)$ and its derivatives. We carried out the computations, required in Steps 1 and 2 of Algorithm 3.5, by means of the sequential quadratic programming software CFSQP [4].

To evaluate our algorithm, we solved the design problem using a two-level optimization algorithm based on CFSQP, which treats the reliability constraint as nonlinear inequality constraints. Every time CFSQP required the evaluation of the reliability index $\beta(p)$ or its gradient, it called CalRel to perform this computations using the modified HL-RF algorithm [6].

Starting from the initial point $(b_1, h_1) = (5.0, 15.0)$, both algorithms converged to $(b, h) = (8.668, 25.0)$. Our algorithm took 14 iterations with 98 evaluations of the limit-state function and 77 evaluations of its gradient. The nested optimization algorithm took 227 evaluations of the limit-state function and 227 evaluations of its gradient.

We note that our implementation of the outer approximations algorithm did not include features such as adaptive precision and constraint dropping schemes (see [2]), which should improve its performance considerably.

6. CONCLUSION

We have presented a new formulation of certain optimal design problems subject to reliability constraints and a new outer approximations algorithm for their solution. Our method does not require the computation of second-derivatives of the limit-state function, nor does it require repeated computation of the first order reliability index. Our preliminary results show that our algorithm outperforms currently used alternatives.

REFERENCES

[1] Enevoldsen, I. and Sorensen, J. D., "Reliability-Based Optimization in Structural Design", Structural Reliability Theory, Paper n. 118, Dept. of Building Technology and Structural Engineering, Aalborg Universitetscenter, August 1993.

[2] Gonzaga, C., Polak, E. and Trahan, R., "An Improved Algorithm for Optimization Problems with Functional Inequality Constraints", IEEE Trans. on Automatic Control, Vol. AC-25, No. 1, pp. 49-54, 1979.

[3] Hettich, R. and Kortanek, K. O., "Semi-Infinite Programming: Theory, Methods and Applications", SIAM Review, Vol. 35, pp. 380-429, 1993.

[4] Lawrence, C., Zhou, J. L., Tits, A. L., "User's Guide for CFSQP Version 2.0: A C Code for Solving (Large Scale) Constrained Nonlinear (Minimax) Optimization Problems, Generating Iterates Satisfying All Inequality Constraints", Electrical Engineering Department, University of Maryland, College Park, 1994.

[5] Liu, P.-L., and Der Kiureghian, A., "Multivariate Distribution models with prescribed marginals and covariances", *Prob. Eng. Mech.*, Vol. 1, pp. 105-112, 1986.

[6] Liu, P.-L., and Der Kiureghian, A., "Optimization Algorithms for Structural Reliability", Structural Safety, 9, pp. 161-177, 1991.

[7] Liu, P.-L., Lin, H.-Z., and Der Kiureghian, A., "Calrel User Manual", Report No. UCB/SEMM-89/18, Department of Civil Engineering, University of California, Berkeley, 1989.

[8] Madsen, H., Krenk, S. and Lind, N., *Methods of Structural Safety*, Prentice-Hall, Englewood Cliffs, New Jersey, 1986.

[9] Polak, E., "On the Mathematical Foundations of Nondifferentiable Optimization in Engineering Design", SIAM Review, Vol. 29, pp. 21-89, 1987.

[10] Thoft-Christensen, P., "151 References in Reliability-Based Structural Optimization", IFIP WG 7.5 Working Conference on Reliability and Optimization of Structural Systems, Munich, September 1991.

14

RELIABILITY ESTIMATION OF TALL RC CHIMNEYS

M.B.Krakovski

Department of Naval Architecture, University of Ulsan, Ulsan P.O.Box 18, Republic of Korea 680-749
[Permanently Leading Researcher, Research Institute of Concrete and Reinforced Concrete, Moscow, Russia]

A RC chimney 330 m high has been designed for a power plant in Kazakhstan. In the course of construction it was discovered that geometrical axis of the chimney had tipped out of vertical. As a result it was decided to reduce the chimney hight down to 300 m. The paper deals with the investigation carried out in order to check whether the chimney 300 m high is sufficiently safe. The methodology of investigation and obtained results are presented.

1. INTRODUCTION

A RC chimney 330 m high has been designed for a power plant in Kazakhstan. In the course of construction it was discovered that geometrical axis of the chimney had tipped out of vertical. The maximum deviations of 65 cm were observed at elevations of 94 and 99.5 m. As a result it was decided to reduce the chimney hight down to 300 m.

The investigation under discussion has been carried out in order to evaluate whether the chimney 300 m high is sufficiently safe. With this aim in view reliabilities of two chimneys 300 and 330 m high were estimated and compared.

The investigation was ordered by a design office of power plant engineering. The office had at its disposal field measurements of concrete strength and of chimney geometrical dimensions. The measurements had been taken to control the quality of construction. These primary data were grouped and represented as histograms, which were taken to be probability density fucntions of corresponding basic variables.

According to the terms of the contract, in the course of reliability analysis all requirements of conventional design should have been met. Therefore it was decided to utilize an available computer program, which designers used for conventional analysis, though it took about 20 minutes to perform one deterministic analysis for specified values of basic variables.

The multitude of failure modes presented additional difficulties: as a result of each deterministic analysis the values of 8 output parameters were determined, i.e., 8 failure modes could occur.

All above peculiarities of the problem make it difficult, if not impossible, to analyse reliability of the chimneys using Level 2 (FORM/SORM) or importance sampling methods (see, e.g., [1-3]). Among other things, both these approaches require probability density functions of basic variables as well as limit state functions to be continuous. But probability density functions in the form of histograms are not continuous and 8 limit state functions can not be defined analytically in closed form: using conventional deterministic analysis it is possible only to check, whether the structure with specified values of basic variables fails or not.

In this connection in order to carry out reliability analysis, Monte Carlo simulation with subsequent approximation of the results by Pearson's curves [4] was used [5]. In the paper the methodology of investigation as well as obtained results are described.

2. STATEMENT OF PROBLEM

For reliability analysis strengths of concrete S and reinforcement R as well as geometrical dimensions of the chimneys (internal radius r and thickness of the walls h) were taken as basic variables. The loads were assumed to be deterministic and equal to their design values. The analysis has shown that under this assumption the calcultions resulted in conservative estimates.

In the course of construction of the chimney 300 m high the following parameters were checked at 26 elevations: strength of concrete, internal radii of the chimney, wall thickness. Strength of concrete was checked in two ways - by compression tests of specimens and by non-destructive ultrasonic tests. The results of geodetic measurements as well as concrete tests were represented as histograms in chimney sections at 26 elevations. To take an example, in Fig. 1 are shown the histograms at an elevation of 130 m for internal radius (a), wall thickness (b) and concrete strength (c). Thus, a number of basic variables representing geometrical dimensions and concrete strength was 78.

As to strength of reinforcement, it was dealt with in the following way. Ten different types of rebars (namely, 12, 16, 18, 20, 22, 25, 28, 32, 36, 40 mm in diameter) were used for chimney reinforcement. Since the rebars of different types are produced separately, they can be regarded as distinct materials. Therefore strengths of different types of rebars were assumed to be normally distributed statistically independent random variables with the same characteristics - mean values, 450 MPa, and standard deviations, 25 MPa. Design strength of reinforcment, 375 MPa, has an exceedance probability of 0.9986. Thus, additional 10 basic variables representing strength of reinforcement were considered, making a total number of basic variables equal to 88.

As a result of each deterministic analysis the following 8 output parameters were determined: stresses in concrete σ_c; stresses in vertical reinforcement σ_v and hoop bars σ_h; vertical a_v^1, a_v^2 and horizontal a_h^1, a_h^2 crack widths; deflection f of the top of the chimney. Here superscripts 1 and 2 are related, respectively, to the upper part of the chimney 1/3H high and to its lower part 2/3H high (H is a total hight of the chimney).

The allowable values of crack widths are: $[a_v^1] = [a_h^1] = 0.1$ cm; $[a_v^2] = [a_h^2] = 0.2$ cm. The allowable deflections $[f]$ of the top of the chimneys 300 and 330 m high are, respectively, 6.03 and 6.58 m. The allowable values for stresses in concrete and reinforcement are discussed below.

All above numerical values are determined in accordance with Russian codes and standards regulating analysis and design of chimneys.

a

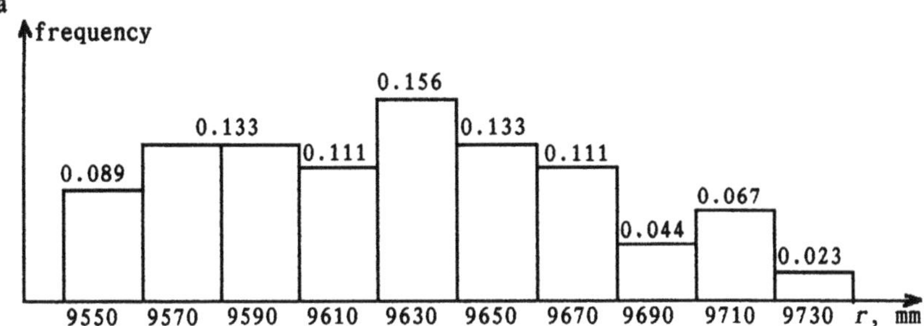

b

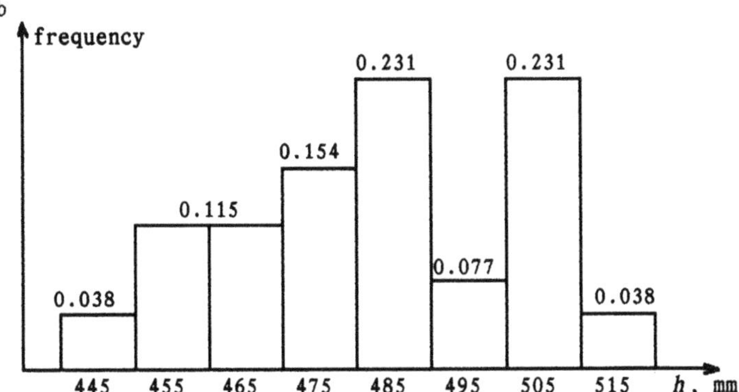

c

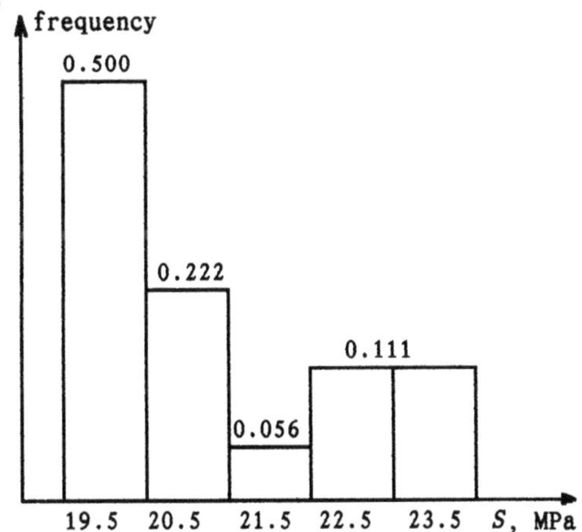

Fig. 1. Histograms

Reliability of the chimneys was assumed to be the probability that output parameter values will be below their allowable values. Reliability was determined with respect to:

- each of the output parameters (σ_c, σ_v, σ_h, $\overset{1}{a_v}$, $\overset{2}{a_v}$, $\overset{1}{a_h}$, $\overset{2}{a_h}$, f);
- strength limit state (failure occurs if at least one of the parameters σ_c, σ_v, σ_h exceeds its allowable value);
- serviceability limit state (failure occurs if at least one of the parameters $\overset{1}{a_v}$, $\overset{2}{a_v}$, $\overset{1}{a_h}$, $\overset{2}{a_h}$, f exceeds its allowable value);
- strength and serviceability limit states (total reliability, failure occurs if at least one of all output parameters exceeds its allowable value);

3. SOLUTION OF PROBLEM

The following algorithm was used to solve the problem.

1. According to specified probability density functions of basic variables obtain m sets of their random realizations (88 realizations in each set).
2. Carry out m deterministic analyses of the structure by the conventional method. As a result obtain m sets of 8 output parameters.
3. Choose one value representing each set of the output parameters. This item is discussed in detail later.
4. For m values, obtained at step 3, choose a suitable probability density function $y(z)$ out of the family of Pearson's curves.
5. Determine reliability of the structure by numerical integration

$$P = \int_{u_1}^{u_2} y(z)dz \tag{1}$$

The ways of specifying the values of u_1, u_2 are described below.

Now let us discuss step 3 of the algorithm, when reliability analysis is carried out with respect to different combinations of the output parameters. Consider first the case of reliability estimation with respect to one parameter σ_v. Ten random realizations R_i^k (i = 1,...,10) of strength of rebars of 10 different above mentioned types are used as input data for kth deterministic analysis. As a result of this analysis 26 values of stresses in vertical reinforcement σ_{vj}^k (j = 1,...,26) are obtained at 26 sections of the chimney. At each section the minimum

$$\min_i \; (R_i^k - \sigma_{vj}^k) \tag{2}$$

is found and then minimum σ_v^k of these minima among all 26 sections is determined

$$\sigma_v^k = \min_j \; \min_i \; (R_i^k - \sigma_{vj}^k) \; (k = 1,...,m) \tag{3}$$

At this point step 3 of the algorithm is completed. At step 4 the values of σ_v^k are approximated by a suitable Pearson's curve $y(z)$. The structure is safe if and only if $\sigma_v^k \geq 0$. Therefore the integration by eqn. (1) is performed with $u_1 = 0$, $u_2 = +\infty$.

In much the same way reliability with respect to each of other output parameters (σ_c, σ_h, a_v^1, a_h^1, a_v^2, a_h^2, f) was determined.

In order to check the calculated values of reliabilities, another approach was used as well. Instead of minima calculated in accordance with eqn. (2), the following maxima were determined:

$$\max_i \frac{\sigma_{vj}^k}{R_i^k} \tag{4}$$

and then maximum σ_{v1}^k of these maxima among all 26 sections was determined - eqn. (5) was used instead of eqn. (3):

$$\sigma_{v1}^k = \max_j \max_i \frac{\sigma_{vj}^k}{R_i^k} \tag{5}$$

Thereupon again a suitable Pearson's curve is chosen and integration by eqn. (1) is performed with $u_1 = -\infty$, $u_2 = 1$.

Now consider reliability evaluation with respect to strength limit states, i.e., three output parameters σ_c, σ_v and σ_h are taken into account. In this case the values σ_{c1}^k and σ_{h1}^k similar to σ_{v1}^k are determined:

$$\sigma_{c1}^k = \max_j \frac{\sigma_{cj}^k}{S_j^k} \qquad (k = 1,...,m) \tag{6}$$

$$\sigma_{h1}^k = \max_j \max_i \frac{\sigma_{hj}^k}{R_i^k} \quad (k = 1,...,m) \tag{7}$$

and then

$$\sigma^k = \max\,(\sigma_{c1}^k,\ \sigma_{v1}^k,\ \sigma_{h1}^k) = \max\left[\left(\max_j \frac{\sigma_{cj}^k}{S_j^k}\right), \left(\max_j \max_i \frac{\sigma_{vj}^k}{R_i^k}\right), \left(\max_j \max_i \frac{\sigma_{hj}^k}{R_i^k}\right)\right] \quad (k = 1,...,m) \tag{8}$$

is determined, representing the outcome of the kth simulation. Here S_j^k is a realization of concrete strength at jth section (j = 1,...,26) in kth simulation (k = 1,....,m). The values of S_j^k are used as input data for each deterministic design. The limits of integration in eqn. (1) are $u_1 = -\infty$, $u_2 = 1$.

Reliability with respect to serviceability limit states as well as total reliability (with respect to all output parameters) is calculated in much the same manner. Take, for example, the total reliability. The outcome of each simulation is represented by the value:

$$t^k = \max\left[\left(\max_j \frac{\sigma_{cj}^k}{S_j^k}\right), \left(\max_j \max_i \frac{\sigma_{vj}^k}{R_i^k}\right), \ldots, \left(\max_j \frac{a_{hj}^k}{[\ a_h^2]}\right), \frac{f^k}{[\ f]} \right] \quad (9)$$

$$(k = 1,\ldots,m)$$

Then a suitable Pearson's curve $y(z)$ is chosen for t^k and integration between limits $u_1 = -\infty$ and $u_2 = 1$ is performed.

As was indicated above, it took about 20 minutes to carry out one deterministic analysis. Therefore several computers working simultaneously were used.

In order to determine necessary sample size for reliability analysis, the values of reliabilities were computed after every 500 simulations. The results for sample sizes 2000, 2500, 3000 appeared to be closely allied. Therefore it was decided to consider results with $m = 3000$ as final ones.

By way of illustration in Table 1 are presented random realizations of basic variables r, h, S at an elevation of 130 m obtained from histograms in Fig. 1, and in Table 2 are given the values of some output parameters at the same elevation. All data are listed for the first 5 simulations. As can be seen from the Tables, it was not clear in advance, which of the chimneys was more reliable.

Below are given some of the Pearson's curves, approximating results of Monte Carlo simulations. For values of σ_c^k

$$\sigma_c^k = \min_j (S_j^k - \sigma_{cj}^k) \quad (k = 1,\ldots,m) \quad (10)$$

in chimneys 300 and 330 m high Pearson's curves turned out to be, respectively, of types 7 eqn. (11) and 4 eqn. (12):

$$y(z) = 0.2498\left[1 + \left(\frac{z - 3.402}{0.2315}\right)^2\right]^{-2.882} \quad (11)$$

$$y(z) = 0.2753\left[1 + \left(\frac{z - 1.460}{0.2096}\right)^2\right]^{-11.04} \times \exp\left[-11.04 \arctan\left(\frac{z - 1.460}{0.2069}\right)\right] \quad (12)$$

For values σ_v^k (3) in chimneys 300 and 330 m high Pearson's curves turned out to be of type 2:

$$y(z) = 0.1418\left[1 - \left(\frac{z - 157.7}{107.75}\right)^2\right]^{-2.111} \quad (13)$$

$$y(z) = 0.1307\left[1 - \left(\frac{z - 158.0}{61.31}\right)^2\right]^{-1.227} \quad (14)$$

Table 1
Realizations of basic variables at an elevation of 130 m

Simulation No	r, mm	h, mm	S, MPa
1	9690	475	19.5
2	9570	465	19.5
3	9570	495	19.5
4	9630	505	20.5
5	9610	505	25.5

Table 2
Results of Monte Carlo simulation at an elevation of 130 m

Simu-lation No	Stresses, MPa						Crack width $\times 10^3$, cm			
	σ_c		σ_v		σ_h		a_v^2		a_h^2	
	Chimney hight H =									
	300m	330m	300m	330m	300m	330m	300m	330m	300m	330m
1	11.0	9.7	193	192	253	276	75	73	74	72
2	11.2	8.9	190	189	250	253	72	71	71	70
3	10.6	8.6	197	197	260	259	79	87	78	83
4	10.7	8.6	200	218	263	263	83	66	80	85
5	10.9	10.3	200	217	263	222	83	78	80	77

The results of the investigation were as follows. Probabilities of failure of both chimneys with regard to each of the output parametets σ_c, σ_v, σ_h as well as with regard to strength limit state were less than 10^{-7}. Probabilities of failure of both chimneys with regard to each of the output parameters a_v^1, a_v^2, a_h^1, a_h^2, f as well as with regard to serviceability limit state were less than 10^{-5}. Reliabilities estimated according to equations of type (3) and (5) with respect to the same output parameter were in close agreement. The total failure probabilities of chimneys 300 and 330 m high were, respectively, 8.184×10^{-7} and 3.166×10^{-6}. All failure probabilities are small enough. Failure probabilites of chimney 300 m high did not exceed those of chimney 330 m high. On the basis of these results it was concluded that the chimney 300 m high was sufficiently safe. The chimney is in service now.

4. CONCLUSION

A rather complex practical problem was solved: reliabilites of two RC chimneys - 300 m high (with deviations of geometrical axis) and 330 m high (without the deviations) - were estimated. Results of field measurements in the form of histograms were used as probability density functions of basic variables. Eight failure modes of the chimneys could not be defined analytically in closed form. Therefore it was very difficult, if not impossible, to use Level 2 (FORM/SORM) or importance sampling methods for reliability analysis. To solve the problem Monte Carlo simulation with subsequent approximation of the results by Pearson's curves was applied. The methodology of investigation was developed and checked for accuracy. Calculations were performed, reliabilities of two chimneys with respect to different combinations of output parameters were determined and compared with each other. On the basis of the obtained results it was concluded that the chimney 300 m high was sufficiently safe. The chimney is in service now.

REFERENCES

1. P.Thoft-Christensen and M.J.Baker, Structural Reliability Theory and its Applications, Springer, New York, 1982.
2. M.Hohenbichler, S.Soltwitzer, W.Kruse and R.Rackwitz, New light on first- and second-order reliability methods, Structural Safety, 4, 1987, 267.
3. S.Engelung and R.Rackwitz, A benchmark study on importance sampling technique in structural reliability, Structural Safety, 12, 1993 255.
4. W.P.Elderton and N.L.Johnson, Systems of Frequency Curves, Cambridge University Press, 1969.
5. M.B.Krakovski, Improvement of Russian Codes for RC Structures Design on the Basis of Reliability Theory, Proceedings of IFIP WG7.5 Fifth Working Conference on Reliability and Optimization of Structural Systems, Japan, March, 1993, North-Holland, 141.

15

Discussion on response surface approximations for use in structural reliability

J. Labeyrie[a] and F. Schoefs[b]
[a] IFREMER , Post Box 70 , 29280 Plouzané - FRANCE
[b] Laboratoire de Mécanique, Université de Nantes, 2 rue de la Houssinière
44072 Nantes Cedex 03

ABSTRACT

Response Surface Methodology (RSM) has been largely discussed by the past, from a biometric viewpoint and also for use in nuclear reactor safety. The main concern comes to perform some approximations as a surrogate for the original model in subsequent uncertainty and sensitivity studies.

We propose in this paper to focus attention on RSM for use in structural reliability analysis. The requalification of offshore platforms through the process of re-assessment confirms that such questions merit to be more investigated.

Obviously there is a large amount of basic statistical techniques as stochastic differential or Monte Carlo analyses which remain a common background. But thinking is that the criteria for constructing response curves are drastically depending of how the models will be managed through the computational procedure and for what physical meanings mainly.

1 - INTRODUCTION

Response Surface Methodology (RSM) is a formal aid-tool for most of the quantitative investigations where the required response $\mathbf{y}$ can be considered as the output of a system and which varies in response to the changing levels of several input variables [1][2]. The system is to be modelled by some mathematical function of the random variables $\mathbf{x}$ (or processes) involved in the system, and characterised by statistical informations $\boldsymbol{\Theta}$ (moments, free or parametric distribution functions, fourier series, ...),

$$y = f(x/\Theta) \tag{1}$$

The exact form of f being unknown generally, different targets which may conflict are in considerations :

- ➔ it is wished to fit mainly a function in order to describe the data with reduction in storage.
- ➔ the function is required to be meaningful in the sense that it is based as large as possible on some understanding of the underlying mechanism.
- ➔ the function is to be used for inference purposes.
- ➔ the function can be fitted without too complex computational requirements.

Mathematically, assuming that f is a continuous function and the vector $\mathbf{x}$ varies in a finite range, the Weierstrass approximation theorem ensures that one can approximate f by a polynomial function to any desired accuracy.
The most frequently used response curves are linear, quadratic or cubic, as they are very often obtained by least squares estimates of the parameters without too complex calculations. This can be extended by applying the same result to $\Phi(f)$ where Φ is a specified basic scale function (inverse, power, logarithmic, exponential, ...).

It is to notice that these polynomial or derived functions seem to be used as simplest readily-available smoothing curves, without any appeal to their theoretical properties as approximations to the true response function. In particular there is no discussion and even an eclipse of the issue of a polynom order to be admissible as an asymptotic property.

There is no hope to define a general procedure but when a response curve is chosen, then this choice has to be well argued. Otherwise the results are to be presented as conditional on the model being corrrect and consequently with loss of generality.

By means of an example (wave actions for structural reliability) we suggest to illustrate application of techniques for validating some response curve following two essential criteria given by physical meanings and statistical distribution effects. Complexity reduction in modelling and computational tractability are introduced at a second stage as increasing use and progress of electronic computer and algorithms offer important potentialities.

2 - WAVE ACTIONS FOR STRUCTURAL RELIABILITY

Since some decades, the Morison equation remains a robust means of predicting wave forces on slender cylinders [3]. The usual form can be already viewed as a response curve by itself,

$$f(t)=\frac{1}{2}\rho D\, C_d\, u_t|u_t| + \frac{1}{4}\,\pi\rho D^2\, C_m\, u'_t \tag{2}$$

where $f(t)$ is the force per unit span separated into drag and inertial components, ρ is water density, D is cylinder diameter and u_t is the instantaneous flow velocity. The drag and inertia coefficients C_d and C_m closely related to the Reynolds and Keulegan-Carpenter numbers are obtained by applying least squares procedures to measured force and velocity data.

2.1 - Physical meanings

The equation (2) generally predicts the main trends in measured data quite well, once an appropriate joint distribution function for C_d and C_m, depending of sea-state parameters, can be provided [4].
Measurements of Morison coefficients were performed also in many other investigations with particular interest in realistically fouled and roughened cylinders, and in conditions where waves and currents act simultaneously.

Nevertheless some interesting characteristics of the flow are not represented with enough accuracy (e. g. high frequency content, gross vortex shedding effects, ...). So when applying eq.(2) we have to get in mind that some sources of response problems for an offshore platform are missing.

As an example, extensions of eq.(2) are to be evaluated, using the NARMAX modelling technique (Non linear Auto-Regressive Moving Average with eXogenous inputs) [5] [6]. It is obtained from the frequency domain analysis that some systematic structure is present in higher order which is not predicted by Morison's equation. But they don't allow for instance a direct physical interpretation of the additional term. There is probably no hope to propose a general model consistent whatever the application but it is relevant to validate extension of the physics "governed" by eq.(2) and specially for non linear effects.

2.2 - Polynomial approximations

When predicting the stochastic response of offshore platforms under Morison type non linear random wave loading, a lost of works suggest a response curve of $f(t)$ in the form of linear, quadratic or cubic expansion as a function of the instantaneous flow velocity. From this tractable form, the resulting non linear equivalent system can be solved using the Volterra series method or some other integration technique. Then response moments are computed.

We propose to examine this procedure under the designation distribution effect, e. g. the interrelationships between probability distributions functions through the response curves.

Let $Y=\varphi(X)$ be a random variable defined as a function of a random variable X with probability density f_X, where φ is differentiable and bijective over the supports of the distributions. We can obtain the probability density f_Y by :

$$f_Y(y) = f_X\left(\varphi^{-1}(y)\right) \cdot \left|\frac{1}{\varphi'}\left(\varphi^{-1}(y)\right)\right| \tag{3}$$

Considering φ as a response curve and let P be a polynomial approximation. From eq.(3) and assuming P closed enough of φ, a consistent approximation of φ with respect of the response density f_Y needs also a good approximation of the derivative of φ. Taking P as linear, quadratic or cubic expansions then the derivative P' is constant, linear or quadratic.

The risk of discrepancy between these three last functions is very high and their effects in changing the upper and lower tails of the density f_X are quite different.

The choice of a polynom of low order, so tractable for the computational procedure, although apparently relevant for an approximation of the response curve, can lead to an erroneous probability density output.

In particular this explains results presented in different papers from simulation studies, which show that equivalent systems deduced from quadratization and cubicization methods give inappropriate response skewness and kurtosis coefficients.

This simple example illustrates that the distribution effects can significantly influence response curve construction as giving criteria. It is very often important to diagnose the possible ramifications of these effects. The issue of RSM is an area of sensitivity analysis where interaction among model builders and those conducting the sensibility analysis is essential.

There is no generally applicable technique for determining whether the distribution effect is affecting the computational procedure through the RSM. But there is a clear scope for models of higher order properties than those frequently used for simplification mainly.

This problem increases in complexity, in the case of several variables to be selected. Usual criteria are based on minimization bias and/or variance of the deviation from the true response, integrated over some region. As an extension of eq.(3), there is a need for an additional criterion such as good fitting of the Jacobian matrix,

$\left(\frac{D(x_1,\ldots,x_n)}{D(y_1,\ldots,y_m)}\right)$ in order to prevent the main distribution effects.

2.3 - Formal geometrical modelling

The eq.(2) is to be quite relevant to compute quasi-static loads under extreme environmental conditions. In the following, the guiding idea is to maintain f as a response curve without any more approximation in order to avoid inherent distribution effects and to focus attention in providing a suitable formal geometrical modelling of the instantaneous flow velocity [7]. At a first stage we consider wave actions for simplicity of presentation only.

Let Oxyz be a canonical repair in the euclidean space R^3. The origin O is taken on the mean sea level. The axis Oz is vertical and increasing. We note $\overrightarrow{OU}$ the directional vector of waves ($\overrightarrow{OU} = Ox\cos\theta + Oy\sin\theta$ where θ is the wave direction).

The instantaneous kinematics field of waves being a random vector in R^3, we propose to evaluate this vector field as resulting of geometrical transformations (similarity, rotation, ...) of $\overrightarrow{OU}$.

The Navier Stokes equations are basic for the most mathematical models in fluid mechanics. They are second-order differential equations due to the friction terms, and non linear because of the convective inertia terms.

The small amplitude plane harmonic waves known as Airy waves can be derived from the potential ϕ,

$$\phi(x,z,t)=\frac{H}{T}\frac{\pi}{k}\frac{\cosh(kz)}{\sinh(kd)}\sin(kx-\Psi_t) \quad (4)$$

H : wave height, T : wave period,

k : wave number, Ψ_t: pseudo phase angle,

and d : water depth.

After some algebraic operations the associated gradient field is expressed from geometrical transformations (rotation and similarity product),

$$\vec{V}_{airy} = H(O,\lambda_1) \circ \Re\left(\overrightarrow{OV},\alpha\right) \overrightarrow{OU} \tag{5}$$

where the similarity ratio λ_1, and the rotation angle α are random functions,

$$\alpha = \operatorname{atan}\left[\tanh(kz)\tan(kx - \Psi_t)\right] \text{ and,} \tag{6}$$

$$\lambda_1 = \frac{H}{2}\sqrt{\frac{g}{d}}\sqrt{\frac{\cosh kz}{\cosh kd}}\sqrt{\frac{kd}{\tanh kd}}\sqrt{\cos^2(kx - \Psi_t) + \tanh^2(kz)\sin^2(kx - \Psi_t)} \tag{7}$$

Considering now the Stokes 2nd order expansion, there is an additional term to eq.(5), written in a similar form,

$$\vec{V}_s = H(O,\lambda_s) \circ \Re\left(\overrightarrow{OV},\alpha_s\right) \overrightarrow{OU} \tag{8}$$

where $\alpha_s \approx 2\alpha$ and λ_s is a random fonction which varies also with (H,k).

This gives arguments in discussing the level of complexity to be introduced in kinematics field modelling of waves for structural reliability analyses.

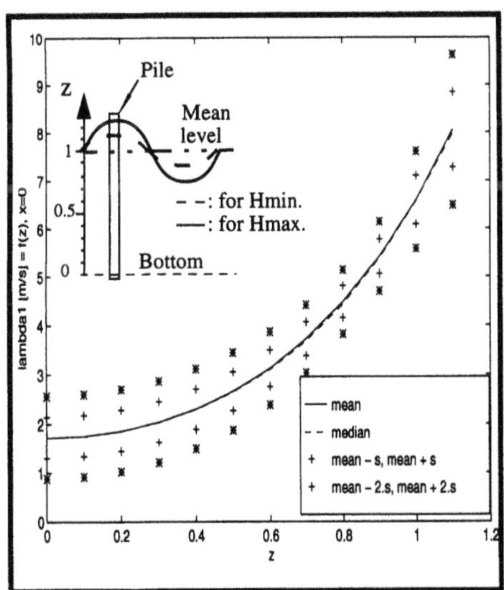

figure 1 Statistical variations of λ_1 with depth (North sea)

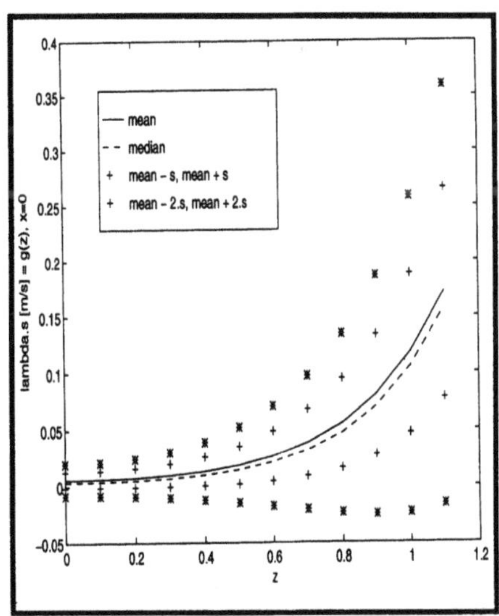

figure 2 Statistical variations of λ_s with depth (North sea)

In fact the two similarity ratios λ_1 and λ_s have been compared as functions of the two random variables H and k of which ranges and distributions were performed from North Sea in situ measurements (see figures 1 and 2). The influence of λ_s appears quite of minor importance. It confirms that among Stokes models family, the Airy wave model contains the essential structure of basic random variables to be analysed for structural reliability under quasi-static loads.

The angle rotation α is a random function with a specified trajectory (see eq. (6)) of which the stochastic fluctuations are governed by the random variable k. In the case of severe conditions, the range and the statistics of k are well adapted for use of first order taylor's expansion around the mean wave number expectation $\bar{k}$, when developing the angle rotation function.

Denoting by $\underline{\alpha} = \alpha(\bar{k})$, $\underline{\alpha}' = \alpha'(\bar{k})$, $\lambda_0 = (k - \bar{k}).\underline{\alpha}'$, $\lambda_2 = \frac{1}{\sqrt{1+\lambda_0^2}}$ .Then after some mathematical operations in the euclidean space R^3, and considering a repair connected with waves $(O, \overrightarrow{OU}, \overrightarrow{OV}, \overrightarrow{OZ})$, we obtain a sharp approximation among geometrical operators,

$$\Re\left(\overrightarrow{OV}, \alpha\right) \approx H(O, \lambda_2) \circ \left\{ I_d + H(O, \lambda_0) \circ \Re\left(\overrightarrow{OV}, \frac{\pi}{2}\right)\right\} \circ \Re\left(\overrightarrow{OV}, \underline{\alpha}\right) \tag{9}$$

Introducing this operator approximation inside eq.(5), the kinematics field deduced from the Airy model has the very convenient form,

$$\vec{V}_{airy} = \mu \vec{A} + \nu \vec{B} \tag{10}$$

where $\vec{A}$ and $\vec{B}$ are deterministic orthonormal vectors defined by,

$$\vec{A} = \Re\left(\overrightarrow{OV}, \underline{\alpha}\right)\overrightarrow{OU} \quad \text{and} \quad \vec{B} = \Re\left(\overrightarrow{OV}, \frac{\pi}{2}\right)\vec{A}.$$

The coefficients μ and ν are random scalar functions with specified trajectories depending of the couple of random variables (H, k).

The non linear term $u|u|$ (see eq. (2)) is then written easily $\sqrt{\mu^2 + \nu^2}\,(\mu \vec{A} + \nu \vec{B})$.

Finally f(t) is also approximate in the closed form of a linear combination of deterministic vectors ; the coefficients of which are random functions which depend of a set of basic random variables H, k, C_d, C_m and possibly D_e the number's effective diameter to take account for estimating the marine growth thickness. For another wave direction θ_1, then the response curve is obtained by changing $\vec{A}$ and $\vec{B}$ only through a rotation of axis Oz and angle $\theta_1 - \theta$.

This formal geometrical modelling can be extended to study wave/current interactions [8]. The maximum current during a storm rarely occurs at the same time or in the same direction as the maximum waves. A one third decrease in current design value is to be expected into the site under consideration when considering concomitants of order statistics (e.g. distribution of current speed given that significant wave height is at its maximum). Due to the respective ranges of the wave kinematics and current speeds, this implies no important decrease in the near-surface zone (≈ 5%) where loadings are the most severe. Obviously the decreasing effect is very sensitive to depth parameter and when the current profile is very uneven, the current/wave interaction becomes a crucial item.

At a second stage this formal approach allows to identify the external equivalent concentrated nodal forces by a finite linear combination of the form,

$$\Sigma \lambda_i \vec{F}_N^{(i)} \tag{11}$$

where the λ_i are random multipliers which contain the same basic random variables as mentioned previously, enlarged with λ_w, λ_c wind and current velocities measured at reference levels. The $\vec{F}_N^{(i)}$ are deterministic vectors which depend essentially of the structure topology, and also of the different deterministic profile functions of actions and of wave location with respect to the structure [9].

A computer program was developed and implemented as a module of the general package ARPEJ (Evaluation of Jackets from a Probabilistic Redundant Analysis). It gives the opportunity to perform sensitivity studies and it indicates when necessary, where to concentrate efforts in terms of probabilistic modelling of environmental conditions.

3 - CONCLUSION

The Response Surface Methodology is clearly a relevant scope for Reliability and Optimization of structural system. There is no doubt that response curves for limit state functions, ranking of input variables or sensitivity studies are to be provided.

This paper is briefly discussing the main criteria for response surface construction. It is suggested to start with functions based on some understanding of the underlying mechanism.

It is showned that usual polynomial approximations of low order are very sensitive to distribution effect.

A formal geometrical approach through an illustration to wave actions modelling, comes to response curves in a very convenient form for reliability analysis, and allows to prevent the main distribution effects.

The efficiency of RSM for use in the requalification process depends largely on ability in estimating realistic ranges and distributions for the input variables under consideration.

Finally existing data bases enhanced by satellite measurements of wave and wind, experience feedback provided by in situ measurements at platform or a neighbouring location, are to be worked with a dedicated attention to this extension.

ACKNOWLEDGEMENTS

The work reported here had been carried out within the Clarom R&D project "Requalification process of existing offshore structures", and funded by its participants (namely : Bureau Veritas, Ifremer, Geodia, EDF, Bouygues Offshore and Principia). Their support is gratefully acknowledged.

References

1.Mead, R. and Pike, D.J. a review of response surface methodology from a biometric viewpoint. Biometrics 31, 803-851 (1975)

2.Iman, R.L. An approach to sensitivity analysis of computer models :Part **II**. Ranking of input variables, response surface validation, distribution effect and technique synopsis, J. of Quality Technology, Vol.13, n°4, 232-240 (1981)

3.Morison, J.R.; O'Brien, M.P., Johnson, J.W. and Schaff, S.A. the forces exerted by surfaces waves on piles, Petroleum trans., 189, 149-154 (1950)

4.Nerzic, R. and Lebas, G. Uncertainties in wave loading from full scale measurements, in proc. BOSS'88, 1399-1412 (1988)

5.Leontaritis, I.J. and Billings, S.A. Input-output parametric models for non linear system. Part II : stochastic non linear systems. International Journal of Control 41, 329-344 (1985)

6.Worden, K., Stansby, P.K. and Tomlinson, G.R. Identification of non linear wave forces, Journal of fluids and structures, 8, 19-71 (1994)

7.Schoefs, F. and Labeyrie, J. A formal geometrical modelling of the wave actions for structural reliability, to be published.

8.Labeyrie, J. and Olagnon, M. Stochastic sensitiveness to combined extreme environmental loads in structural reliability, proc. 12th OMAE, Vol II, 107-116 Glasgow, (1993)

9.Labeyrie, J., Stochastic load models for marine structure reliability, 3rd IFIP WG 7.5, Berkekey, California (1990)

16

Site-specific and Component-specific Bridge Load Models

Jeffrey A. Laman[1] and Andrzej S. Nowak[2]

[1] Research Assistant, Department of Civil and Environmental Engineering, University of Michigan, Ann Arbor, Michigan, 48109.

[2] Professor of Civil Engineering, University of Michigan, Ann Arbor, Michigan, 48109.

Bridge reliability analysis is based on load and resistance models. In particular, load spectra vary depending on site and component. The objective of this paper is to present the results of measurements of truck loads on girder bridges. The study included several structures located on interstate highways and surface streets.

The study confirms that truck loads are strongly site specific. There is a negative correlation between law enforcement effort and occurrence of overloaded trucks. Illegally loaded trucks are observed on roads not controlled by truck weigh stations. A comparison of the truck citation data and weigh-in-motion (WIM) measurements obtained in this study confirms this observation.

Load spectra are measured for various components, including each girder. The results indicate that the stress spectra are component specific. Girders located directly under the traffic lanes support the largest portion of the load. This observation is important for fatigue analysis considerations. Measured loads are presented in terms of strains and gross vehicle weight (GVW). Cumulative distribution functions (CDF's) of the data are plotted for enhanced interpretation.

1. INTRODUCTION

Presented are measurement results of truck GVW's and strain loads on girder bridges. A study of four bridge structures was performed at locations of different average daily truck traffic (ADTT) and proximity to stationary weigh stations. Measurements were conducted using a bridge weighing system and a data acquisition system. Approximately 22,000 truck files are included in the data base of WIM. In addition, 1985 citation data from the Michigan State Police Motor Carrier Division was processed. The recorded truck data was processed to develop cumulative distribution functions (CDF's) of GVW for each of four sites. Stress histories are collected at mid span of each girder bottom flange for one week periods and processed using the rainflow algorithm. This data is then converted to an equivalent (root mean cube) stress as a convenient method of comparison between components. Load spectra, particularly fatigue loads, differ depending on ADTT, proximity of industrial facilities, and position of the girder. The results indicate that live load is strongly site specific and component specific.

Bridge live load is a dynamic load which may be considered as a sum of static and dynamic forces. This study is concerned with the both the static portion of the load and the resulting dynamic response. Actual truck axle weights and spacing, GVW, ADTT, and the load effects of the trucks such as moments, shears, and stresses are important parameters used in the effective evaluation of a bridge. Truck data is available from highway weigh station logs, citation data, and also through the use of WIM technology. The stationary weigh scales at weigh stations are biased and will not reflect accurately the distribution of truck axle weights, spacing, and gross vehicle weights due to avoidance of scales by illegally loaded trucks. WIM measurements of trucks and strain measurements can be taken discretely, resulting in unbiased data for a statistically accurate sample of truck traffic crossing a particular bridge. Strain measurements can be collected at various locations on the bridge to sample the stress effects of the measured truck traffic. Presented in this study are WIM gross vehicle weight (GVW) for the bridge site and strains in the main girders.

2. SELECTED BRIDGES

During 1991, 1992, and 1993 WIM and strain measurements were conducted at seven bridge sites. These sites are located in southeast Michigan on US, Interstate, Michigan, and city roadways. The location and description of the bridges are as follows:

1. The bridge on US-23 over the Huron River (US-23/Huron Rr.) east of Ann Arbor, Michigan is constructed as a two lane, five simple span, 6 steel composite girder structure and carries northbound (NB) traffic. (See figure 1)

2. The bridge on US-23 over the Saline River (US-23/Saline Rr.) south of Milan, Michigan is constructed as a two lane, three simple span, 10 steel composite girder structure and carries southbound (SB) traffic between the metropolitan Detroit area and Toledo Ohio. There is no weigh station on the route. (See figure 2)

3. The bridge on I-94 over Pierce Road (I-94/Pierce Rd.) west of Ann Arbor, Michigan and east of Jackson, Michigan is constructed as a three simple span, composite steel, ten girder bridge and carries eastbound (EB) traffic between Detroit and Chicago. The site is 4 miles east of a weigh station. (See figure 3)

4. The bridge on US-23/M-14 over the New York City Railroad (M14/New York RR) north of Ann Arbor, Michigan is constructed as a two lane, three simple span, 8 steel composite girder structure and carries eastbound (EB) traffic. (See figure 4)

3. TRUCK WEIGHT MEASUREMENTS

Statistical data is presented in the form of cumulative distribution functions (CDF's). CDF's are used to present and compare the critical extreme values of the data and are plotted on normal probability paper.

Most states allow a maximum GVW of 356 kN (80 kips), however in Michigan the legal limit for an eleven axle truck is over 712 kN (160 kips), depending on axle configuration. From the tables and graphs it can be seen that there were a number of illegally loaded trucks measured during data collection at several of the sites.

GVW results are plotted in figure 5. Measurements of GVW taken at US-23/Huron Rr NB indicate the upper tail is dominated by 11 axle vehicles. This bridge is not on a route near or between major industrial areas and the data reflects this as there are almost entirely legally loaded vehicles, despite the absence of a weigh station in the vicinity. The US-23/Saline Rr SB bridge site is along a north-south route between the major industrial areas of metropolitan Detroit and Toledo, Ohio with no weigh station along the route. It can be expected, as the GVW data reveals, that heavily and illegally loaded trucks will be motivated to travel this route. The heaviest vehicle weighed was a 1100 kN (248 kip) 11 axle truck. Several vehicles weighed over 900 kN (200 kips) and many exceeded legal limits.

Measurements taken at I-94/Pierce Rd. in 1993 were conducted when the nearby stationary weigh station was closed for repairs. As a comparison, WIM data was collected in 1991 with the stationary weigh station in operation . It is suspected that illegally loaded trucks avoid stationary weigh scales causing the data to be biased. The data presented supports this notion. The 1993 maximum GVW of 800 kN (181.5 kips) is considerably larger than the 1991 maximum GVW of 590 kN (133.3 kips). GVW measurements collected at M-14/New York RR EB indicate very heavy vehicles with the upper tail dominated by the 11 axle vehicles. This bridge, while not in an industrial area, may be a route used between northern Detroit suburbs and the metropolitan Chicago area

4. STRAIN MEASUREMENTS

All strain measurements were taken in the bridge girder bottom flange at mid span. The strain data was collected using the rainflow algorithm enabling one week of continuous data collection at each bridge site. Equivalent stress (root mean cube) results of the strain data collection are presented in figure 10a through 10d. For orientation, girder 1 is located at the extreme right of the right lane of the bridge.

5. CONCLUSIONS

From the presented data of GVW and lane moments, load spectra is highly site-specific and is dependent upon a number of factors. These factors include proximity to stationary weigh stations, weigh station hours of operation, proximity to industrial areas, and desirability of the route for traffic between major metropolitan areas. The distribution of moments caused by the truck traffic is related to the same factors as GVW as well. In some locations the level of the lane moments may be exceedingly high. Michigan State Police Citation data confirms that very heavy trucks should be expected when data is collected without observation, which is possible with WIM equipment.

Strain spectra is component dependent and varies greatly from girder to girder. The tests consistently indicated that girders located nearest the right lane left wheel track experienced the highest stains, typically girder 3 to 5. Girders supporting the left lane experienced moderate strains, and outer girders low levels of strain.

ACKNOWLEDGMENTS

The research presented in this paper has been sponsored by the Michigan Department of Transportation and the Great Lakes Center for Truck Transportation Research, (GLCTTR), which is gratefully acknowledged.

REFERENCES

Nowak, A.S. and Laman, J.A., *Effect of Truck Loads on Bridges*, Research Report UMCE 94-22, Department of Civil and Environmental Engineering, The University of Michigan, Ann Arbor, Michigan, September 1994.

FIGURES

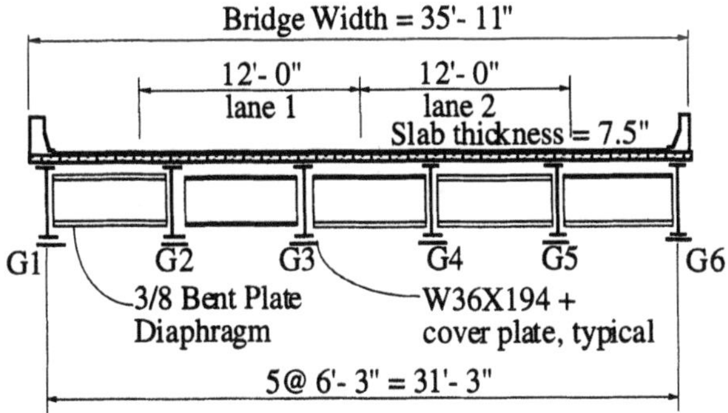

Fig. 1, US23/Huron River Northbound Section Looking South.

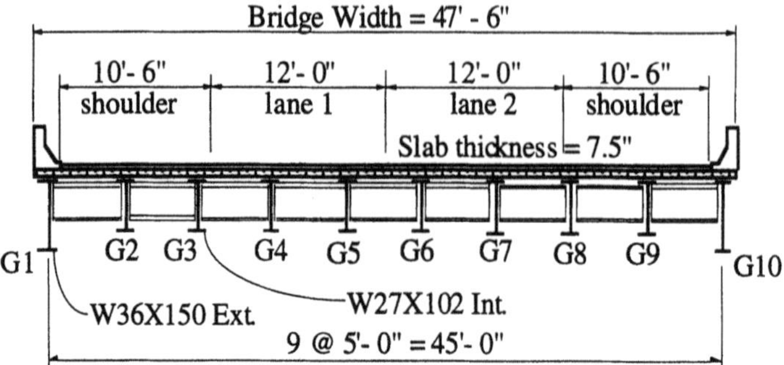

Fig. 2, US23/Saline River Southbound Section Looking North.

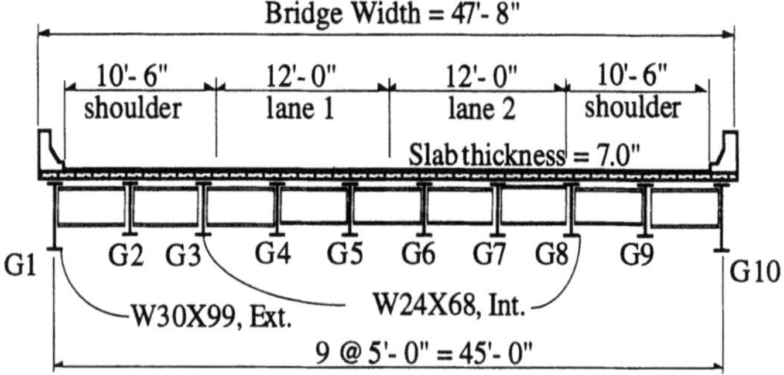

Fig. 3, I-94/Pierce Road Eastbound Section Looking West.

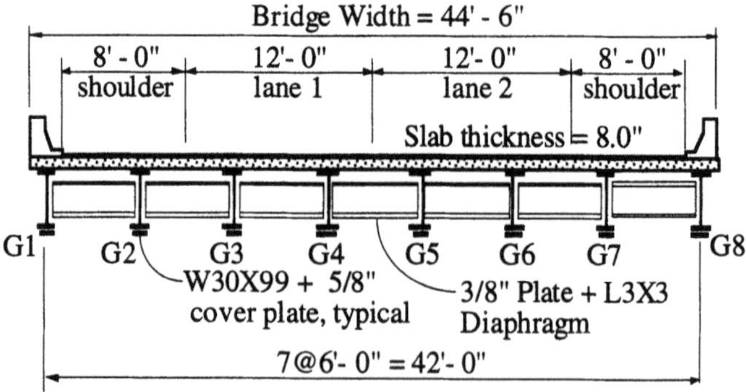

Fig. 4, M14/NYC Railroad Section Looking West.

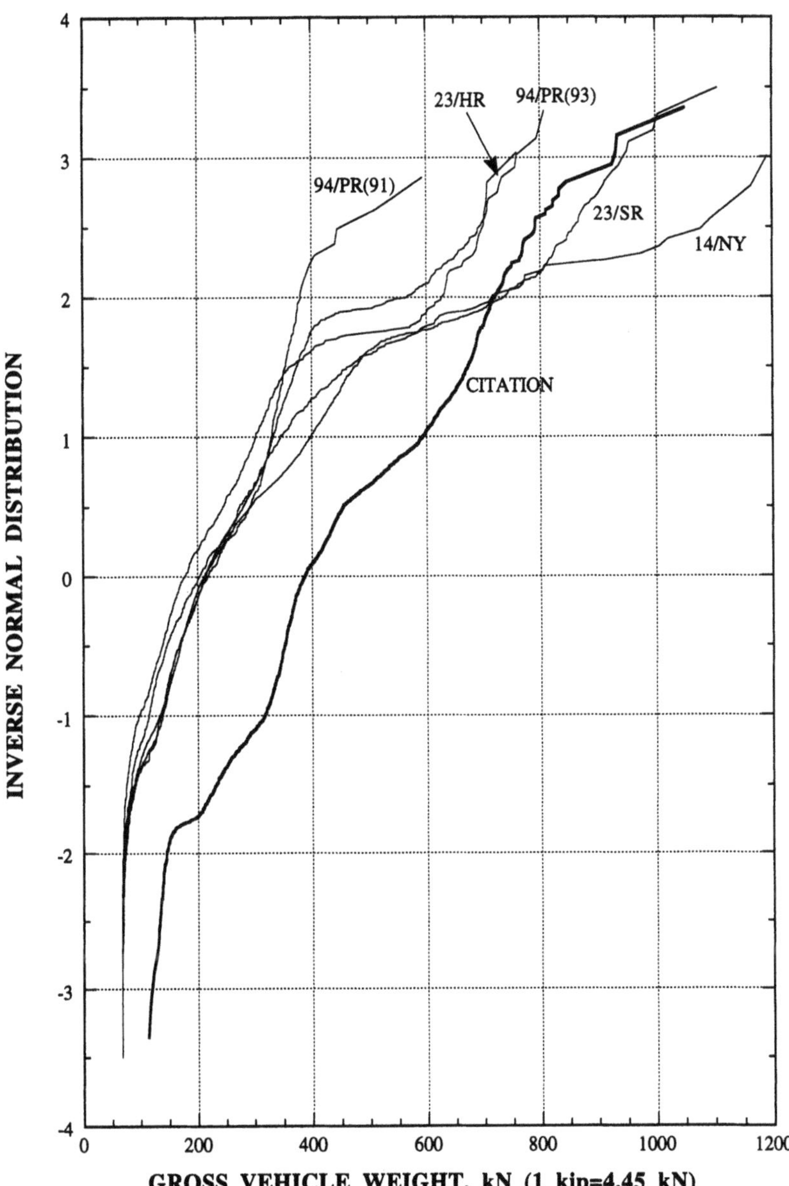

Fig. 5, Gross Vehicle Weight CDF's of All Trucks > 65 kN with Citation Data.

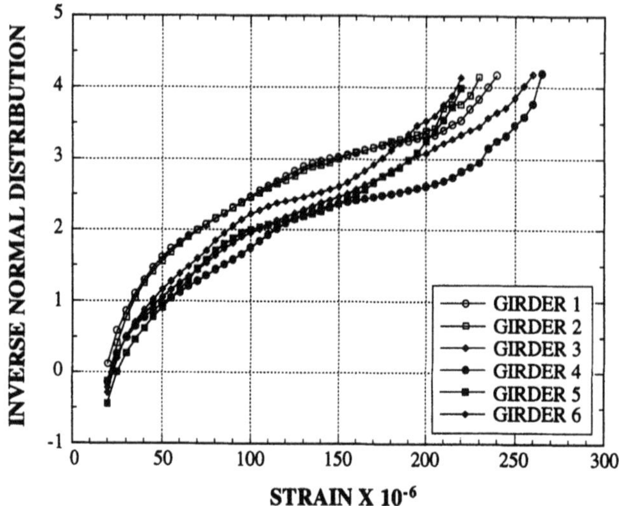

Fig. 6, US23/Huron Rr NB Rainflow Strain CDF's for G1-G6.

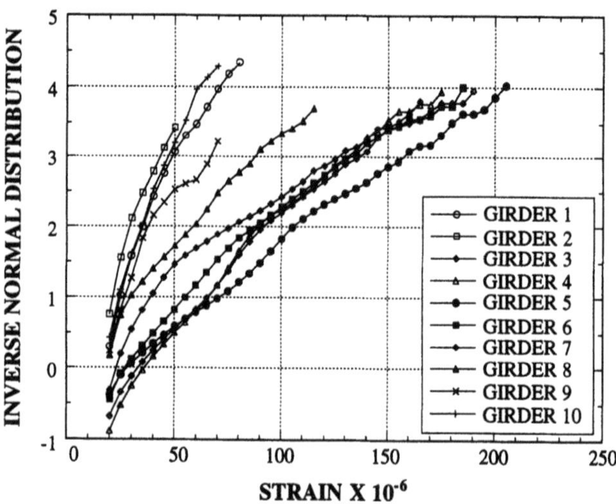

Fig. 7, US-23/Saline Rr SB, Rainflow Strain CDF's for G1-G10.

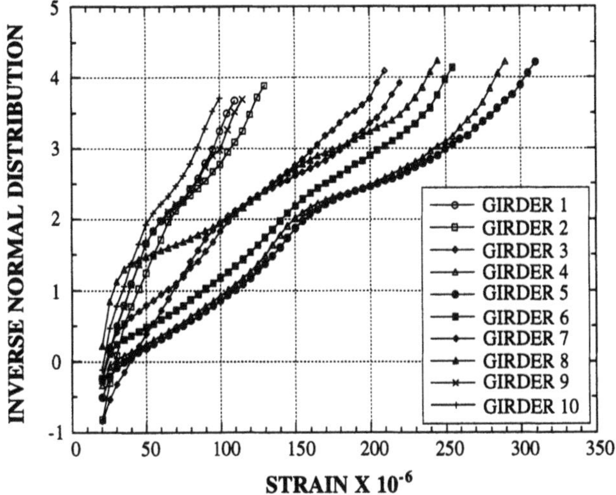

Fig. 8, I-94/Pierce Rd EB Rainflow Strain CDF's for G1-G10.

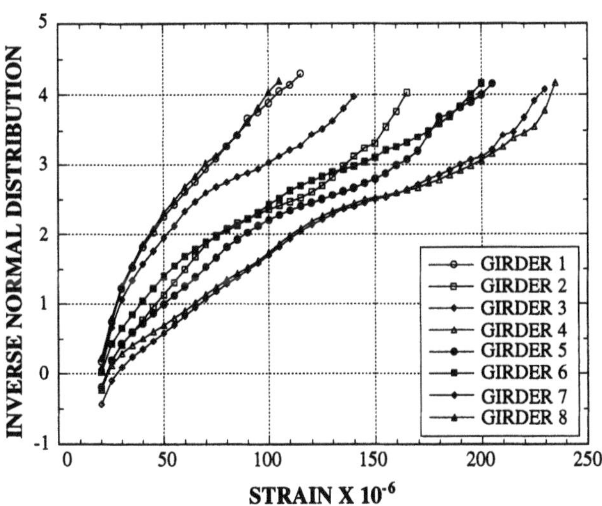

Fig. 9, M-14/New York City RR EB, Rainflow Strain CDF's for G1-G8.

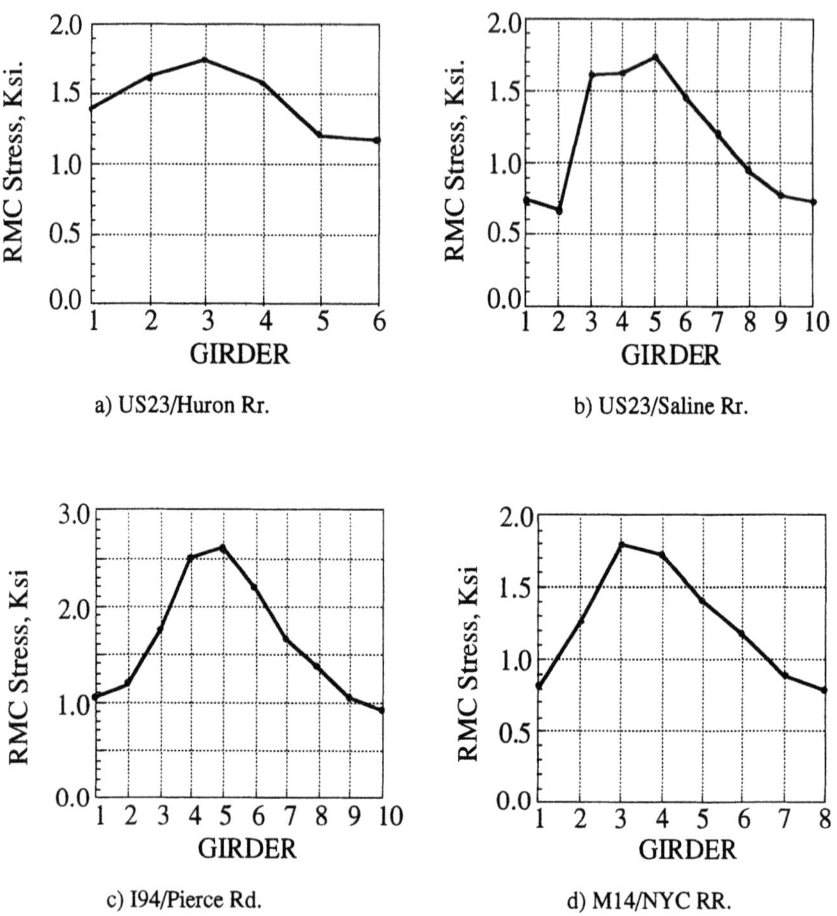

Fig. 10, Equivalent (Root Mean Cube) stress for each Girder of the Selected Bridges. (1 ksi = 6.9 MPa)

17

Functionals of Stochastic Processes and Random Fields in Reliability Analysis

Claus Lange
Hochschule für Technik und Wirtschaft Dresden
Fachbereich Informatik/Mathematik
D-01069 Dresden, Friedrich-List-Platz 1
Germany

Abstract
Probability theory methods for calculating approximations of first excursion and fatigue failure probabilities in the case of arbitrary differentiable stochastic process models are discussed. The discrete random variable of upcrossings of a level by a differentiable stochastic process is used to consider the functionals of stochastic processes which are of practical interest in reliability analysis.

1. Introduction

Many problems in reliability analysis of structural systems can be solved only by the theory of stochastic processes and random fields. Random excitations like wind, earthquake and sea waves, random loads like snow and vehicles, random temperature influences and random material or damping properties are only a few examples where methods of that theory are required to estimate the reliability of structural systems.

The approximation of first excursion or fatigue failure probabilities and of other characteristics of reliability bases on the distribution of different functionals of stochastic processes and random fields. The maximum and the minimum of a stochastic process, the distance between a local maximum and the following local minimum of a stochastic process, the time a process spends above or below a level and local maxima and minima of a random field are examples of such functionals.

It is given a method which approximates the distributions of all those functionals by the number of upcrossings (or downcrossings) of an arbitrary level by a stochastic process. The theory can be used to calculate the failure probability in the case where both types of failure for a structural system, that means fatigue failure as well as first excursion failure, are thinkable to occur.

2. First excursion failure models

The reliability of structural systems is to determine in many engineering applications. It is assumed that the mathematical models of the structural systems, the random loads and the critical levels are given. The failure type is first excursion failure. An arbitrary differentiable (with probability one) stochastic process $X(t), t\epsilon T \subset [0, \infty)$ is introduced for first excursion failure study. Typical functionals for reliability analysis are the random variables $X_1 = \max X(t)$ or sometimes $X_2 = \min X(t)$. Let be $S = [s_A, s_E] \subset T$, c the critical level and $X(s_A) < c$ with probability one. In Lange [9] is discussed the approximation of probabilities

$$P(\max_{t\epsilon S} X(t) < c) = P(X(t) < c \ \forall t\epsilon S) \tag{2.1}$$

by the discrete random variable N of upcrossings of the level c by $X(t), t\epsilon T$:

$$P(N = 0) = p_0, \tag{2.2}$$

especially with moments

$$m_i = EN^i = \sum_{k=0}^{\infty} k^i p_k,\ m_0 = 1,\ i = 1, 2, ...;\ P(N = k) = p_k$$

and factorial moments

$$m_{(i)} = EN(N-1)...(N-i+1) = \sum_{k=0}^{\infty} k(k-1)...(k-i+1)p_k,\ i = 1, 2, ...$$

follow estimations like for example

$$p_0 \geq 1 - m_i \tag{2.3}$$

and series developments like

$$p_0 = 1 + \sum_{i=1}^{\infty} \frac{(-1)^i}{i!} m_{(i)} \tag{2.4}$$

as well as Gram-Charlier-Series developments like

$$p_0 = e^{-m_1} \sum_{i=0}^{\infty} q_i (-1)^i \tag{2.5}$$

with

$$q_0 = 1,\ q_1 = 0,\ q_i = \sum_{j=0}^{i-1} (-1)^j \frac{m_1^j}{j!(i-j)!} (m_{(i-j)} - m_1^{i-j}),\ i = 2, 3, ...$$

The formulas (see Lange [9]) are valid for arbitrary differentiable stochastic processes and arbitrary low and high constant levels, level functions and random levels , can be used for stochastic vector processes (see Engelund, Rackwitz, Lange [4]) and can be generalized to random fields.

Here is the approximation of moments and factorial moments of the discrete random number N very important (see for example Cramer, Leadbetter [3], Friedrich, Lange [5] , Lange [9]), especially basing on random counting measures.

3. Fatigue failure models

Now the failure type is fatigue failure. A random measure $D(t)$ as a functional of the given arbitrary differentiable stochastic process $X(t), t\epsilon T \subset [0,\infty)$ is introduced for modelling the fatigue in the following way:

$$D(t) = \int_0^t G(D(\tau), S(\tau))d\tau. \tag{3.1}$$

$S(\tau)$ is the stress amplitude. $D(t)$ is normalized, that means at time $t = 0$ is $D(t) = 0$ with probability one and at time $t = L$ (L is the life time, a continuous random variable) is $D(L) = 1$ with probability one. It is not necessary that $D(t)$ is a monotone increasing function.

Here is $G(\cdot,\cdot)$ a special integrable function depending on the practical problem (for the deterministic case see Hennig, Friedrich, Heinrich, Mauersberger [8]) with $G(D(\tau); S(\tau)) > 0$, if $S(\tau) > S_D$ (S_D is the endurance limit) and $G(D(\tau), S(\tau)) = 0$, if $S(\tau) \leq S_D$. Formula (3.1) is a generalization for the deterministic Palmgren-Miner-Theory (see Bolotin [2]; Heinrich, Hennig [7]). For the stochastic process $X(t), t\epsilon T$ (which is now assumed to be a centralized narrow-band process) the random measure $D(t)$ is given (in generalization from Heinrich, Hennig [7]) for example by

$$D(t) = -\int_0^t \left[\ddot{X}(s)\delta(\dot{X}(s))\chi(X(s) - S_D)\right] G(D(s), X(s))ds. \tag{3.2}$$

In (3.2) are $\delta(.)$ the Dirac delta distribution and

$$\chi(X(s) - S_D) = \begin{cases} 1 \text{ if } X(t) > S_D \\ 0 \text{ if } X(t) \leq S_D \end{cases}.$$

The calculation of the probability density function of the random variable L from (3.2) and $D(L) = 1$ is in general not possible. In the following the special case

$$G(D(s), X(s)) = \frac{1}{b_2}(X(s))^{b_1} = g(X(s)) \tag{3.3}$$

is considered, where b_1, b_2 are material constants. Let $ED(t)$ be the mean function of $D(t), t\epsilon [0, L]$ and $\sigma_D(t)$ the corresponding standard deviation function. The mean life time EL can be approximated, if

$$\lim_{t\to\infty} \frac{\sigma_D(t)}{ED(t)} = 0 \tag{3.4}$$

can be shown, because in this case the probability that an arbitrary realization of $D(t)$ at time $t = EL = \tilde{L}$ is equal one, if $ED(\tilde{L}) = 1$, tends to one with increasing

t. So the mean life time $\tilde{L}$ can be calculated from $ED(\tilde{L}) = 1$. Formulas (3.2) and (3.3) give now

$$ED(\tilde{L}) = \int_0^{\tilde{L}} (\int_{S_D}^{\infty} (\int_{-\infty}^{0} \ddot{x}\, g(x) f(x, 0, \ddot{x}, t) d\, \ddot{x}) dx) dt = 1, \tag{3.5}$$

where f is the probability density function of the stochastic process $X(t), t\epsilon T$, its first and its second order derivatives. For stationary narrow band processes Renger, Mohr [12] proved the validity of (3.4). So further follows

$$\tilde{L} \int_{S_D}^{\infty} \int_{-\infty}^{0} \ddot{x}\, g(x) f(x, 0, \ddot{x}) d\, \ddot{x}\, dx = \tilde{L}\, ED = 1 \tag{3.6}$$

For a stationary narrow band Gaussian process $X(t), t\epsilon T$ follows from (3.6)

$$ED = \frac{1}{b_2} \frac{1}{2\pi} \frac{\sigma_{\dot{X}}}{\sigma_X} (\sqrt{2}\sigma_X)^{b_1} \Gamma(\frac{b_1 + 2}{2}; \frac{1}{2}(\frac{S_D}{\sigma_X})^2), \tag{3.7}$$

with the gamma function

$$\Gamma(x, y) = \int_y^{\infty} s^{x-1} \exp\{-s\}\, ds.$$

A special case of (3.6) is the Miles formula (see Miles [11], Heinrich, Hennig [7]), which can be obtained with $S_D = 0$ and $\Gamma(x, 0) = \Gamma(x)$:

$$ED = \frac{1}{b_2} \frac{1}{2\pi} \frac{\sigma_{\dot{X}}}{\sigma_X} (\sqrt{2}\sigma_X)^{b_1} \Gamma(\frac{b_1 + 2}{2}). \tag{3.8}$$

4. First excursion or fatigue failure

In the following the theory is used to calculate the failure probability in the case where both types of failure for a structural system, that means fatigue failure as well as first excursion failure, are thinkable to occur. The probabilities of no first excursion failure

$$P(X(t) < c\ \forall t\epsilon S) \tag{4.1}$$

and of no fatigue failure

$$P(D(t) < 1\ \forall t\epsilon S) \tag{4.2}$$

were discussed in ch.2 and ch.3 separately. If both types of failure are thinkable to occur, it has to be considered the probability

$$P(\{D(t) < 1\} \cap \{X(t) < c\}\ \forall t\epsilon S). \tag{4.3}$$

Functional $D(t), t\epsilon T$ and stochastic process $X(t), t\epsilon T$ are dependent. So for the calculation of (4.3) has to be used the conditional probability:

$$P(\{X(t) < c\}\ \forall t\epsilon S) \cdot P(\{D(t) < 1\} \mid \{X(t) < c\}\ \forall t\epsilon S). \tag{4.4}$$

The first factor in (4.4) is equal $P(N=0)$ corresponding ch.2. A new construction for the functional $D(t), t \epsilon S$ in generalization of Miles [11] and Heinrich, Hennig [7] is now derived to make possible the calculation of the second factor in (4.4):

Let $N_{c_i}(t_1, t_2)$ be the number of upcrossings of the level c_i by $X(t), t \epsilon [t_1, t_2) \subset T$ and $D_{c_i}(t_1, t_2)$ the growth of damage in an interval $[c_i, c_{i+1})$ for $t \epsilon [t_1, t_2)$. It is clearly, that the growth of damage depends on the difference of the numbers of local maxima and local minima, the material function g (see (3.3)) and the corresponding interval . The random variable "number of local maxima minus number of local minima in an interval $[c_i, c_{i+1})$" is equal $N_{c_i}(t_1, t_2) - N_{c_{i+1}}(t_1, t_2)$. Then is defined

$$D_{c_i}(t_1, t_2) = \left[N_{c_i}(t_1, t_2) - N_{c_{i+1}}(t_1, t_2)\right] \cdot g(c_i) \tag{4.5}$$

With $c_1 = 0$ and $c_{i+1} = c_i + \Delta$, $\Delta > 0$ the damage $D(t_1, t_2)$ is now

$$D(t_1, t_2) = \sum_{i=1}^{\infty} D_{c_i}(t_1, t_2) = \sum_{i=1}^{\infty} g(c_i) \left[N_{c_i}(t_1, t_2) - N_{c_{i+1}}(t_1, t_2)\right]$$

$$= \sum_{i=1}^{\infty} N_{c_i}(t_1, t_2) \left[g(c_{i+1}) - g(c_i)\right] = \sum_{i=1}^{\infty} N_{c_i}(t_1, t_2) \left[\frac{g(c_i + \Delta) - g(c_i)}{\Delta}\right] \Delta.$$

For $\Delta \to 0$ follows

$$D(t_1, t_2) = \int_0^{\infty} N_c(t_1, t_2) \frac{\partial g(c)}{\partial c} dc. \tag{4.6}$$

The formula for the mean value $ED(t_1, t_2)$ of the damage

$$ED(t_1, t_2) = \int_0^{\infty} EN_c(t_1, t_2) \frac{\partial g(c)}{\partial c} dc \tag{4.7}$$

is valid for an arbitrary differentiable stochastic process and by this formula can be approximated under certain simple assumptions the mean life time. Formula (4.7) is a generalization of the Miles formula.

Let $X(t), t \in T$ be a stationary Gaussian process with the probability density function f of the stochastic process and its first order derivative:

$$f(x, \dot{x}; t) = \frac{1}{2\pi \sigma_X \sigma_{\dot{X}}} \exp\left\{-\frac{x^2}{2\sigma_X^2}\right\} \exp\left\{-\frac{\dot{x}^2}{2\sigma_{\dot{X}}^2}\right\}. \tag{4.8}$$

After the calculation of

$$EN_c(t_1, t_2) = \frac{(t_2 - t_1)}{2\pi} \frac{\sigma_{\dot{X}}}{\sigma_X} \exp\left\{-\frac{1}{2}\left(\frac{c}{\sigma_X}\right)^2\right\}$$

and with

$$g(c) = \frac{1}{b_2} c^{b_1} \; ; \; \frac{\partial g(c)}{\partial c} = \frac{b_1}{b_2} c^{b_1 - 1}$$

and (4.7) follows then

$$ED(t_1,t_2) = \frac{(t_2-t_1)}{2\pi}\frac{\sigma_{\dot{X}}}{\sigma_X}\frac{b_1}{b_2}\int_0^\infty \exp\left\{-\frac{1}{2}(\frac{c}{\sigma_X})^2\right\} c^{b_1-1}dc$$

$$= \dots = \frac{(t_2-t_1)}{2\pi b_2}\frac{\sigma_{\dot{X}}}{\sigma_X}(\sqrt{2}\sigma_X)^{b_1}\Gamma(\frac{b_1+2}{2}).$$

For nonstationary Gaussian processes can be obtained from (4.7) the mean

$$ED(t_1,t_2) = \frac{1}{2\pi b_2}\int_{t_1}^{t_2}\sqrt{1-\alpha^2(t)}\frac{\sigma_{\dot{X}}(t)}{\sigma_X^2(t)}I(\Theta,b_1,t)dt$$

with

$$I(\Theta,b_1,t) = \int_0^\infty c^{b_1}\exp\left\{-\frac{1}{2}(\frac{c}{\sigma_X(t)})^2\right\}\cdot$$

$$\cdot\left[\frac{c}{\sigma_X(t)}\exp\left\{-\frac{c^2\Theta^2}{2\sigma_X^2(t)}\right\} - \sqrt{\frac{\pi}{2}}\Theta(1-\frac{c^2}{\sigma_X^2(t)})(1+\Phi(\frac{c\Theta}{\sqrt{2}\sigma_X(t)}))\right] dc,$$

$$\alpha(t) = \frac{\rho_{X\dot{X}}(t)}{\sigma_X(t)\sigma_{\dot{X}}(t)};\ \Theta = \Theta(t) = \sqrt{\frac{\alpha^2(t)}{1-\alpha^2(t)}}$$

and $\Phi(\cdot)$ as the (Gaussian) $N(0,1)$-distribution function.

Now the calculation of probability (4.3) will be considered. To simplify the notation is used without any restriction: $S = [0,t]$; $t_1 = 0$; $t_2 = t$ and $N_c(0,t) = N_c(t)$; $D(0,t) = D(t)$:

$$p_0 = p_0(t) = P(\{D(t) < 1\} \cap \{N_c(t) = 0\})$$

$$= P(\{N_c(t) = 0\}) \cdot P(\{D(t) < 1\} \mid \{N_c(t) = 0\}).$$

With (4.6) is then

$$p_0(t) = P(\{N_c(t) = 0\}) \cdot P(\left\{\int_0^\infty N_c(t)\frac{\partial g(c)}{\partial c}dc < 1\right\} \mid \{N_c(t) = 0\}) \tag{4.9}$$

Formula (4.6) makes it possible to calculate the conditional probability in (4.9):

$$p_0(t) = P(\{N_c(t) = 0\}) \cdot P(\left\{\int_0^c N_c(t)\frac{\partial g(c)}{\partial c}dc < 1\right\}). \tag{4.10}$$

The failure probability in the case where both types of failure for a structural system are thinkable to occur can be calculated on the basis of formula (4.10) for arbitrary differentiable stochastic processes. A lower bound for $p_0(t)$ is given by application of Markov inequality:

$$p_0(t) \geq (1 - EN_c(t))(1 - \int_0^c EN_c(t)\frac{\partial g(c)}{\partial c}dc) \tag{4.11}$$

The distribution function for the discrete random variable $N = N_c(t)$ is approximately a Poisson distribution, if the critical level c is "high enough" (see for example Cramer, Leadbetter [3] for Gaussian processes). So follows the lower bound

$$p_0(t) \geq \exp\{-EN_c(t)\}\,(1 - \int_0^c EN_c(t)\frac{\partial g(c)}{\partial c}dc) \tag{4.12}$$

At last an example is considered. Let $X(t), t\epsilon T$ be a stationary Gaussian process with mean function $EX(t) \equiv 0$, standard deviation function $\sigma_X(t) \equiv \sigma_X = 1$ and standard deviation function of $\dot{X}\ (t), t\epsilon T$ (the derivative with probability one) $\sigma_{\dot{X}}(t) \equiv \sigma_{\dot{X}}$. Then it is easy to calculate

$$p_0(t) \geq \exp\left\{-t\frac{\sigma_{\dot{X}}}{2\pi}\cdot\exp\left[-\frac{c^2}{2}\right]\right\}(1 - \frac{t\sigma_{\dot{X}}}{2\pi}\frac{b_1}{b_2}\cdot 2^{\frac{b_1-2}{2}}\Gamma_c(\frac{b_1}{2})) \tag{4.13}$$

with the incomplete gamma function

$$\Gamma_c(x) = \int_0^c y^{x-1}\exp\{-y\}\,dy.$$

A generalization of the results for nonconstant critical levels is possible.

The following figure shows lower bounds for $p_0(t)$ corresponding formula (4.13) with $\sigma_X(t) = \sigma_{\dot{X}}(t) \equiv 1$; $b_1 = 2$; $b_2 = 40$; $c = 3$ and $c = 4$. It is easy to see, that the influence of a possible first excursion failure is small for high critical levels.

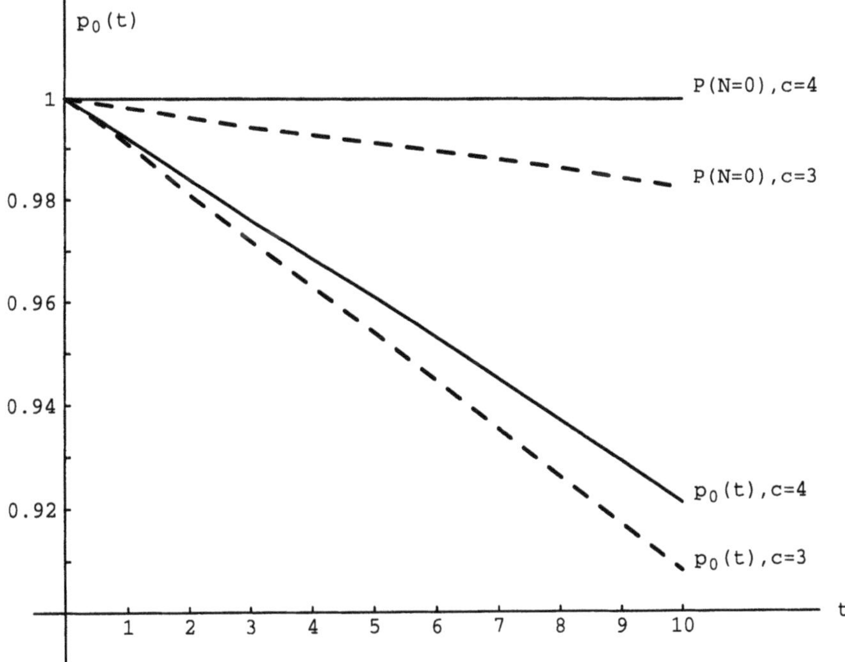

5. References

[1] **Beljajev, Y.K.**
Distribution of the maximum of a random field and its application to problems of reliability theory (in Russ.), Isv. Akad. Nauk SSSR, Techn. Kib. 2 (1970) 44-56
[2] **Bolotin, V.V.**
Wahrscheinlichkeitsmethoden zur Berechnung von Konstruktionen, Verlag für Bauwesen, Berlin 1981
[3] **Cramer, H., Leadbetter, M.R.**
Stationary and related processes, Wiley, New York 1969
[4] **Engelund, S., Rackwitz, R., Lange, C.**
Approximations of first-passage times for differentiable processes, publication in Prob. Eng. Mech. 1994
[5] **Friedrich, H., Lange, C.**
Niveauüberschreitungswahrscheinlichkeiten bei stochastisch belasteten mechanischen Systemen, Report R-Mech-04/81, Berlin 1981, 230 pp.
[6] **Hasofer, A.M.**
The mean number of maxima above high levels in Gaussian random fields, Journ. of Appl. Prob. 13 (1976) 377-379
[7] **Heinrich, W., Hennig, K.**
Zufallsschwingungen mechanischer Systeme, Akademieverlag, Berlin 1977
[8] **Hennig, K., Friedrich, H., Heinrich, W., Mauersberger, G.**
Studie zur Arbeit auf ausgewählten Gebieten der Sicherheit und Zuverlässigkeit im Maschinenbau und Bauwesen, AdW, ZIMM, Berlin 1979
[9] **Lange, C.**
First excursion probabilities for low threshold levels by differentiable processes, Proc. of the 4th WG 7.5 IFIP conference on reliability and optimization of structural systems 1991, Munich, Springer 1992, pp 261-275
[10] **Lange, C., Friedrich, H.**
Zuverlässigkeitsuntersuchungen für mechanische Systeme auf der Grundlage zufälliger Felder, FMC-Series No.43, Chemnitz 1989, 90 pp
[11] **Miles, J.W.**
On structural fatigue under random loading, Journal Aero. Sci. 21 (1954) 753-762
[12] **Renger, A., Mohr, H.**
Eine Studie über experimentielle Forschungen auf dem Gebiet der Zufallsschwingungen mechanischer Systeme, Forschungsbericht der HfR Festkörpermechanik, Berlin 1979

18

Analysis of Ignorance Factors in Design Criteria Subject to Model Uncertainty

Marc A. Maes
Civil Engineering Department, The University of Calgary
Calgary, Alberta, Canada T2N 1N4

1. INTRODUCTION

The effect of model uncertainty on both structural reliability analysis and reliablity-based design has been discussed in a number of articles. Der Kiureghian (1989) and Der Kiureghian and Liu (1986) formulate the basic framework for analysis; they develop several measures of structural safety given imperfect models, and focus on a complete description of the different types and sources of model uncertainty, together with appropriate estimation and analysis methods. Also, several measures of imprecision are proposed.

In a design context, rules and specifications should encourage the gathering of information and the use of more refined models to reduce model uncertainty : this has been investigated by Der Kiureghian (1989) and by Maes (1991), through the use of optimal reliability metrics on the one hand, and through the use of ignorance factors on the other.

In obtaining ignorance factors, an essential objective is to obtain as much information about the behaviour of ignorance factors with varying degrees of uncertainty, on the basis of the smallest possible number of calibration steps. In Maes (1991) a two-step calibration scheme is discussed. The objective of Step 1 is to fine-tune all the partial factors (applicable to the basic random variables) using a reliability-based optimization scheme. This analysis is performed assuming perfect models (zero model uncertainty). Step 2 focuses on determining numerical values of the ignorance factors for varying degrees of model quality, *given* that the previously derived partial factors remain constant. Ideally, just one analysis would be required to yield optimal design check equations: this is the subject of the present paper.

Winterstein et al. (1994) develop an interesting approach to deriving reliability-based design criteria for uncertain models. The idea is to "correct" results based on a median response (which requires an analysis with fixed model uncertainty parameters), based on FORM omission factors. Omission sensitivity factors (Madsen, 1988) give the relative error in the reliability index when a basic random variable is replaced by a deterministic number. In this paper, the opposite problem is examined : what is the effect of expanding the analysis by replacing a fixed parameter by a random variable ? Inverse FORM is used by Winterstein et al. (1994) to define contours corresponding to a specified level of reliability. This type of analysis is discussed in more detail in Der Kiureghian et al. (1994) and some aspects of it are retained and/or generalized in the present analysis.

2. MODEL EXPANSION FACTORS

In this section, we consider a limit state model $g(\mathbf{x}, \overline{\boldsymbol{\theta}})$ for the basic random variables $\mathbf{X}$, formulated using a set of *constant* model parameters $\overline{\boldsymbol{\theta}}$. These parameters typically originate from a statistical analysis performed in the process of building or fitting the model; or, they could simply represent an empirical estimate or an expert's best opinion. For simplicity, we will assume that $\overline{\boldsymbol{\theta}}$ are the mean values of the random variables $\boldsymbol{\Theta}$ to be considered next. This assumption is not restrictive : a simple scaling of the subsequent results needs to be performed if they are different from the means of $\boldsymbol{\Theta}$.

Let P_0 be the failure probability associated with the failure region $\{g < 0\}$ where we keep in mind that g is a deterministic model with constant parameters $\overline{\boldsymbol{\theta}}$:

$$P_0 = \Pr(g(\mathbf{X}, \overline{\boldsymbol{\theta}}) < 0) \tag{1}$$

This result is now contrasted with an "expanded" structural reliability analysis which includes all model errors and parameters in the set of random variables $\boldsymbol{\Theta}$:

$$P_m = \Pr(g(\mathbf{X}, \boldsymbol{\Theta}) < 0) \tag{2}$$

The basic objective of the "expansion" problem considered here is to estimate P_m using information from the P_0 analysis *only*, i.e. the analysis *without* model uncertainty. This problem may be contrasted with the "omission" problem (Madsen, 1988), where the relative error on the reliability index is estimated when one or more variables are replaced by deterministic number(s) : this would correspond with the converse problem of finding P_0 based upon a full P_m-analysis.

It is clear that $P_m \to P_0$ as $\boldsymbol{\Theta}$ converges in distribution to the fixed set $\overline{\boldsymbol{\theta}}$. Furthermore, we have :

$$P_m(\mathbf{X}, \boldsymbol{\Theta}) = \int_\theta P(\mathbf{X} \mid \boldsymbol{\theta}) f_\theta(\boldsymbol{\theta})\, d\boldsymbol{\theta} \tag{3}$$

where $P(\mathbf{X} \mid \boldsymbol{\theta})$stands for $\Pr(g(\mathbf{X}, \boldsymbol{\theta}) < 0)$ conditional upon $\boldsymbol{\Theta} = \boldsymbol{\theta}$, and $f_\theta(\boldsymbol{\theta})$ is the joint density of $\boldsymbol{\theta}$. A Taylor expansion of the integrand about the mean vector $\overline{\boldsymbol{\theta}}$, followed by integration, yields a "model expansion factor" ξ_θ, approximately equal to:

$$\xi_\theta = \frac{P_m}{P_0} \cong \left[1 + \tfrac{1}{2}\sum_i\sum_j \frac{1}{P_0}\left(\frac{\partial^2 P}{\partial\theta_i\partial\theta_j}\right)_{\theta=\bar{\theta}} \sigma_{\theta_i\theta_j}\right] \tag{4}$$

where $\sigma_{\theta_i\theta_j}$ are the elements of the covariance matrix $\Sigma_{\theta\theta}$ of the model uncertainties Θ. Exact values of the second order sensitivities may be hard to obtain under general conditions. However, as shown in the next section, excellent asymptotic estimates can readily be obtained as a by-product of the basic P_0-analysis.

3. ASYMPTOTIC EXPRESSIONS FOR THE EXPANSION FACTOR

Asymptotic expressions for first-order parameter sensitivities of $P(\mathbf{X}, \boldsymbol{\theta})$can be found in Breitung (1994). Breitung's analysis covers both distributional parameters as well as model parameters. Here, only the latter are required, namely :

$$\frac{\partial P(\mathbf{X}, \boldsymbol{\theta})}{\partial\theta_i} \sim -P_0\left[\frac{\partial g}{\partial\theta_i}\frac{|\nabla l|}{|\nabla g|}\right]_{\theta=\bar{\theta}, \mathbf{x}=\mathbf{x}^*} \tag{5}$$

where the gradient ∇ is taken with respect to $\mathbf{x}$, and where $\mathbf{x}^*$ represents the coordinates (in the original variable space) of the point of maximum likelihood (PML), as described in Breitung (1994) and Maes et al.(1993); the PML $\mathbf{x}^*$ can readily be obtained as the solution of the basic optimization problem needed to solve the structural reliability analysis problem for the P_0-case, i.e. to maximize $l(\mathbf{x})$ subject to $g(\mathbf{x} \mid \bar{\boldsymbol{\theta}}) = 0$, where $l(\mathbf{x}) = \ln f_{\mathbf{X}}(\mathbf{x})$ is the loglikelihood function of the basic random variables. Breitung's (1994) sensitivity factor analysis is based on a generalization of Leibnitz' rule for the derivative of parameter-dependent integrals, and on asymptotic expansions for multivariate Laplace type integrals. This analysis can easily be extended (Breitung, 1994b) to second-order sensitivities, which yields the asymptotic approximation of the "expansion" ratio (4):

$$\frac{P_m}{P_0} \sim 1 + \tfrac{1}{2}\left[\left(\frac{\nabla l}{\nabla g}\right)^2 (\nabla_\theta g)^T \Sigma_{\theta\theta} \nabla_\theta g\right]_{\bar{\theta}, \mathbf{x}^*} \tag{6}$$

It should be stressed that the ratio $|\nabla l| / |\nabla g|$ is equal to the Lagrange multiplier associated with the above optimization problem.

4. IGNORANCE FACTORS

4.1. Comparing Basic Design Checks With and Without Model Uncertainty

First, consider the perfect model without model uncertainty. Assume that the mathematical expression used for the deterministic design check (e.g. in LRFD format) is the same as that of the limit state model (They do not strictly need to be the same for the subsequent analysis to apply, but the notation simplifies considerably).

Therefore, in order to determine the (minimum) required resistance, denoted here by some capacity-related design parameter r_0, the following DCE needs to be solved :

$$g(\mathbf{x}^*, \overline{\boldsymbol{\theta}}, r_0) = 0 \tag{7}$$

where $\mathbf{x}^*$ are the (input) design values of the basic variables X. This DCE entirely defines the corresponding limit state model $g(\mathbf{X}, \overline{\boldsymbol{\theta}}, r_0)$. For this model, of course, we wish to achieve a desired reliability level, i.e.

$$P(\mathbf{X}, \overline{\boldsymbol{\theta}}, r_0) = P_T \tag{8}$$

where the same abreviation is used for P as in (3), and where P_T denotes the target failure probability. If the model without model uncertainty is perfectly calibrated, $\mathbf{x}^*$ may, without loss of generality, be considered to correspond to the PML on the surface (7).

The second step is now to include model uncertainty in the reliability analysis. The model g is mathematically the same. A larger resistance r_m will now be required to meet the same target reliability level :

$$P(\mathbf{X}, \boldsymbol{\Theta}, r_m) = P_T \tag{9}$$

At the design level, however, it makes sense to keep things simple; the approach is :

- to keep the *same* design values of the basic variables; the load and resistance factors, the specified probability levels, etc..., used in (7) remain unchanged.
- to use ignorance factors $\boldsymbol{\theta}^*$ to encapsulate the effect of model uncertainty (Maes, 1991).

This results in the following DCE :

$$g(\mathbf{x}^*, \boldsymbol{\theta}^*, r_m) = 0 \tag{10}$$

If (10) is linearized with respect to $\boldsymbol{\theta}$ and r in the neighbourhood of $\overline{\boldsymbol{\theta}}$ and r_0, and after inserting (7), it follows that :

$$(r_m - r_0) \cong \left[-\left(\frac{\partial g}{\partial r}\right)^{-1} (\boldsymbol{\nabla}_{\theta} g)^T (\boldsymbol{\theta}^* - \overline{\boldsymbol{\theta}}) \right]_{\mathbf{x}^*, \overline{\theta}, r_0} \tag{11}$$

The essential aspect of the design rule (10) is that $\mathbf{x}^*$ is unchanged from (7). Consequently, the original DCE (7) can be used to achieve (9), provided ignorance factors are used in (7) rather than mean values. The following section shows that approximate $\boldsymbol{\theta}^*$ can be determined solely on the basis of a P_o-analysis.

4.2. Inverse Reliability

Given that a properly calibrated model (8) is available using the fixed parameters $\overline{\boldsymbol{\theta}}$, the next step is now to ensure that, by adjusting the ignorance factors in (10), the inclusion of model uncertainty also results in a model having the desired level of reliability (9). A Taylor expansion of P_m yields :

$$P(\mathbf{X}, \boldsymbol{\Theta}, r_m) \sim P(\mathbf{X}, \boldsymbol{\Theta}, r_0) + (r_m - r_0)\frac{\partial P(\mathbf{X}, \boldsymbol{\Theta}, r_0)}{\partial r} \tag{12}$$

The "expansion" result (4) may now be used to link the P_m and P_0 analyses at $r = r_0$; when higher order derivatives are neglected, and when $(r_m - r_0)$ is replaced by (11), together with the condition

$$P(\mathbf{X}, \Theta, r_m) = P(\mathbf{X}, \overline{\boldsymbol{\theta}}, r_0) = P_T \tag{13}$$

then the asymptotic version of (12) can be derived based on (6) and (5):

$$(\boldsymbol{\nabla}_{\theta} g)^T(\boldsymbol{\theta}^* - \overline{\boldsymbol{\theta}}) \sim -\tfrac{1}{2}\frac{|\nabla l|}{|\nabla g|}(\boldsymbol{\nabla}_{\theta} g)^T \Sigma_{\theta\theta} \boldsymbol{\nabla}_{\theta} g \tag{14}$$

and, for the special case of just one ignorance factor:

$$\theta^* \sim \ \overline{\theta} - \tfrac{1}{2}\frac{|\nabla l|}{|\nabla g|}\frac{\partial g}{\partial \theta}\sigma_\theta^2 \tag{15}$$

with all of the derivatives evaluated at the PML of the P_0-problem. In the single parameter case, the magnitude of $(\theta^* - \overline{\theta})$ is thus seen to be proportional to the variance of the model uncertainty (see, for example, Maes, 1991).

The previous approach can be extended to multiplicative uncertainties. This leads to a design format which is quite pervasive in all areas of civil engineering. The asymptotic expression for an ignorance factor ψ^* associated with a (single) multiplicative model uncertainty Ψ, for which :

$$\Psi > 0 \qquad \text{and} \qquad \boldsymbol{E}\ (\Psi) = 1\ , \tag{16}$$

can most conveniently be derived from the previous results using a logarithmic transformation $\varphi = \ln \psi$, together with an adjustment for $\mathbf{E}\ (\ln \psi)$ in the above equations :

$$\psi^* = 1 - \tfrac{1}{2} v_\psi^2 \left(\frac{|\nabla l|}{|\nabla g|}\frac{\partial g}{\partial \psi} + 1 \right)_{\psi=1, \mathbf{x}=\mathbf{x}^*} + o(v_\psi^2) \tag{17}$$

where v_ψ is the coefficient of variation (COV) of the model uncertainty parameter Ψ; the PML $\mathbf{x}^*$ is obtained for the model $g(\mathbf{x} \mid \psi = 1)$ with mean model uncertainty 1.

5. EXAMPLE APPLICATION

Collapse of downhole oil and gas casing and tubing structures occurs when a pipe is accidentally or intentionally evacuated of internal fluids. As a result, the thick-walled tubular is exposed to the full external pressure induced by the formation pore pressure. The sensitivity of the collapse failure mode to imperfections, especially when the onset of plasticity precedes instability, makes it difficult to "predict" collapse loads. In a major development of reliability-based criteria for casing and tubing pipes (Gulati et al., 1994), the ultimate capacity of moderately thick and thick tubes loaded by external pressure is, therefore, calibrated based on the results of well executed experiments. Several data sets are available and they show different degrees of uncertainty depending on manufacturer, type of use, grade, age, geographical and geological context, etc...

Regression allows the COV of the multiplicative model uncertainty Ψ to be determined on the basis of a comparison of each series of test results with the analytical expression developed by Timoshenko and Gere(1961); the objective is then to determine ignorance factors for this model in order to allow for easy consideration of any degree of model quality and variability. The idea is thus to compensate for increasing model error by means of "reducing" the nominal collapse capacity using appropriate ignorance factors $\psi^* \leq 1$. The limit state model contains 5 basic random variables and one model uncertainty variable Ψ :

$$g(\sigma_y, E, t, \xi, \Delta p, \Psi) = \Psi \cdot p_c(\sigma_y, E, t, \xi) - \Delta p \tag{18}$$

where Δp is the net internal-external pressure difference at a point along the casing string, and p_c is the Timoshenko collapse capacity :

$$p_c(\sigma_y, E, t, \xi) = \frac{1}{2}\left[p_Y + p_e(1 + \frac{3\xi D}{t}) - \sqrt{\left(p_Y + p_e(1 + \frac{3\xi D}{t})\right)^2 - 4p_Y p_e}\right] \tag{19}$$

where D is the outer diameter, and t is the wall thickness. The ovality ξ is defined as $\xi = \frac{2(D_{\max} - D_{\min})}{D_{\max} + D_{\min}}$. The elastic buckling pressure is $p_e = \frac{2E}{1-\nu^2}\left(\frac{t}{D}\right)^3$ and the yield pressure is $p_Y = 2\sigma_y \frac{t}{D}$, where σ_y is the yield stress, E is Young's modulus, and ν is Poisson's ratio.

The effect of Ψ depends critically on the variability of the "loading term" Δp in (18). Several load domains were considered in the study (Gulati et al., 1994) as part of the overall zonation scheme of the design set. As an example, we only consider salt loading (direct contact with a flowing salt formation), which imposes the most severe collapse load on casing. Also, the present analysis is restricted to just one nominal D/t ratio ($D/t = 12$), one steel grade (L80) (In reality, a wide range of conditions were assumed).

It suffices then to calibrate the nominal values in (18) in such a way that $\Pr(g \leq 0) = P_T = 10^{-3.5}$. This analysis is performed with Ψ set to one, that is, with the probability distributions of the 5 basic variables only. The PML and the Lagrange multiplier $\lambda = |\nabla l| / |\nabla g|$ are obtained automatically, and together with the fact that $\frac{\partial g}{\partial \psi} = p_c$ at the PML, equation (17) can be used directly to determine the required ignorance factor ψ^* as a function of the Timoshenko model error COV, v_ψ.

The resulting ignorance factor is shown in Figure 1. It can be seen that a model uncertainty COV of 10% requires an ignorance factor ψ^* of about 0.87. This indicates that, for the design conditions considered in this example, a 87% reduction of the collapse capacity p_c would be needed to achieve a design product with the same reliability level as that corresponding with the use of a perfect model without experimental error.

For comparison, Figure 1 also shows "exact" values of ψ^*. These were obtained by including a lognormal random variable Ψ with a COV varying in the range 0 to 0.15, and solving each resulting 6-variable inverse reliability problem to the target value $P_T = 10^{-3.5}$; following (18) the "exact" ignorance factor associated with a model error COV equal to v is then given by the ratio $p_c(v_\psi = 0)/p_c(v_\psi = v)$. It can be seen that the values provided by the expansion approach are indeed $o(v_\psi^2)$ and that, in this case, they slightly overestimate the required collapse resistance reduction, which makes for a somewhat conservative design rule.

6. CONCLUSIONS

Model expansion factors are useful in assessing the effect of model uncertainties on a reliability analysis performed on the basis of an imperfect model. They address the question of how the failure probability P_m varies with respect to the base case of a perfect model (P_0), if model uncertainty variables are included in the analysis. It is shown that the ratio P_m/P_0 can be approximated using information from a P_0-analysis only.

Ignorance factors intervene in the inverse problem of reliability-based design. For this second type of problem, the critical question is how the design check equations need to be modified in order to "compensate" for the effect of increasing model uncertainty.

In both cases, the factors can be determined on the basis of an analysis which ignores model uncertainty (i.e. constant mean values of Θ and Ψ). In fact, only the Lagrange multiplier $|\nabla l| / |\nabla g|$ of the constrained loglikelihood maximization problem needs to be determined together with the gradient of g with respect to $\boldsymbol{\theta}$ a the resulting maximum point. The approximations result in accurate failure probability ratio estimates, and in reliable error-inclusive design rules.

REFERENCES :

Breitung, K (1991). "Probability approximations by loglikelihood maximization", *J. Engrg. Mech.*, ASCE, 117(3), 457-477.

Breitung, K (1994). *Asymptotic approximations for probability integrals.* Mathematics Series, Springer Lecture Notes in Mathematics, No. 1592.

Breitung, K (1994b). "Parameter dependent integrals : some mathematical tools", *6th IFIP Working Conference on Reliability and Optimization of Structural Systems*, Assisi (Perugia), Italy.

Der Kiureghian A. and Liu, P.-L. (1986). "Structural reliability under incomplete probability information", *J. Engrg. Mech.*, ASCE, 112(1), 85-104.

Der Kiureghian A. (1989). "Measures of structural safety under imperfect states of knowledge", *J. Struct. Engrg.*, ASCE, 115(5), 1119-1140.

Der Kiureghian A., Zhang, Y., and Li, C.-C. (1994). "Inverse reliability problem", *J. Engrg. Mech.*, ASCE, 120(5), 1154-1159.

Gulati, K.C., McKenna, D.L., Maes, M.A., Brand, P.R., Johnson, R.C., Lewis, D.B., Riekels, L. and Maute, R.E. (1994). "Reliability-based design and application of drilling tubulars", *1994 Offshore Technology Conference*, Houston, No. 7557, 4, 423-430.

Madsen, H.O. (1988). "Omission sensitivity factors", *Structural Safety*, 5, 35-45.

Maes, M.A. (1991). "Codification of design load criteria subject to modeling uncertainty", *J. Struct. Engrg.*, ASCE, 117(10), 2988-3007.

Maes, M.A., Breitung, K. and Dupuis, D.J. (1993). "Asymptotic importance sampling", *Structural Safety*, 12, 167-186.

Timoshenko, S.P. and Gere, J.M. (1961). "*Theory of elastic stability.* McGraw Hill Brook Cy., New York.

Winterstein, S.R., Ude T.C., Cornell, C.A., Bjerager, P., and Haver, S. (1994). "Environmental parameters for extreme response : inverse form with omission factors", *Structural Safety & Reliability*, Ed. by Schüeller, G.I., Shinozuka, M. and Yao, J.I.P., Vol.I, 551-557.

Figure 1: Ignorance Factor For The Timoshenko Collapse Limit State

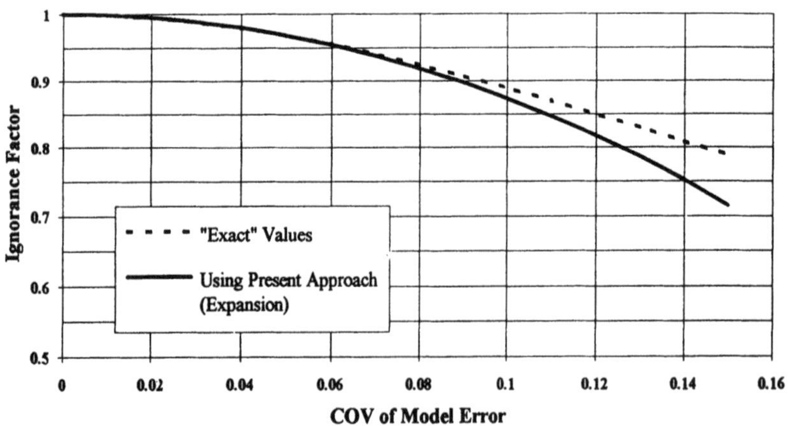

19

AN EFFICIENT *FORM*-BASED STRATEGY FOR THE RELIABILITY ANALYSIS OF MARINE STRUCTURES

L. Manuel[a] and C.A.Cornell[b]

[a]Jack R. Benjamin & Assoc., Inc., 444 Castro St., Suite 501, Mountain View, CA 94041

[b]Dept. of Civil Engineering, Stanford University, Stanford, CA 94305-4020.

1. ABSTRACT

In structural reliability analyses, one often finds that response computations needed to examine limit state(s) can be very expensive. Within a first- or second-order reliability method (FORM or SORM) framework which involves an iterative search for the design point, these limit state functions (and their gradients) need to be repeatedly evaluated. Clearly, it would be desirable to minimize the number of such evaluations while still retaining the same rate of convergence and the overall analysis framework.

Using a marine structure for illustration, we propose a FORM-based strategy that achieves this objective. It does so by making only approximate computations for gradients of the limit state function. These approximations are based on localized exact response estimates and we do not perform any additional response computations for them. The approximate gradients are computed by employing user-defined response transfer functions and response surfaces. The proposed strategy is shown to be efficient and accurate when compared to conventional FORM analyses.

2. INTRODUCTION

In most structural reliability applications, the pertinent random variables can be broadly divided into three classes: load, capacity, and response. Of these, response analyses required to establish the distribution of the random response are often the most expensive. For example, in the reliability of marine structures, load and capacity random variables may be thought to make up the long-term reliability problem. Dynamic response analyses conditional on load and capacity realizations form the short-term reliability problem and are the most expensive part of the overall reliability analysis, often requiring time-domain structural response simulations.

When FORM analyses are employed, inherent in the methodology is a need to iteratively evaluate the performance function and its gradients (with respect to all random variables) at each new point during the search for the design point. Because each such evaluation often involves expensive short-term response analyses, one would like to minimize the number of these evaluations but at the same time retain the general framework of the first-order method and not significantly degrade the efficiency of the search for the design point.

We propose a strategy which might be thought of as deviating only slightly from a conventional FORM approach: in each iteration, gradients of the limit state function are obtained by simpler approximate methods (not full-blown short-term response analyses). The precise nature of these approximations is intimately tied to aspects of the problem such as the importance of dynamics and the extent to which the response is non-Gaussian. We will show how fairly accurate gradients may be computed when gross (up to second order) statistics are captured by employing response transfer functions and higher order statistics of the response are captured by employing adaptive response surfaces.

The manner in which the marine structural reliability analysis problem is usually formulated, namely with a clear distinction between short-term (response) and long-term (load and capacity) random variables, makes this problem a good candidate for demonstration of our proposed strategy. The proposed strategy may, however, be generalized to include a broader class of problems than ones associated with marine structure reliability. Essentially, any reliability problem where limit state function evaluations can be expensive would benefit from a strategy such as the one proposed here where response transfer functions and response surfaces are employed and defined entirely within the FORM framework.

In our numerical studies, we will use a jack-up rig subjected to random hydrodynamic loading. We choose this structure because of its many interesting aspects. Jack-up rigs are dominated by non-Gaussian drag loads and are flexible, and thus dynamically sensitive especially when used in deep water. Several recent studies have addressed issues dealing with the behavior and reliability of jack-up rigs (see, for example, Karunakaran *et al.* [2], Kjeøy *et al.* [3], Løseth *et al.* [4], Manuel [5], McDonald and Bea [6]). We will study two different jack-ups and will use a long-term environmental description representative of the North Sea. We will define required response transfer functions and response surfaces (for our proposed strategy) in two different ways for each structure. We will compare our results with those obtained using the conventional FORM approach.

3. FORMULATION

3.1. Conventional Approach

The conventional first-order reliability method (FORM) algorithm, in its most common implementation, involves the following set of steps that need to be performed iteratively after defining a vector X made up of N random variables, and a limit state function, $g(X)$ (see, for example, Rackwitz and Fiessler [7])

Step 1 Guess x_0 (of size N) as an estimate of the (unknown) design point, x^*
Step 2 Evaluate $g(x_0)$
Step 3 Obtain N gradients of $g(x_0)$
Step 4 Improve estimate of x^* using $g(x_0)$ and gradients;
return to Step 1 with improved estimate of x^* if no convergence yet.

It is immediately clear from the above algorithm that in Steps 2 and 3 above, a total of $N+1$ evaluations of the limit state function are required within each FORM iteration.

3.2. Proposed Strategy and Modification to Conventional Approach

In order to describe our proposed strategy, it is most appropriate to compare the steps in the approach we will take with those involved in the conventional algorithm. The essential differences are a direct result of our objective which is to minimize the number of limit state function evaluations per iteration (which, we have seen above, is $N+1$ in the conventional approach). Very simply stated, we will achieve this by doing away with Step 3 in its conventional sense. We will thus require only *one* conventional limit state function evaluation (for Step 2); however, we will evaluate the N required gradients (for Step 3) only in an approximate manner taking advantage of information we have about the limit state function locally (from Step 2 of the present iteration) by defining response transfer functions there and by also building (and improving adaptively) response surfaces based on a specified number of preceding iterations in our search for the design point. Computational savings are a result of our use of "approximate" gradients which will not require response analyses of the same level of accuracy as we would employ in Step 2.

In order to describe our proposed strategy in detail, it is convenient to think of the random variable vector, $\boldsymbol{X}$, as being made up of two parts: a load/system variable vector, $\boldsymbol{W}$, and a response random variable, Z. Long-term joint distributions for $\boldsymbol{W}$ will generally be available from environmental descriptions and structural system properties. The function of short-term response analyses is to establish the conditional distribution of Z given $\boldsymbol{W}$. The random variable Z is usually a critical response measure (for example, a response extreme or a cumulative damage measure) of a time-varying underlying response process, $Y(t)$. Evaluations of the limit state function and its gradients implicitly involve establishing (in each iteration) the conditional distribution of Z given $\boldsymbol{W}$. When this conditional distribution is represented by a standard distribution type, this implies establishing the parameters of such a distribution.

Assume that a vector of parameters, $\boldsymbol{\theta}$, needs to be estimated in order to establish the conditional distribution of Z given $\boldsymbol{W}$ and evaluate the limit state function. Assume also, that in Step 2 of iteration j, we have estimated this parameter vector, $\boldsymbol{\theta}_j$. In Step 3 of the same iteration, we will then need to estimate $\boldsymbol{\theta}_{j,\Delta i}$ in order to obtain the gradient in the direction i (where $\boldsymbol{\theta}_{j,\Delta i}$ is our shorthand notation for the parameter vector resulting from an incremental change in random variable W_i in iteration j). It is to obtain $\boldsymbol{\theta}_{j,\Delta i}$ that we will employ our response transfer functions and response surfaces.

In general, a clear distinction will be possible among (1) the entire parameter vector, $\boldsymbol{\theta}$, (2) those parameters, $\boldsymbol{\Omega}$, that retain gross statistical information (up to second order moments, i.e., r.m.s. level) and (3) other parameters, $\boldsymbol{\Psi}$, that will have to capture any higher order statistics. In a problem such as the one chosen for illustration here (the reliability analysis of a jack-up rig to random wave loading), the role of the former sub-set, $\boldsymbol{\Omega}$, might also be thought of as capturing the dynamic characteristics of the response while that of the latter sub-set, $\boldsymbol{\Psi}$, might be thought of as preserving the non-Gaussian nature of the loading or capturing quasi-static characteristics of the response. With a defined distinction of this type, we have that $\boldsymbol{\theta} = \{\boldsymbol{\Omega}, \boldsymbol{\Psi}\}$ and we will next describe procedures for establishing $\boldsymbol{\Omega}_{j,\Delta i}$ and $\boldsymbol{\Psi}_{j,\Delta i}$, the former using response transfer functions and the latter using response surfaces. This will, then, enable us to establish $\boldsymbol{\theta}_{j,\Delta i}$ and thus, gradients in iteration j.

3.3. Establishing Response Transfer Functions for $\boldsymbol{\Omega}$ in iteration j

If the parameter vector, $\boldsymbol{\Omega}$, is to include response statistics only up to second order, we can *select* suitable load measures, L_μ and L_σ, with respect to which we will define response transfer functions $R(\mu, \boldsymbol{w}_j)$ and $R(\sigma^2, \boldsymbol{w}_j, f)$ (in terms of realizations $\boldsymbol{w}_j$ of the random variable vector, $\boldsymbol{W}$ in Step 2 of iteration j, where the variable f corresponds to frequency) that will enable computation (in Step 3) of the mean and standard deviation (μ and σ) of the underlying response process, $Y(t)$ as follows:

$$R(\mu, \boldsymbol{w}_j) = \frac{E[Y(\boldsymbol{w}_j)]}{E[L_\mu(\boldsymbol{w}_j)]} \tag{1a}$$

$$R(\sigma^2, w_j, f) = \frac{S_Y(w_j, f)}{S_{L_\sigma}(w_j, f)} \tag{1b}$$

where $E[Y(w_j)]$ and $S_Y(w_j, f)$ are the mean value and the power spectral density of $Y(t)$ as obtained in Step 2; $E[L_\mu(w_j)]$ and $S_{L\sigma}(w_j, f)$ are the mean value and the power spectral density, respectively, of the two selected load measures. Note the dependence on the current values of the load/system variables w_j, i.e., the system is presumed to be linear only locally and only for small deviations (gradients).

3.4. Establishing Response Surface for Ψ in iteration j

In Step 2 of iteration j, the parameter vector Ψ of size P retained from the last N_I iterations (up to and including the jth iteration) is used to establish the matrix A required to define an adaptive response surface as follows:

$$\underset{P \times M}{A^{(j)}} \; \underset{M \times 1}{G(w_k)} = \underset{P \times 1}{\Psi(w_k)} \quad \text{where } k = j, j-1, \ldots j-(N_I-1) \quad (N_I \text{ normal equations}) \tag{2}$$

It is important to note that the user selects the number N_I of iterations to be used in building the response surface. The user also selects the functional form of the vector $G(w)$ and the size M of this vector. Regarding these selections, it is necessary that $N_I \geq M$; also, a judicious choice of $G(w)$ (based on the user's preliminary judgement of a functional dependence of Ψ on W) can be important in improving the rate of convergence. This is especially true in early iterations. N_I, M and the form of $G(w)$ can be changed as the search progresses; indeed, early in the process, N_I can clearly not be large since there will not have been a sufficient number of iterations already carried out. Then, M and $G(w)$ must be such that they permit simpler functional forms for $G(w)$ that are needed. In general, the procedure for building a response surface as defined above may either be based on a least squares minimization using the N_I normal equations when $N_I > M$ or on collocation when $N_I = M$.

3.5. Approximate Gradients in iteration j

In order to evaluate gradients of the limit state function in iteration j, the parameter vector $\theta_{j,\Delta i}$ must be defined for each random variable W_i. It is convenient to think of the limit state function in iteration j as g(w_j, $Z(w_j, \theta_j)$). Then, the gradient in the direction of W_i is given as follows:

$$\left.\frac{\partial g}{\partial w_i}\right|_{X = x_j} = \frac{g(w_{j,\Delta i}, Z(w_{j,\Delta i}, \theta_{j,\Delta i})) - g(w_j, Z(w_j, \theta_j))}{\Delta w_i} \tag{3}$$

where the unknown parameter vector, $\theta_{j,\Delta i} = \{\Omega_{j,\Delta i}, \Psi_{j,\Delta i}\}$, can now be estimated as follows:

$$\Omega_{j,\Delta i} = \left[R(\mu, w_j) \cdot E[L_\mu(w_{j,\Delta i})] \, , \int_0^\infty R(\sigma^2, w_j, f) \cdot S_{L_\sigma}(w_{j,\Delta i}, f)\, df \right] \tag{4a}$$

$$\Psi_{j,\Delta i} = A^{(j)} \; G(w_{j,\Delta i}) \tag{4b}$$

where $R(\mu, w_j)$ and $R(\sigma^2, w_j, f)$ are defined in Equation (1), $A^{(j)}$ is defined in Equation (2), and the incremental change in the direction of W_i together with the definitions of L_μ, L_σ, and $G(w)$ make it possible to evaluate $\theta_{j,\Delta i}$. This in turn enables us to compute $g(w_{j,\Delta i}, Z(w_{j,\Delta i}, \theta_{j,\Delta i}))$ and the desired gradients using Equation (3).

As can be seen from the above formulation of our proposed strategy, response analyses carried out in Step 2 are followed by additional computations in order to define "new" local response transfer functions and to "update" response surfaces. These additional computations are, however, extremely easy ones and make it unnecessary to perform N additional response analyses in order to compute required gradients.

4. NUMERICAL STUDIES

We now address the reliability of a jack-up rig subjected to random wave loading. Additional non-random wind and current are also present. Salient aspects of the geometry, the material properties, and the loading are described in Figure 1.

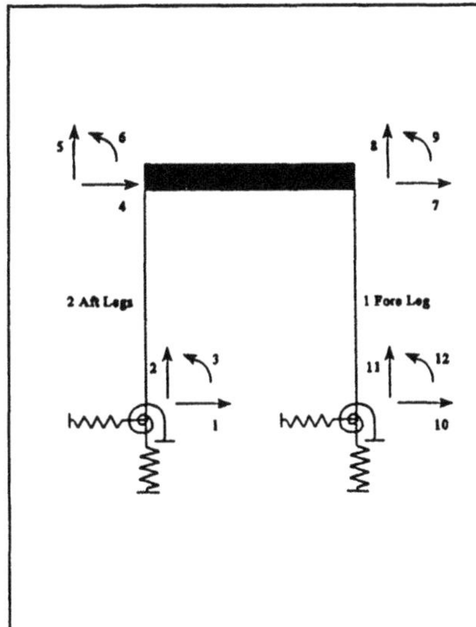

Wave Spectrum: JONSWAP (γ = 3.3)
Depth of water, d = 75.0 m
Steady Wind Force, F_{wind} = 0.27 x 10^6 N
Current, $u_0(z)$ = 0.8 m/s (-50 m < z < 0 m)
0.4 m/s (-75 m < z < -50 m)
Length of each leg = 100.0 m
Spacing between fore and aft legs = 51.0 m
Spacing between two aft legs = 55.0 m
Weight of each leg = 1.03 x 10^7 N
Weight of hull = 8.53 x 10^7 N
Leg Flexural Rigidity = 1.56 x 10^{12} N-m^2
Area of cross-section of each leg = 0.48 m^2
P-δ effects included
Airy wave theory with Wheeler stretching
Relative Morison hydrodynamics
Structural damping: 2 percent of critical.
For pinned model: displacements 1-2, 10-11 = 0
For fixed model: displacements 1-3, 10-12 = 0

Figure 1. Jack-up Rig Geometry, Material Properties, Loading, and Modeling Assumptions.

The prototype is a three-legged jack-up rig that is triangular hull in plan. We employ an idealized two-dimensional structural frame model to represent it. In this model, the dominant loading direction is assumed to be from aft to fore of the rig. The left leg of the frame model represents the two aft legs and the right leg represents the single fore leg of the prototype. Fixity conditions at the soil-structure interface will be varied in order to determine the reliability of jack-ups with different flexibilities; pinned and fixed end conditions will be compared.

In addition to the random wave loading, a steady current velocity profile collinear with the waves is adopted. A steady wind force also in the direction of the dominant wave loading is applied at the level of the deck. In order to obtain hydrodynamic forces due to waves and current, we employ Airy wave theory along with Wheeler stretching corrections. Full response analyses in our examples consist of time-domain load and response simulations necessary to obtain stable estimates of response extremes for the duration of a seastate.

Consider a first-excursion limit state function where $Y(t)$ represents the instantaneous deck displacement of our jack-up rig and Y_m represents an extreme value of $Y(t)$ in a seastate of duration

T (in our examples, $T = 3$ hours). We will assume that our random variable vector, $\boldsymbol{X}$, is made up of two load-related random variables (significant wave height, H_s, and spectral peak period, T_p) and one response random variable, Y_m. In our previous notation, $\boldsymbol{X} = \{\boldsymbol{W}, Z\}$ where $\{\boldsymbol{W}\} = \{H_s, T_p\}$ and $Z = Y_m$. Y_c is a safe response threshold or a permissible deck displacement which we will assume to be deterministic. The limit state function may then be written as follows:

$$g(\boldsymbol{W}, Z) = Y_c - Y_m \tag{5}$$

In the limit state function defined above, Y_m is an implicit function of $\boldsymbol{W}$. In our examples, the joint probability distribution $F_{\boldsymbol{W}}(\boldsymbol{w})$ for the load random variables H_s and T_p and associated parameters are assumed to be those given by Bitner-Gregersen and Haver [1] for the Haltenbanken site off mid-Norway. The conditional distribution of Z given $\boldsymbol{W}$ must be an extreme value distribution type; we employ a Hermite transformation model (see Winterstein [7]) whose parameters may be shown to be related to the zero upcrossing rate, $\nu_Y(0)$, and the first four moments of the underlying deck displacement, $Y(t)$. Our vector of parameters, $\boldsymbol{\theta}$, is needed to establish the conditional distribution, $F_{Z|\boldsymbol{W}}(z)$. Only after this parameter vector is estimated, can the limit state function be evaluated. In our example studies, we have $\boldsymbol{\theta} = \{\boldsymbol{\Omega}, \boldsymbol{\Psi}\}$ where $\boldsymbol{\Omega} = \{\mu, \sigma^2\}$ and $\boldsymbol{\Psi} = \{\alpha_3, \alpha_4, \nu_Y(0)\}$ where α_3 and α_4 are the coefficients of skewness and kurtosis respectively of $Y(t)$.

For each jack-up structure (pinned and fixed), we will perform reliability analyses in three different ways: (1) using the conventional FORM approach, (2) using the quasi-static base shear V to define our load measure of interest, i.e., $L_\mu = L_\sigma = V$, and (3) using $L_\mu = H_s$, the significant wave height, and $L_\sigma = \eta(t)$, the instantaneous sea surface elevation process. These approaches as listed are in decreasing order of effort required per FORM iteration. As discussed before, the conventional FORM approach involves $N+1$ response analyses (i.e., response simulations in the time domain) per iteration; the approach using base shear as the load measure of interest requires one full load and response simulation for Step 2 and N additional load simulations for each random variable in order to obtain gradients for Step 3; finally, the approach using $L_\mu = H_s$ and $L_\sigma = \eta(t)$ only requires the single load and response simulation for Step 2 while for Step 3 neither load nor response simulations are required. In our example, $N_I = 3$, and we choose $G(\boldsymbol{W}) = \{1, T_p, H_s\}$, i.e., $N_I = M$ in Equation (2).

Jack-up rigs with two different fixity conditions are considered: (1) a pinned model where no rotational restraint is assumed at the bottom of each leg of the jack-up; the period of this structural model is about 5.6 seconds and hence, dynamic effects are fairly important in this case, and (2) a fixed model which is assumed to be fully restrained at the bottom of each leg; the period of this structure is about 2.8 seconds and, as a result, the behavior of the structure is quasi-static, and the response is highly non-Gaussian in this case.

The above two choices lead to contrasting dynamic and non-Gaussian features which, in turn, serve to highlight the importance of distinct aspects of our proposed strategy. In the pinned model where the dynamics are important, the response transfer functions play a vital role in the analysis; whereas, in the fixed model the very non-Gaussian nature of the response requires that the response surfaces capture higher order statistics (beyond a gross r.m.s. level) accurately.

In the following tables, we summarize the results obtained from all the reliability analyses. The two faster approximate methods produce virtually the same answers as the conventional approach without degradation of the search efficiency.

(1) Jack-up Pinned at Base

Start at $H_s = 2.38$ m, $T_p = 9.23$ s, $Y_m = 0.17$ m (in standard normal space, $\boldsymbol{u}_0 = \{0, 0, 0\}$)
$g(X) = Y_c - Y_m$; $Y_c = 1.00$ m

Method	Design Point, x^* H_s(m)	T_p(s)	Y_m(m)	Importance Factors α^2_{Hs}	α^2_{Tp}	α^2_{Ym}	β	P_f	Number of Iterations
Conventional	10.15	13.46	1.00	80.1	17.4	2.5	3.32	4.54×10^{-4}	14
Approximation using *base shear*	10.07	13.18	1.00	77.3	20.6	2.1	3.35	3.99×10^{-4}	13
Approximation using *sea surface elev.*	10.20	13.77	1.00	84.4	12.5	3.1	3.25	5.82×10^{-4}	13

(2) Jack-up Fixed at Base

Start at $H_s = 6.89$ m, $T_p = 12.67$ s, $Y_m = 0.95$ m (in standard normal space, $\boldsymbol{u}_0 = \{2, 0, 0\}$)
$g(X) = Y_c - Y_m$; $Y_c = 0.25$ m

Method	Design Point, x^* H_s(m)	T_p(s)	Y_m(m)	Importance Factors α^2_{Hs}	α^2_{Tp}	α^2_{Ym}	β	P_f	Number of Iterations
Conventional	11.43	16.17	0.25	96.0	0.1	3.9	3.38	3.68×10^{-4}	10
Approximation using *base shear*	11.41	16.14	0.25	95.8	0.2	4.0	3.37	3.70×10^{-4}	9
Approximation using *sea surface elev.*	11.37	16.31	0.25	95.4	0.0	4.6	3.37	3.74×10^{-4}	8

5. DISCUSSION AND CONCLUSIONS

A FORM-based strategy that uses response transfer functions and response surfaces has been employed to analyze two different jack-up rigs. Both base shear and sea surface elevations were found to perform adequately as load measures from which to infer locally linear response transfer functions.

The building of a response surface within the FORM search is adaptive (i.e., it requires no additional effort *a priori*). In the examples, a simple linear functional form was employed for the response surface and only the three preceding iterations were utilized to update the response surface in a given FORM iteration.

The contrasting nature of the structures studied revealed that in situations where the dynamics are significant (as in the response of the pinned jack-up model), the response transfer function definitions employed in the proposed strategy are very important. On the other hand, when the response is very non-Gaussian (as with the fixed jack-up model), the response surface definitions prove to be very important.

It was found that roughly the same number of iterations are required with the proposed strategy as with the conventional FORM method. However, the cost per iteration for the proposed strategy can be significantly lower. In the examples studied, we employed two different approximate strategies where the base shear and the sea surface elevation, respectively, were defined as important load measures. The base shear approximation resulted in 40 percent savings in cost per iteration while the sea surface elevation approximation resulted in 67 percent savings relative to the conventional FORM approach. If a larger number of random variables are involved, the savings realized from use of the proposed strategy can be expected to be far greater.

6.ACKNOWLEDGMENTS

The authors gratefully acknowledge the support of the following sponsors of the Reliability of Marine Structures (RMS) program at Stanford University: Amoco, Chevron, Conoco, DNVI, Exxon, Mobil, Norsk Hydro, Saga, Shell, Statoil, and Texaco.

REFERENCES

1. E.M.Bitner-Grigersen and S.Haver, Joint Long Term Description of Environmental Parameters for Structural Response Calculation, *Proc. 12th Intl. Workshop on Wave Hindcasting and Forecasting*, Vancouver, Canada, April 1989.
2. D.Karunakaran, B.J.Leira, S.Haver, and T.Moan, Parametric Influence of Extreme Dynamic Response of Drag-Dominated Platforms, *Proc. Second Intl. Offshore and Polar Eng. Conf., Vol III*, pp. 463-471, San Francisco, California, June 1992.
3. H.Kjeøy, N.G.Bøe, and T.Hysing, Extreme Response Analysis of Jack-Up Platforms, *Marine Structures*, 2:305-334, 1989.
4. R.Løseth, O.Mo, and I.Lotsberg, Probabilistic Analysis of a Jack-Up Platform with respect to the Ultimate Limit State, Proc. 1st European Offshore Mechanics Symp., pp. 322-330, Trondheim, Norway, August 1990.
5. L.Manuel, A Study of the Nonlinearities, Dynamics, and Reliability of a Drag-Dominated Marine Structure, *Tech. Report RMS-12, Stanford Univ.*, Reliability of Marine Structures Program, Stanford, California, December 1992.
6. D.McDonald and R.G.Bea, Reliability Evaluation of a Jack-Up Drilling Unit, *Trans. Soc. of Naval Architects and Marine Engineers, Annual Meeting*, pp. 169-186, San Francisco, California, October 1990.
7. R.Rackwitz and B.Fiessler, Structural Reliability under Combined Random Load Sequences, *Computers and Structures*, 9:489-494, 1978.
8. S.R.Winterstein, Nonlinear Vibration Models for Extremes and Fatigue, *J. of Eng. Mech., ASCE*, 114(10), pp. 1772-1790, October 1988.

20

DIFFERENTIATION OF PROBABILITY FUNCTIONS ARISING IN STRUCTURAL RELIABILITY

K. Marti

Federal Armed Forces University Munich, Aero Space Engineering and Technology, D-85577 Neubiberg / München, Germany

In reliability-oriented design and optimization of engineering systems one needs various derivatives of the probability of systems survival $P(x) = P(y_\ell \le y(a(\omega),x) \le y_u)$. Here, $y=y(a,x)$ denotes the vector of response or output variables depending on the design or input vector x and the vector of random system parameters $a=a(\omega)$; we assume that $a(\omega)$ has a given probability density function $f=f(a)$. Furthermore, y_ℓ, y_u are the vectors of given lower and upper bounds for y. There is shown that in many cases derivatives $D_\ell P(x)$ of arbitrary order ℓ can be obtained by applying an integral transformation T_x to the integral representation of P(x) such that the transformed domain of integration becomes independent of x. The derivatives result then by interchanging differentiation and integration. Based on this mean value representation of $D_\ell P(x)$, estimations of the derivatives can be obtained by using sampling techniques. Furthermore, having the mentioned integral representation, $D_\ell P(x)$ can be computed approximatively by writing $D_\ell P(x)$ first as a Laplace integral and applying then the asymptotic expansion techniques known for Laplace integrals.

1. INTRODUCTION

The reliability of a technical stochastic system is measured [1-4] usually by probability functions of the type

$$P(x) = P(y_{\ell i}<(\le)y_i(a(\omega),x)<(\le)y_{ui},\ i=1,\dots,m). \tag{1}$$

Here

$$y_{\ell i}<(\le)y_i(a,x)<(\le)y_{ui},\ i=1,\dots,m, \tag{2}$$

are the basic operating conditions or behavioral constraints of the underlying system, where

$$y = (y_1,\dots,y_i,\dots,y_m)' \tag{3}$$

are certain response variables. The response variables are functions

$$y_i = y_i(a,x),\ i=1,\dots,m, \tag{3.1}$$

of a decision r-vector $x = (x_1,\dots,x_k,\dots,x_r)'$ and a parameter ν-vector

$a = (a_1,\ldots,a_j,\ldots,a_\nu)'$, where x_k, $k=1,\ldots,r$, are the **deterministic (nominal)** design variables or deterministic system coefficients; moreover, $a_j = a_j(\omega)$, $j=1,\ldots,\nu$, are the **random** system parameters or coefficients. We assume that the random ν-vector

$$a(\omega) = (a_1(\omega),\ldots,a_\nu(\omega))' \tag{3.2}$$

has a given probability density f. Finally, $y_\ell = (y_{\ell 1},\ldots,y_{\ell i},\ldots,y_{\ell m})$, $y_u = (y_{u1},\ldots,y_{ui},\ldots,y_{um})'$ are the m-vectors of given lower, upper bounds (margins) $y_{\ell i} < y_{ui}$, $i=1,\ldots,m$, for the response vector y. Working through the relevant literature, one finds relatively few papers, see e.g. [5-7], containing analytical results on the differentiation of probability functions and parameter-dependent integrals. In [8] several new **constructive differentiation methods** were presented: Differentiation of probability functions by the I) Transformation Method, II) Stochastic Completion and Transformation Method, III) Orthogonal Function Series Expansion, IV) Combinations of (I)-(III). A preliminary, short description of Methods (I) and (II) can be found in [9], see also [10,11]. In the following, the derivatives - of any order - of probability functions arising in structural reliability analysis are obtained by the Transformation Method.

2. THE DIFFERENTIATION OF STRUCTURAL RELIABILITIES

In structural reliability and design the probability of survival (safety) of a structure (structural system) can be represented by the equation

$$P(x) = P(g(u(a(\omega),x))\geq 0). \tag{4}$$

Here, $u=u(a,x)$ denotes the m-vector of the basic displacement variables $u_i=u_i(a,x)$, $1\leq i\leq m$, depending on the r-vector x of **nominal** design variables x_k and the random parameter vector

$$a = a(\omega) := \binom{p(\omega)}{F(\omega)}, \tag{4.1}$$

where $F=F(\omega)$ is the random load m-vector of the structure and $p=p(\omega)$ is a ν_1-vector of further stochastic structural parameters such as elastic moduli, manufacturing errors. Moreover, g is a given vector function selected such that the inequality

$$g(u(a,x)) \geq 0 \tag{4.2}$$

describes the relevant behavioral constraints.

From structural mechanics we know that the displacement vector $u(a,x)$ is given by

$$u(a,x) := K(p,x)^{-1}F, \tag{4.3}$$

where $K=K(p,x)$ denotes the stiffness $m\times m$ matrix of the structure. Assuming in the following - without restrictions - that $p(\omega),F(\omega)$ are stochastic independent random vectors having probability densities $f_1=f_1(p)$, $f_2=f_2(F)$, we find

$$P(x) = P(g(u(a(\omega),x))\geq 0)$$
$$= \int_{g(K(p,x)^{-1}F)\geq 0} f_1(p)f_2(F)dpdF. \tag{5}$$

Hence, we consider - for given vector x - the transformation T_x given by

$$\binom{p}{F} = T_x(q) := \binom{q_1}{K(q_1,x)q_2},\ q = \binom{q_1}{q_2}. \tag{6}$$

Since K(p,x) is positive (semi)definite, the absolute value of the functional determinant of T_x reads

$$|\det(\frac{\partial T_x}{\partial q}(q))| = \det(K(q_1,x)). \tag{6.1}$$

Applying (6),(6.1) to the intgegral in (5), we find

$$P(x) = \int_{\tilde{B}} f_1(q_1)f_2(K(q_1,x)q_2)\det(K(q_1,x))dq_1dq_2, \tag{7}$$

where

$$\tilde{B} := \{\binom{q_1}{q_2} \in \mathbb{R}^{\nu_1} \times \mathbb{R}^m:\ g(q_2)\geq 0\}. \tag{7.1}$$

Since the domain of integration $\tilde{B}$ in (7) does not depend on the design vector x, under some additional weak assumptions we may differentiate (7) by interchanging differentiation and integration. Since

$$\frac{\partial}{\partial x_k}\det(K(q_1,x)) = \det(K(q_1,x))\operatorname{tr}(K(q_1,x)^{-1}\ \frac{\partial K}{\partial x_k}(q_1,x)), \tag{7.2}$$

where "tr" designates the trace of a matrix, we set

$$h_k(q_1,q_2;x) := f_1(q_1)\{\nabla f_2(K(q_1,x)q_2)'\frac{\partial K}{\partial x_k}(q_1,x)q_2$$
$$+ f_2(K(q_1,x)q_2)\operatorname{tr}(K(q_1,x)^{-1}\ \frac{\partial K}{\partial x_k}(q_1,x))\}\det(K(q_1,x)); \tag{7.3}$$

clearly, the existence of the gradient $\nabla f_2(F)$ of $f_2(F)$ is presupposed here for all arguments under consideration.

Theorem 2.1. For a given fixed integer k, $1\leq k\leq r$, and given components $x_\ell^o, \ell\neq k$, define $x=x(t):=(x_1^o,\ldots,x_{k-1}^o,t,x_k^o,\ldots,x_r^o)'$. Suppose that for a given interval $I\subset\mathbb{R}$ the multiple integral (5) or (7) exists and is finite for each $x=x(t)$ with $t\in I$. Furthermore, suppose that function $h_k(q_1q_2;x(t))$, defined by (7.3), exists for all $t\in I$ and has an integrable majorant, i.e. a nonnegative measurable function $H_k=H_k(q_1,q_2)$, defined at least on $\tilde{B}$, such that

$$|h_k(q_1,q_2;x(t))| \leq H_k(q_1,q_2) \text{ for all } t\in I \tag{7.4}$$

and

$$\int_{\tilde{B}} H_k(q_1,q_2)dq_1dq_2 < +\infty. \tag{7.5}$$

Then $\frac{\partial P}{\partial x_k}(x)$ exists for each $x \in (x(t):t \in I)$ and is given by

$$\begin{aligned}
\frac{\partial P}{\partial x_k}(x) &= \int_{\tilde{B}} h_k(q_1,q_2;x)dq_1dq_2 \\
&= \int_{g(K(p,x)^{-1}F)\geq 0} f_1(p)(\nabla f_2(F)'\frac{\partial K}{\partial x_k}(p,x)K(p,x)^{-1}F \\
&\qquad + f_2(F)\mathrm{tr}(K(p,x)^{-1}\frac{\partial K}{\partial x_k}(p,x)))dpdF \\
&= -\int_{g(u(a,x))\geq 0} \mathrm{div}_F(f(a)(\frac{\partial u}{\partial F}(a,x))^{-1}\frac{\partial u}{\partial x_k}(a,x))da,
\end{aligned} \tag{8}$$

where $a = \binom{p}{F}$, $f(a)=f_1(p)f_2(F)$, and $u=u(a,x)$ is the vector of displacement variables defined by (4.3).

Note

By iteration of the above procedure also the higher order partial derivatives $\partial^{\underline{\ell}}P(x)$, $\underline{\ell}=(\ell_1,\ldots,\ell_s)$, of $P(x)$ with respect to $x_{\ell_1},\ldots,x_{\ell_s}$ can be obtained.

3. COMPUTATION OF PROBABILITIES AND ITS DERIVATIVES BY ASYMPTOTIC EXPANSIONS OF INTEGRALS OF LAPLACE TYPE

The computation of reliabilities of mechanical structures is a well established method, see e.g. [12,13]. In the following we examine the potential of this technique to yield also the corresponding sensitivities, i.e. the derivatives of the probabilities of survival with respect to certain design variables or deterministic system parameters x_k.

According to (7) and (8) the probability function $\partial^{(o)}P(x):=P(x)$ and its partial derivative $\partial^{(k)}P(x):=\frac{\partial P}{\partial x_k}(x)$ can be represented jointly by the formula

$$\partial^{(\ell)}P(x) = \int_{\tilde{B}} c_x^{(\ell)}(q_1,q_2)f_1(q_1)dq_1dq_2, \quad \ell=0,1,\ldots,r, \tag{9}$$

where $c_x^{(\ell)}=c_x^{(\ell)}(q_1,q_2)$ denotes the multiplier of $f_1(q_1)$ in the integrand of (7), the multiplier of $f_1(q_1)$ in (7.3), respectively. Note that for the higher order partial derivatives $\partial^{\underline{\ell}}P(x)$ we get - under corresponding assumptions -

$$\partial^{\underline{\ell}}P(x) = \int_{\tilde{B}} c_x^{\underline{\ell}}(q_1,q_2)f_1(q_1)dq_1dq_2 \tag{9.1}$$

with a more complicated function $c_x^{\underline{\ell}}=c_x^{\underline{\ell}}(q_1,q_2)$.

In order to return to the primary variables $a = \binom{p}{F}$ without involving the argument x, we consider in (9) the integral transformation

$$\binom{q_1}{q_2} = S\binom{p}{F} := \binom{p}{K(p_o,x_o)^{-1}F}, \tag{10}$$

where $(p_o,x_o) \in \mathbb{R}^{\nu_1} \times \mathbb{R}^m$ is any **fixed pair** of vectors such that $K(p_o,x_o)$ is positive definite. Application of (10) to (9) yields

$$\begin{aligned} \partial^{(\ell)}P(x) &= \int_{S^{-1}(\tilde{B})} c_x^{(\ell)}(p,K(p_o,x_o)^{-1}F)\frac{1}{\det(K(p_o,x_o))}f_1(p)dpdF \\ &= \int_{S^{-1}(\tilde{B})} C_x^{(\ell)}(p,F)f_1(p)f_2(F)dpdF, \end{aligned} \tag{11}$$

where

$$C_x^{(\ell)}(p,F) := c_x^{(\ell)}(p,K(p_o,x_o)^{-1}F)\frac{1}{\det(K(p_o,x_o))}\cdot\frac{1}{f_2(F)}, \tag{11.1}$$

$$S^{-1}(\tilde{B}) := \{\binom{p}{F} \in \mathbb{R}^{\nu_1} \times \mathbb{R}^m:\ g(K(p_o,x_o)^{-1}F) \geq 0\}. \tag{11.2}$$

Clearly, for $\partial^{\ell}P(x)$ we find the related formula

$$\partial^{\ell}P(x) = \int_{S^{-1}(\tilde{B})} C_x^{\ell}(p,F)f_1(p)f_2(F)dpdF, \tag{11a}$$

where $C_x^{\ell}(p,F)$ follows from (11.1) by replacing $c_x^{(\ell)}$ by c_x^{ℓ}, see (9.1).

Representing the random vector $a(\omega) = \binom{p(\omega)}{F(\omega)}$ by

$$\binom{p(\omega)}{F(\omega)} = \Gamma(z(\omega)) = \binom{\Gamma_p(z_p(\omega))}{\Gamma_F(z_F(\omega))}, \tag{12}$$

where

i) $\Gamma(z) = \binom{\Gamma_p(z_p)}{\Gamma_F(z_F)}$ is a certain 1-1-transformation of $\mathbb{R}^{\nu_1} \times \mathbb{R}^m$, and

ii) $z(\omega) = \binom{z_p(\omega)}{z_F(\omega)}$ is a $N(0,I)$-normal distributed random (ν_1+m)-vector,

we may represent $\partial^{(\ell)}P(x)$ by

$$\begin{aligned} \partial^{(\ell)}P(x) &= E1_{S^{-1}(\tilde{B})}(a(\omega))C_x^{(\ell)}(a(\omega)) \\ &= E1_{S^{-1}(\tilde{B})}(\Gamma(z(\omega)))C_x^{(\ell)}(\Gamma(z(\omega))) \\ &= \int_{\Gamma^{-1}(S^{-1}(\tilde{B}))} C_x^{(\ell)}(\Gamma(z))(2\pi)^{-(\nu_1+m)/2}\exp(-\tfrac{1}{2}||z||^2)dz, \end{aligned} \tag{13}$$

where 1_M denotes the characteristic function of a set M and

$$\Gamma^{-1}(S^{-1}(\tilde{B})) = \{\binom{z_p}{z_F} \in \mathbb{R}^{\nu_1} \times \mathbb{R}^m:\ g(K(p_o,x_o)^{-1}\Gamma_F(z_F)) \geq 0\}; \tag{13.1}$$

of course, a corresponding representation holds also for $\partial^{\ell}P(x)$, see (11a).

Let $z^* = \binom{z_p^*}{z_F^*}$ denote the projection of the origin $0 = \binom{0_p}{0_F}$ in $\mathbb{R}^{\nu_1} \times \mathbb{R}^m$ onto $\Gamma^{-1}(S^{-1}(\tilde{B}))$. According to (13.1) we find $z_p^* = 0$, and z_F^* is the projection of 0_F onto

$$Z_F := \{z_F \in \mathbb{R}^m:\ g(K(p_o,x_o)^{-1}\Gamma_F(z_F)) \geq 0\}. \tag{13.2}$$

Moreover, let

$$\beta^* := ||z^*|| = ||z_F^*||. \tag{13.3}$$

In the following we assume that

$$\beta^* \text{ is "large" (i.e. } P(||z(\omega)|| \leq \beta^*) \text{ is close to 1)}. \tag{14}$$

Remark 3.1

If $0_F \notin Z_F$, i.e. $\beta^* > 0$, then condition (14) can be fulfilled in many cases by an appropriate selection of the vectors p_o, x_o in the stiffness matrix $K(p_o,x_o)$. If $0 \in Z_F$, i.e. $\beta^* = 0$, then condition (14) can be reached by replacing first $\tilde{B}$ by its complement $\tilde{B}$ and considering therefore $P_1(x) := 1 - P(x)$ instead of $P(x)$.

Under the assumption (14) we apply to (13) the integral transformation in $\mathbb{R}^{\nu_1+m}$ defined by

$$z := \beta^* w. \tag{15}$$

We obtain

$$\partial^{(\ell)}P(x) = (2\pi)^{-(\nu_1+m)/2}\beta^{*\nu_1+m} I(\beta^{*2}), \tag{16}$$

where the integral $I = I(\lambda)$ of **Laplace type** [14] is given by

$$I(\lambda) := \int_{w \in \frac{1}{\beta^*}\Gamma^{-1}(S^{-1}(\tilde{B}))} c_x^{(\ell)}(\Gamma(\beta^* w)) \exp(\lambda(-\tfrac{1}{2}||w||^2))\,dw. \tag{16.1}$$

Note that a corresponding representation can be derived also for $\partial^{\ell}P(x)$, cf. (11a),(13).

Since β^* is large by assumption (14), the integral (16.1) can be evaluated now **analytically** by using the theory of asymptotic expansions [14]:

Theorem 3.1. If the above assumptions hold, then we have the expansion (being asymptotically exact for $\beta^* \longrightarrow +\infty$)

$$\partial^{(\ell)}P(x) \approx \frac{1}{(2\pi)^{1/2}\beta^*} C_x^{(\ell)}(\Gamma(z^*))\frac{1}{|J(w^*)|^{1/2}} \exp(-\frac{1}{2}\beta^{*2}), \tag{17}$$

where $J(w^*)$ is given by

$$J(w^*) = w^{*\prime}Q(w^*)w^*. \tag{17.1}$$

Here $Q(w^*)$ is the matrix of cofactors of the matrix

$$\tilde{Q}(w^*) := -(I+\mu^*\nabla^2\psi(w^*)), \tag{17.2}$$

where $\psi(w)=\psi(w;w^*)$ is a function such that the hypersurface of $\frac{1}{\beta^*}\Gamma^{-1}(S^{-1}(\tilde{B}))$ in a neighbourhood $U(w^*)$ of w^* is given by

$$\{w\in U(w^*): \psi(w;w^*) = 0\}, \tag{17.3}$$

and μ^* is determined by the equation

$$1 - ||w^*|| = \mu^*||\nabla\psi(w^*)||. \tag{17.4}$$

Remark 3.2

i) If $\frac{1}{\beta^*}\Gamma^{-1}(S^{-1}(\tilde{B}))$ has a plane hypersurface in a certain neighbourhood of w^*, then $|Q(w^*)|=1$.

ii) A corresponding asymptotic expansion holds - of course - also for $\partial^{\underline{\ell}}P(x)$.

For the **numerical computation of $\partial^{(\ell)}P(x)$** we have now the (asymptotically exact) expansion (17) with formulas (11.1) and (17.1)-(17.4). On the other hand, because of the **mean value representation** (13) of $\partial^{(\ell)}P(x)$ and (11a) of $\partial^{\underline{\ell}}P(x)$, these derivatives of order (ℓ),$\underline{\ell}$, resp., can be estimated, of course, also by ordinary sampling techniques.

REFERENCES

1. Melchers, R.E.: Structural reliability: Analysis and prediction. Ellis Horwood Ltd., Chichester (England) 1987
2. Rackwitz, R., Cuntze, R.: Formulations of reliability-oriented optimization. Engineering Optimization 11 (1987), 69-76
3. Schuëller, G.I.: A critical appraisal of methods to determine failure probabilities. J. Structural Safety 4 (1987) 4, 293-309
4. Thoft-Christensen, P., Baker, M.J.: Structural Reliability Theory and its Applications. Springer Verlag, Berlin 1982
5. Breitung, K.: Parameter Sensitivity of Failure Probabilities. In: A. Der Kiureghian, P. Thoft-Christensen (eds.): Reliability and Optimization of Structural Systems '90. Springer Verlag, Berlin-Heidelberg-New York 1990; Lecture Notes in Engineering, Vol.61, 43-51
6. Streeter, V.L., Wylie, E.B.: Fluid mechanics. McGraw-Hill, New York 1951 (first edition)

7. Uryas'ev, St.: A differentiation formula for integrals over sets given by inclusion. Numer. Funct. Anal. and Optimiz. 10 (1989), 827-841
8. Marti, K.: Differentiation of Probability Functions. Lecture Notes, IIASA Workshop on "Approximations of Stochastic Optimization Problems", Laxenburg (Austria), July 12-22, 1993
9. Marti, K.: Stochastic Optimization Methods in Structural Mechanics. ZAMM 70 (1990), T742-T745
10. Marti, K.: Approximations and Derivatives of Probabilities in Structural Design. ZAMM 72 (1992) 6, T575-T578
11. Marti, K.: Approximations and derivatives of probability functions. In: G. Anastassiou, S. Rachev (eds.): Approximation, Probability and Related Fields. Plenum Publishing Corporation, New York 1994, pp. 367-377
12. Breitung, K., Hohenbichler, M.: Asymptotic approximations for multivariate integrals with an application to multinormal probabilities. J. of Multivariate Analysis 30 (1989), 80-97
13. Hohenbichler, M., Rackwitz, R. et al.: New light on first and second order reliability methods. Structural Safety 4 (1987), 267-284
14. Bleistein, N., Handelsmann, R.: Asymptotic Expansions of Integrals. Holt, Rinehart and Winston, New York 1975

21

Relation Between Parameters Sensitivities and Dimensional Invariance on Stochastic Materials Design of Fibrous Composite Laminates

H. Nakayasu[a]

[a]Department of Industrial Management, Osaka Institute of Technology
5-16-1, Omiya, Asahi-ku Osaka 535, Japan
E-mail: nakayasu@nak101.dim.oit.ac.jp

This paper deals with the relation between parameters sensitivities and dimensional invariance on stochastic materials design of fibrous composite laminates. For the materials design of composite materials, there are several cases of the definition of the nominal safety factor. In this materials design, though the safety measures such as safety margins are identical with that by another definition of safety factor in deterministic design, it is not always verified that these consistencies are kept in stochastic fields because of the differences of the numbers of basic random variables. Therefore it is important problem in stochastic materials design whether there is consistency for any definition of safety factor or not.

1. INTRODUCTION

In the materials design of composite materials, there are several cases of the definition of safety factor in order to gain the margin of safety design for strength and modulus[1]. Each definition is essentially derived because of the intrinsic design code or calibration to practical specification. However, though the safety measure such as safety margin is coincided with that by another definition of safety factor in deterministic design, it is not always verified that these consistencies keep in stochastic fields because of the discrepancy of the number of random variables. This design problem arises from the dimensional invariability of the model in stochastic field[2],[3].

As the actual material design of composite laminates has many parameters with uncertainty such as load, strength, modulus, dimensions, and so on, all of these parameters must be evaluated in the probabilistic fields. However it has not been verified that the safety margin obtained by the nominal safety factor under the specified strain conditions in the deterministic analysis coincides with those under the strain conditions in the probabilistic field.

In this paper it is discussed that the differences of the definition of safety factor with different random parameters yield how sensitive to the final measure of reliability analysis. For the purpose of the practical stochastic materials design of composite materials, the sensitivity study on reliability of uni-directional fiber reinforced composite laminates are performed based on the present first order reliability method(FORM)[4]. Two kinds of sensitivity measures such as physical and stochastic sensitivity are defined and the relation between these sensitivities.

2. LIMIT STATE AND SENSITIVITIES OF COMPOSITE LAMINATE

2.1. Limit state function by quadratic polynomial failure criteria

When the coordinate system of unidirectional composite laminate is defined as Figure 1

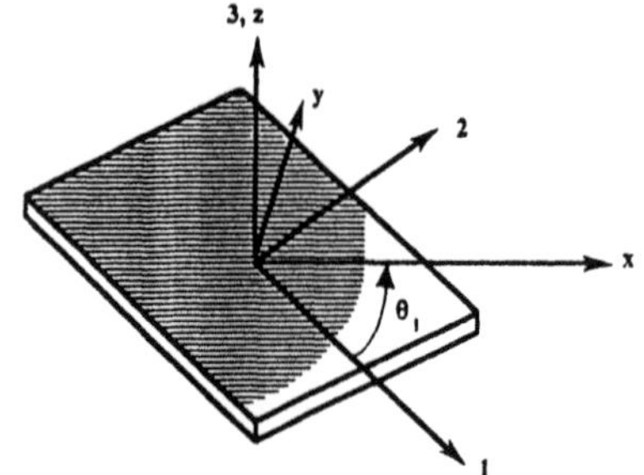

Figure 1. Coordinates system and relation stress and strain components of unidirectional composites laminates.

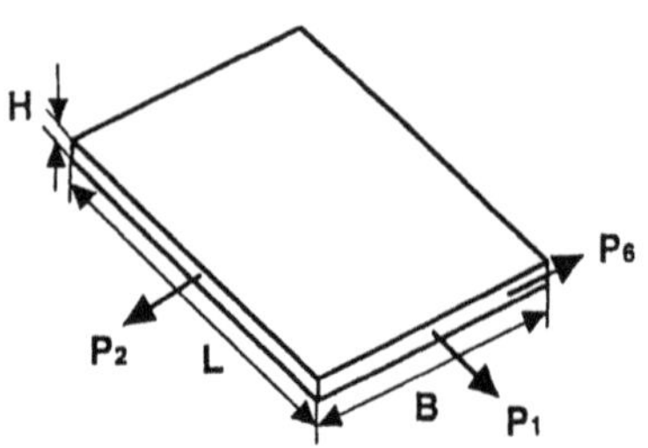

Figure2. Loads and dimensional between parameters.

and applied loads and dimensional parameters are shown in Figure 2, the typical failure criteria for materials design by Tsai-Wu[5], which is based on the mechanical behavior of composite laminates, are formulated as follows:

(1) On stress space

$$g(\mathbf{X}) = 1 - \sigma_x^T \mathbf{F}_{A,x}\, \sigma_x - \mathbf{F}_{B,x}^T \tag{1}$$

where $\mathbf{F}_{A,x}$ and $\mathbf{F}_{B,x}$ are strength parameter matrix and vector as

$$\mathbf{F}_{A,x} = \begin{bmatrix} F_{xx} & F_{xy} & 0 \\ & F_{yy} & 0 \\ \text{sym.} & & F_{ss} \end{bmatrix}, \quad \mathbf{F}_{B,x} = \{ F_x\ F_y\ 0 \}^T \tag{2}$$

$$F_{xx} = (\sigma_{L,t} \cdot \sigma_{L,c})^{-1}, F_{yy} = (\sigma_{T,t} \cdot \sigma_{T,c})^{-1}, F_{xy} = k_{12}(F_{xx} \cdot F_{yy})^{1/2},$$

$$F_{ss} = (\sigma_{LT})^{-2}, F_x = (\sigma_{L,t})^{-1} - (\sigma_{L,c})^{-1}, F_y = (\sigma_{T,t})^{-1} - (\sigma_{T,c})^{-1}$$

$\sigma_{L,t}$:tensile strength on fibrous direction x
$\sigma_{L,c}$:compressive strength on fibrous direction x
$\sigma_{T,t}$:tensile strength on transverse direction y
$\sigma_{T,c}$:compressive strength on transverse direction y
σ_{LT} :shear strength on x-y
$\varepsilon_{L,t}$:ultimate tensile strain on fibrous direction x
$\varepsilon_{L,c}$:ultimate compressive strain on fibrous direction x
$\varepsilon_{T,t}$:ultimate tensile strain on transverse direction y
$\varepsilon_{T,c}$:ultimate compressive strain on transverse direction y

From the definitions of limit state function of i-th element, the state of i-th element is divided into three categories:

$$V = \{ g_i(\mathbf{X}) \le 0 \}, \quad S = \{ g_i(\mathbf{X}) > 0 \}, \quad G = \{ g_i(\mathbf{X}) = 0 \} \tag{3}$$

V, S, and G expresses the failure domain, safety domain, and failure surface, respectively.

(2) On strain space

$$g(\mathbf{X}) = 1 - \boldsymbol{\varepsilon}_x^T \mathbf{G}_{A,x} \boldsymbol{\varepsilon}_x - \mathbf{G}_{B,x}^T \boldsymbol{\varepsilon}_x \tag{4}$$

where $\mathbf{G}_{A,x}$ and $\mathbf{G}_{B,x}$ are strength parameter matrix and vector which are obtained by

$$\mathbf{G}_{A,x} = \mathbf{Q}_x^T \mathbf{F}_{A,x} \mathbf{Q}_x \tag{5}$$

$$\mathbf{G}_{B,x}{}^T = \mathbf{F}_{B,x}{}^T \mathbf{Q}_x \tag{6}$$

2.2. Measures of sensitivity to random parameters

If the failure domain V is rewritten by

$$V = \{\mathbf{u} : \alpha^T \mathbf{u} + \beta \le 0\}, \tag{7}$$

the safety index which is a common measure of structural safety instead of safety factor is obtained by the relations.

$$P_f = P[\mathbf{U} \in V] = \Phi(-\beta) \tag{8}$$

In Eq.(8), standardized normal random vector has mean and standard deviation vectors

$$\mathbf{m} = \{m_1, m_2, m_3, \ldots, m_n\}^T = \{0\ 0\ 0\ \ldots\ 0\}^T = \mathbf{0}, \tag{9}$$

$$\boldsymbol{\sigma} = \{\sigma_1, \sigma_2, \sigma_3, \ldots, \sigma_n\}^T = \{1\ 1\ 1 \ldots 1\}^T = \mathbf{1}, \tag{10}$$

respectively.

Now the equivalent safety index β_E should be defined as well as Eq.(8)

$$\beta_E = -\Phi^{-1}(P[\mathbf{U} \in V]) = -\Phi^{-1}(P_f) \tag{11}$$

which is corresponding to P_f equivalently, where $\Phi^{-1}(.)$ indicates the inverse function of cumulative function of standard normal distribution function.

In order to evaluate the sensitivity of X_i to β_E, it is considered that the derivatives of β_E on m_i and σ_i are useful. Therefore, two kinds of measure of sensitivity

$$s_1 = \left.\frac{\partial \beta_E}{\partial m_i}\right|_{\substack{m \to 0 \\ \sigma \to 1}} \tag{12}$$

$$s_2 = \left.\frac{\partial \beta_E}{\partial \sigma_i}\right|_{\substack{m \to 0 \\ \sigma \to 1}} \tag{13}$$

should be derived from the relation

$$\beta_E = -\Phi^{-1}(P_f) = -\Phi^{-1}(P[\alpha^T \mathbf{u} \le -\beta]) = -\Phi^{-1}(P[\sum_{i=1}^{n} \alpha_i(U_i\sigma_i + m_i) + \beta \le 0])$$

$$= -\Phi^{-1}(P[\sum_{i=1}^{n} \alpha_i\sigma_i U_i \le -\beta + \sum_{i=1}^{n}\alpha_i m_i]) \tag{14}$$

which can be expressed by the Hesse Form

$$\beta_E = -\frac{-\beta + \sum_{i=1}^{n}\alpha_i m_i}{(\sum_{i=1}^{n}(\alpha_i\sigma_i)^2)^{1/2}} \quad . \tag{15}$$

Thus the measure of sensitivity are finally expressed by the safety index and gradients of failure surface on transformed U-space.

$$s_1 = \left.\frac{\partial\beta_E}{\partial m_i}\right|_{\substack{m\to 0\\ \sigma\to 1}} = \alpha_i \tag{16}$$

$$s_2 = \left.\frac{\partial\beta_E}{\partial\sigma_i}\right|_{\substack{m\to 0\\ \sigma\to 1}} = -\beta\alpha_i^2 \tag{17}$$

The s_1 and s_2 mean the perturbation of β_E on m_i and σ_i. The former is a sensitivity on mean of X_i which is called physical sensitivity. On the other hand, the latter is a sensitivity on standard deviation of X_i which is called stochastic sensitivity.

3. NUMERICAL ANALYSIS

3.1. Definition of the three kinds of safety factor

Consider the load condition of a unidirectional carbon/epoxy laminate (T300/5208) as shown in Fig.2 with three kinds of the definitions of safety factor for materials design as:

1) Definition of safety factor on ultimate strength (Type I)

$$SF1_1 = \sigma_{L,t}/\sigma_1^*(\sigma_1^* \ge 0) \text{ or } \sigma_{L,c}/\sigma_1^*(\sigma_1^* < 0)$$
$$SF1_2 = \sigma_{T,t}/\sigma_2^*(\sigma_2^* \ge 0) \text{ or } \sigma_{T,c}/\sigma_2^*(\sigma_2^* < 0)$$
$$SF1_3 = \sigma_s/\sigma_6^* \tag{18}$$

2) Definition of safety factor on in-plane load (Type II)

$$SF2_1 = P_1/P_1^*, SF2_2 = P_2/P_2^*, SF2_3 = P_6/P_6^* \tag{19}$$

3) Definition of safety factor on dimensional factor (Type III)

$$SF3_1 = B^*/B, SF3_2 = L^*/L, SF3_3 = H^*/H \tag{20}$$

In the above equations, (.)* means the nominal value of each factor in every definition of safety factor. Since the load condition subjected to composite laminate is shown in Fig. 2, the mean stress on principle axis will be calculated from the basic relation

$$\sigma_1 = \frac{P_1}{BH}, \sigma_2 = \frac{P_2}{LH}, \sigma_6 = \frac{P_6}{BH} \tag{21}$$

which is in correspondence to these definition of safety factor. These mean stresses by Eq.(21) is transformed into in-plane strain through stiffness matrix or compliance matrix which is composed of modulus factors. In the stochastic materials designs of composite materials, the parameters such as strength factors, modulus factors, load factors, lamination factors, and dimensional factors must be taken into account of as random variables.

Many research works on reliability based materials design of composite materials usually start from the Type I definition of nominal safety factor. Unfortunately there are differences in the number of random variables as shown in Table 1 which is dependent on the types of the definition of safety factor. Because the limit state function which is dependent on the selective failure law on composite laminates requires several kind of random parameters whose number is shown in Table 1. If the value of the safety index does not coincide with that of another definition, the designer must be modify the value of the design variables in practical design to keep equivalent safety margin according to the selection of the definition of safety factor. This design problem arises from the lack of dimensional invariability. Therefore the degree of the differences of the value of safety index should be evaluated for Eqs.(18)-(20) and the relation to the sensitivity of random parameter should be also discussed.

3.2. Evaluation results and some considerations

Numerical analysis of the graphite epoxy unidirectional composite laminates subjected to off axis loads was examined for the verification of the existence of the lack of dimensional invariability and the consideration of the relation to the sensitivity. Table 2 shows the values of

Table 1 Number of random variables corresponding to the definition of safety factor.

Definition of SF	Material strength parameter	Material constant parameter	Lamination parameter	Dimensional parameter	Load parameter	Total number of random variables
Type. I SF1(·)	6	4	1	0	0	11
Type. II SF2(·)	6	4	1	3	3	17
Type. III SF3(·)	6	4	1	3	3	17

Table 2 Characteristic properties of random variables.

(T300/5208)

Parameter	Expectation	Standard deviation	CV
$\sigma_{L,T}$	1500 (MPa)	150	0.10
$\sigma_{L,c}$	1500 (MPa)	180	0.12
$\sigma_{T,t}$	40 (MPa)	4.40	0.11
$\sigma_{T,c}$	246 (MPa)	19.68	0.08
σ_{LT}	68 (MPa)	4.08	0.06
k_{12}	−0.5	0.005	0.01
E_x	181000 (MPa)	9050	0.05
E_y	10300 (MPa)	515	0.05
E_s	7170 (MPa)	358.5	0.05
ν_x	0.28 (MPa)	0.0028	0.01
B	30 (mm)	1.5	0.05
L	150 (mm)	7.5	0.05
H	2 (mm)	0.1	0.05
P_1	90000 (N)	4500	0.05
P_2	12000 (N)	600	0.05
P_6	4080 (N)	204	0.05

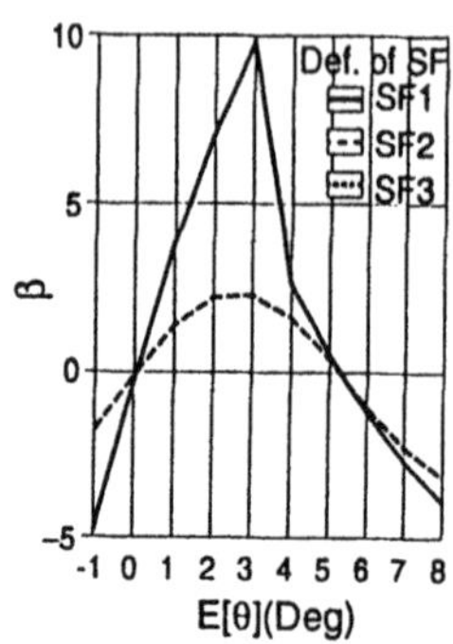

Figure 3 Relation between β and E[θ]in correspondence to the definition of SF.

stochastic features of parameters and design variables for T300/5208, whose distribution types are respectively

$$\sigma_{L,t}\,,\ \sigma_{L,c}\,,\ \sigma_{T,t}\,,\ \sigma_{T,c}\,,\ \sigma_{LT} \rightarrow \text{Log-normal distribution,}$$

$$k_{12}\,,\ E_x\,,\ E_y\,,\ E_s\,,\ \nu_x\,,\ B,\ L,\ H \rightarrow \text{Normal distribution,}$$

$$P_1\,,\ P_2,\ P_6 \rightarrow \text{Weibull distribution (2 parameters).}$$

Figure 3 shows one of the typical differences of the value of safety indices on some value of expectation of lamination angles, where each safety factor are decided that

$$SF1_{(.)}=1.790, \quad SF2_{(.)}=1.810, \quad SF3_{(.)}=1.346$$

as the nominal mean stresses are equal each other on principle axis by Eq.(21). Despite of the same behavior of in-plane strain for any lamination angles, it is seen that there are great differences between the results among $SF1_{(.)}$, $SF2_{(.)}$, and $SF3_{(.)}$. Especially the distinctions become to be remarkable around the maximum point of safety index on lamination angle. However the results by $SF2_{(.)}$ are same to those by $SF3_{(.)}$ whose number of random variables are equal to $SF2_{(.)}$.

For the consideration of this fact, the sensitivity study on random parameters which construct the limit state function was performed. Figure 4 shows the physical and stochastic sensitivities for the expectation degree of the lamination angle. In the figures it is clarified that the load factors such as P1, P2, P3 and the dimensional factors such as B, L, H have some quantities of physical and stochastic sensitivities as well as strength factors. Nevertheless the modulus factors have little sensitivities. Since the definition of Type I safety factor by Eq.(18) does not include these factors, it is suggested that such a differences of final measure of safety indices arises from the effect of sensitivities of load and dimensional factors.

These evaluation results which show the discrepancies of the value of safety indices on the definition of safety factors are in correspondence with the mean values and coefficient of deviations shown in Table 2. For some kinds of combination of the coefficients of variation of random parameters such as load and strength factors, the effectiveness of the lack of dimensional invariance are examined as described in Figure 5. In the figure, it is seen that the differences by the definition of safety factor become to be remarkable in proportion to the increases of the value of coefficient of variation of load factors.

4. CONCLUDING REMARKS

In the practical materials design of composites materials, several definitions are essentially derived because of the intrinsic design specification. However it is verified in this study that

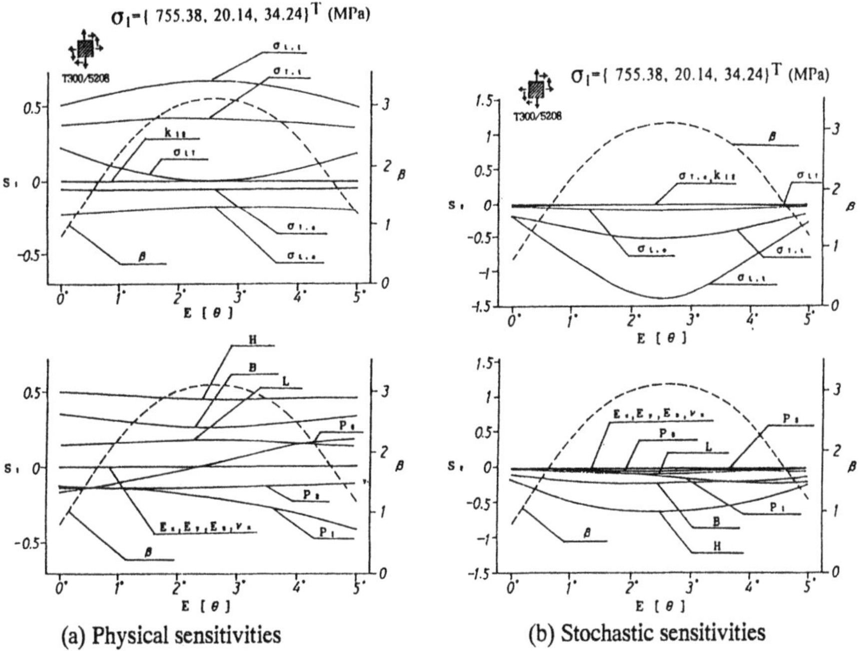

Figure 4 Physical and Stochastic sensitivities of random parameters under off axial load condition.

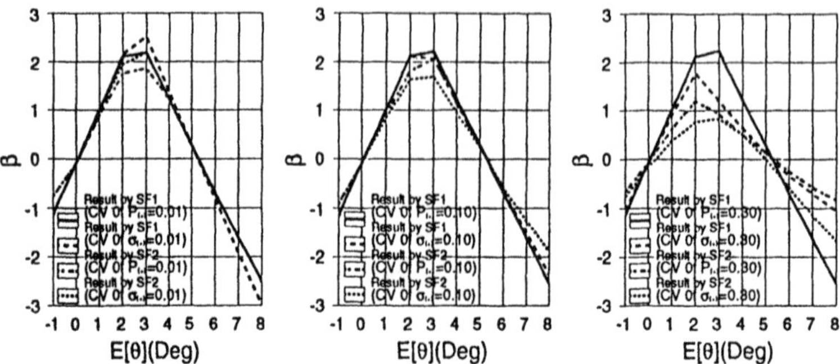

Figure 5 Relation between β and $E[\theta]$ to the combination of CV of load and strength factors.

some differences of the real safety measures considering parameter uncertainties become to be remarkable in dependence how to select the nominal safety factors in design processes. This discrepancy arises from the lack of dimensional invariability. From the previous experiences of the former study[6], there are little discrepancy of the safety measures in the treatment on in-plane stress and strain based limit state functions, though the former has small number of random variables than the latter. The lack of dimensional invariability was negligible problem in the modeling process if the sensitivities of random parameters are small. On the other hand, it should be noted that the lack of dimensional invariability on the design process of stochastic materials design of selecting the nominal safety factor yields great discrepancy of final safety index because of the sensitivities of load and dimensional random parameters. Thus one must be taken into account the differences by the selection of nominal safety factor which arises from the lack of dimensional in variabilities and keep equivalent safety margin whenever any definition of safety factor is specified in the practical materials design of composite materials.

ACKNOWLEDGMENTS

This work was partly supported by the Science Research Fund of the Ministry Education, Science and Culture of Japan, and the Science Research Promotion Fund of Japan Private School Promotion Foundation. The author thanks to Mr. Fukuzumi (Showa Rekisei Co. Ltd.) for his cooperation to this work.

REFERENCES

1. Tsai, S. W., Composite Design (4th ed.), THINCK COMPOSITES, 1988.
2. Cornell, C. A., ACI J. **66**-12 (1969) 974.
3. Hasofer, A. M., and Lind N., Proc. ASCE, J. Engng. Mech.Div., **100** (1974) 111.
4. Hochenbichler, M., Gollwitzer, S., Kruse, K. and Rackwitz, R., Structural Safety, **4** (1987) 267.
5. Tsai, S. W. and Wu E. M., J. Compos. Mater. **5** (1971) 58.
6. Nakayasu, H., Probabilistic Structural Mechanics: Advances in Structural Reliability Methods (Spanos,P.D. and Wu, Y.-T.Eds.),Springer-Verlag, Berlin Heidelberg, 1994, 392.

22

ASSESSMENT OF GLOBAL RELIABILITY FROM LOCAL DAMAGE OBSERVATIONS: APPLICATION TO HIGHWAY BRIDGE DECKS

L. Pardi[a], E. Mancino[b]

[a]AUTOSTRADE SPA, via A.Bergamini 50, 00159 ROMA

[b]PAVIMENTAL, piazza F.De' Lucia 15, 00100 ROMA

ABSTRACT: An attempt is made at assessing the reliability of highway bridges taking into account the local deterioration process with the aid of design sensitivity analysis of commercial F.E. codes.

1. INTRODUCTION

The Italian network of highways consists of approximately 3000 bridges and viaducts which are rapidly deteriorating due to ageing, ever-growing traffic volumes and environmental conditions. For these reasons, the number of bridges in need of repair is expected to increase in the future such that, in view of possible economic constraints, it will be necessary to schedule interventions very precisely, in reference to overall bridge reliability. Reliability is expected to vary with time. Its changes depend on the deterioration of individual bridge components and the consequent modifications of the global structural behaviour.

To describe the process of damage of the individual elements, a deterioration model has been created based on all the annual inspection reports. Compiled since 1967, these consist mostly of visual information, sometimes corroborated by non-destructive tests. All these visual data, appropriately codified, are discretized in a series of possible states, representing different levels of damage, and modelled by Markov chains: predictions are made for individual components only.

To assess global bridge reliability it is necessary to determine all possible failure modes because damage, at different times and at different locations along the structure, will affect the global response in various ways. Of course, the Finite Element Method provides the proper computational framework to model complex systems. In particular recourse has been made to the Design Sensitivity Analysis routines of F.E. commercial codes. The DSA is used to calculate the values of some structural responses (for ex. element forces) whose derivatives are evaluated with respect to some design variables (for ex. element properties). By applying it iteratively , it is possible to identify all the groups of elements which, if they cannot deteriorate together, will result in failure of the structure: the whole system can be suitably represented by means of cut-sets.

The failure probability, associated with each possible cut-set, may be seen as a function of the values of the design variables calculated by means of the DSA (index β). Alternatively, it is possible to express the probability of failure in terms of the probability of each individual component arriving at a certain undesirable visual state, predicted using the damage model.

In order to test this, the entire procedure was applied to some bridge decks of the Italian highway system.

2. DETERIORATION MODEL

The reliability of bridges changes over time as a result of the process of deterioration of its individual component parts. As a consequence, the properties of the structural members are seen as random variables because affected by the deterioration process. Damage is mainly due to the slow but continuous attack of environmental factors such as snow, rain, etc, that modify both the aspect and the behaviour of the bridge elements. All these changes are currently reported in the inspection reports, compiled regularly since 1967, consisting of visual observations and of results of non-destructive tests. These reports form a very long record of damage evolution.

Recently codified and fed into a data-bank, all these data may be interpreted as discrete instants in the continuous life of the structure, and the evolution of the deterioration process may be transformed into a series of successive possible "states", each state being identified by a given set of observations. The main difficulty is that damage does not follow a single path culminating in the most alarming state, but it passes through many different intermediate states: this means that from the same initial visual observation many different subsequent observations may follow. The most suitable mathematical instrument to model the process of deterioration is the Markov process. An important property of Markov models is that the transition probability from state S_i to state Sj only depends on S_i and S_j and not on previous or later states. All the transition probabilities are assembled in the transition matrix that governs the succession of the states. An example of a transition matrix together with the corresponding physical meaning are given in tab.1 and fig.1.

table 1- Transition matrix

	1	2	3	4	5	6	7	8	9	10	11	12	13	14	15	16	17
1	0,975																
2	0,003	0,500															
3	0,016		0,898														
4		0,500		0,909													
5	0,003		0,028		0,909												
6			0,028			0,333											
7	0,016		0,014				0,818										
8			0,028					0,937									
9				0,091					1								
10					0,091		0,181			0,907							
11						0,667					0,862						
12								0,062				1					
13										0,092			0,636				
14											0,138			1			
15													0,363		0,010		
16															0,990	0,926	
17																0,074	1

By iteratively multiplying a vector representative of the state of all the elements by the matrix, we are able to predict the number of elements reaching a given level of deterioration after n intervals of time, and consequently where damage will be located along the structure.

As a conclusion, it can be said that the Markov process permits establishing an exponential law of deterioration which is actually only qualitative: to link this, or the corresponding visual observations, to a probability distribution of strength or other parameters of interest, it is necessary to run in-situ and laboratory tests. In this way it should be possible to assess the variations of reliability over time and the evolution of damage so as to schedule the needed repairs in timely manner.

The deterioration model is a local model, as predictions are made for individual structural elements only. How changes in the element properties will affect the overall bridge behaviour and the global reliability is determined via sensitivity analysis.

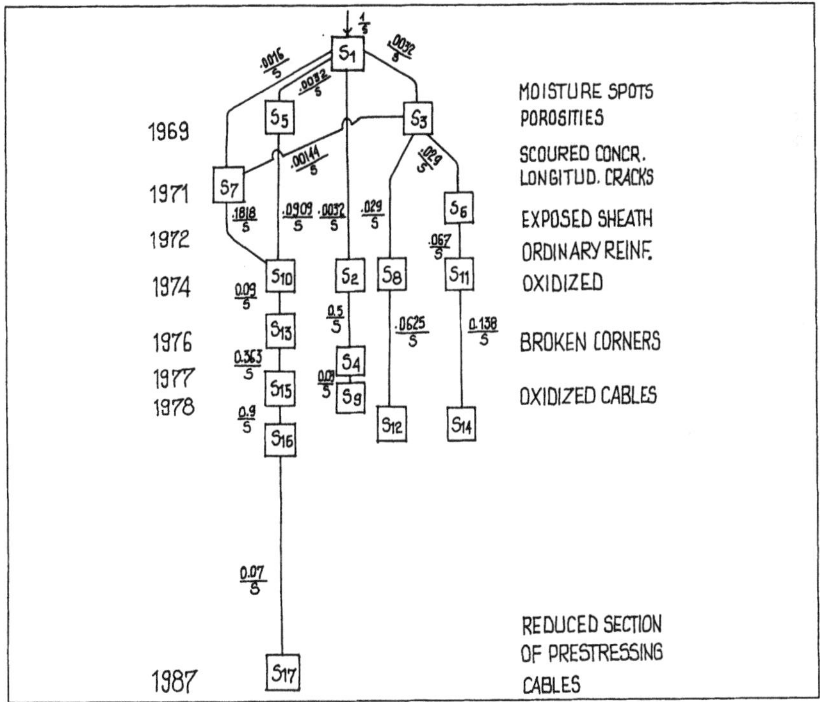

fig. 1 - Graph representation of the transition matrix

3. FAILURE MODES

Bridge decks are complex systems composed of many different elements (beams and cross-beams). Various limit states such as bending action, shear, deflection, etc., may be considered. A system may be subject to many loads in various combinations.

In the following analysis, loads will be considered as deterministic, while the element behaviour will be considered to be elastic-brittle. This means that failure at any point of the structure implies failure of the entire structure: the system could be conveniently represented through a series of possible failure modes.

Within each limit state, for instance bending action, the limit state condition may be reached in any element depending on the number of elements simultaneously deteriorated, and their level of deterioration. The event "attainment of limit condition in element i" can be seen as a failure mode, and therefore each limit state could again be interpreted as a series of events, the deteriorated elements being connected in parallel.

Of course, depending on the level of deterioration and its location along the structure, all groups of 1, 2,....n elements could be possible. To reduce the number of computations, recourse has been made to the Design Sensitivity Analysis (DSA).

The associated overall probability of failure may be computed according to the well-known formula of Ditlevsen which gives the lower and upper bounds of the failure probability of a structural system.

The probability of failure for each cut-set could be very easily computed through the β-index, in the hypothesis that the density probability functions of all the variables of interest were known. If these are not known, the probabilities of arriving at an unacceptable level of visual deterioration can be used.

The two main tasks of computing the probability of failure of complex system are: identifying the cut-sets or failure modes and evaluating the failure probability associated with each failure mode, preferably as a function of the design variables. The second point was studied by the authors in ref.4 where an optimisation numerical procedure was used to find the appropriate combination of the design variable values in each mode. In this paper the attention is focused on the steps to identify the modes themselves. It will be found that the method is a simplification of the well-known truncated enumeration scheme.

4. APPLICATION OF DESIGN SENSITIVITY

The Design Sensitivity Analysis, which is available in many commercial F.E. codes, is used to determine the most important failure modes from the structural importance of individual components. It is used to compute the derivatives of a structural response with respect to design variables. DSA currently outputs the values of the user-defined constraints, and the design sensitivity matrices which are the derivatives of the design constraints with respect to the design variables. The design variables can be material and element properties, while design constraints can be displacements, element forces and element stresses.

The design variables should correspond to model variables that have statistical uncertainty. Every change in a non-dimensional design variable is defined by the formula:

$$\Delta B_i = \frac{b_i}{b_i^0} - \frac{b_i^0}{b_i^0} = B_i - 1$$

where B_i=i[th] non-dimensional design variable

b_i^o=mean value of the design variable

b_i=perturbed design variable

The response of interest is specified by applying a constraint r_{lim} to the model. DSA outputs the constraint value y_i and the sensitivity of this constraint with respect to the non-dimensional design variables (λ_{ij}) so that:

$$\lambda_{ij} = \frac{\Delta\psi_i}{\Delta B_j}$$

therefore it is possible to compute the value of the response due to a change in the jth design variable

$$r_i = r_i^0 \pm \lambda_{ij} |r_{\lim}| \Delta B_j$$

which is the value of the ith response due to a perturbance of the jth design variable. The + or - signs correspond respectively to the upper or lower bounds of the constraint.

In the following paragraph an example is given of how sensitivity analysis is used to find the failure modes of a given structure.

5. APPLICATION TO A BRIDGE DECK

A bridge deck is composed of 4 beams and 4 cross-beams (fig.2), with 2 concentrated forces on the external beam. For each bar element a safety margin is defined as a function of the bending moments as, according to the guidelines of the F.E. code used in this case:

$$r_i = \frac{M_i}{|Limit(i)|} - 1 \bullet SIGN(Limit(i)) \text{.........} if. M_i..(-)$$

$$r_i = 1 \bullet SIGN(Limit(i)) - \frac{M_i}{|Limit(i)|} \text{........} if. M_i...(+)$$

In this way both values approach zero from the same side (fig.3).

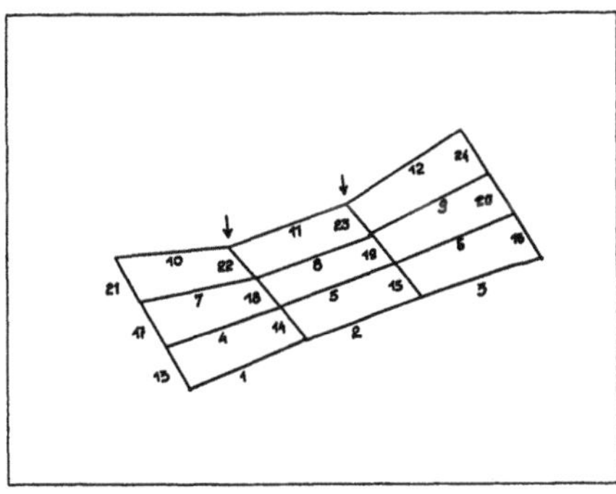

fig.2- Deformed strucuture

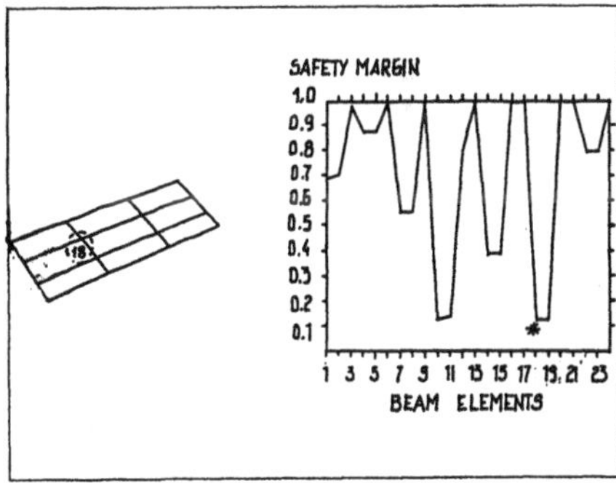

fig.3 - Safety margins - 1st run

In fig.3 we see that the element closest to the limit condition is the 18th. Which are the other elements of the structure whose conditions most affect the performance of element 18? The DSA is run, and the resulting sensitivity coefficients are plotted in fig.4.

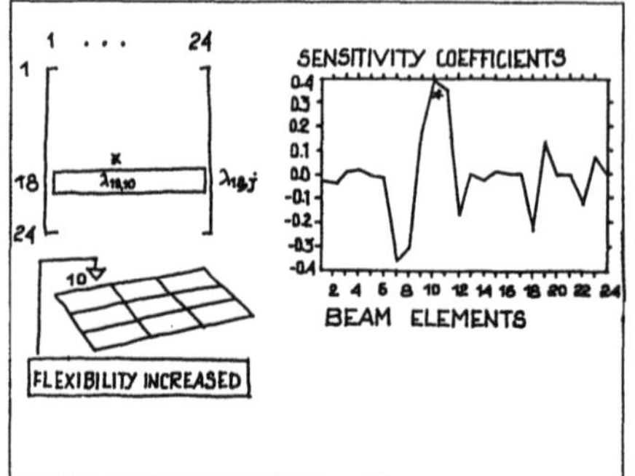

fig.4 - Sensitivity coefficients- 1st run

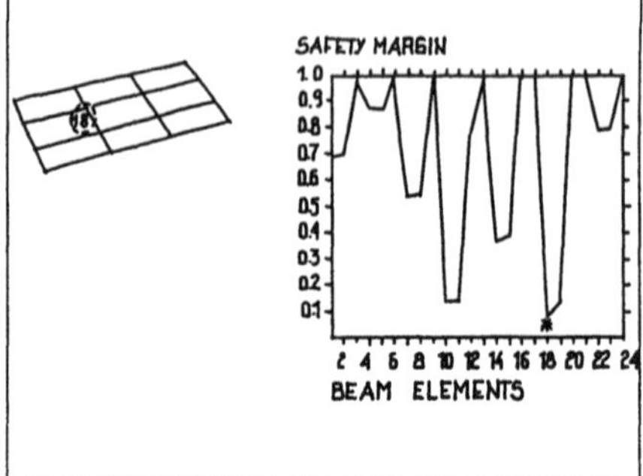

fig.5 - Safety margins - 2nd run

Element 10 has the largest influence on the 18th, so its flexibility is increased by the allowable increment (10%), and a new DSA is run. The safety margin of element 18 has been reduced, but not enough to reach the unsafe domain (fig.5).

The flexibility of the 11th element has to be reduced for the next run (fig.6) and the corresponding values of the safety margin are plotted in fig.7. The 18th element is still the most critical one (fig.8), while the most likely element to be increased is n.9.

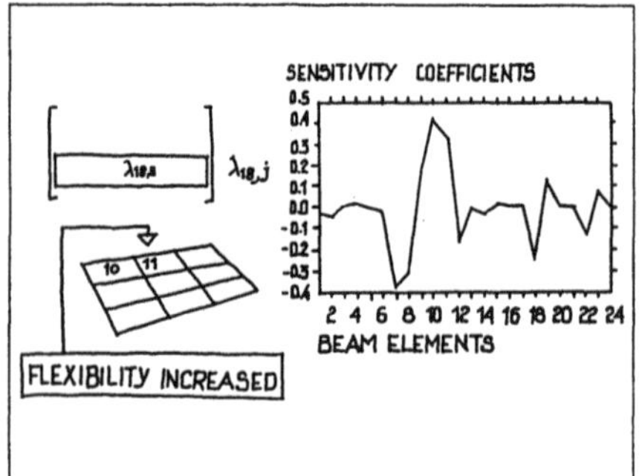

fig.6 - Sensitivity coefficients -2nd run

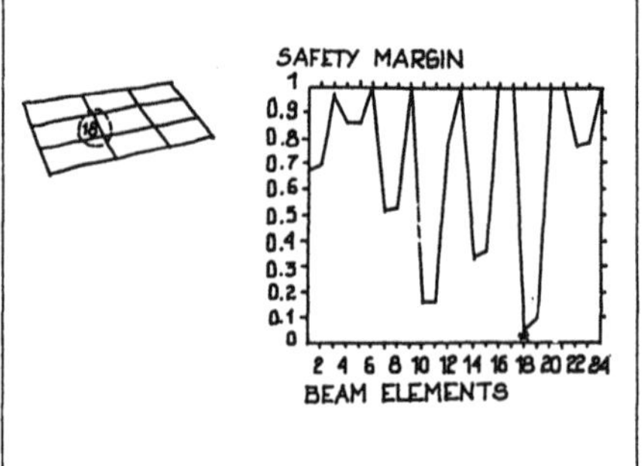

fig.7 - Safety margins - 3rd run

This time (fig.9) the limit condition has been violated, and we may say that elements 9,10 and 11 form a cut set of order 3. The search should go on, checking if element 20 can be included in a set (fig.9). The procedure is run until all the elements are included in at least one set. All the failure modes may be assembled in a series, where each branch is given by a parallel of elements.

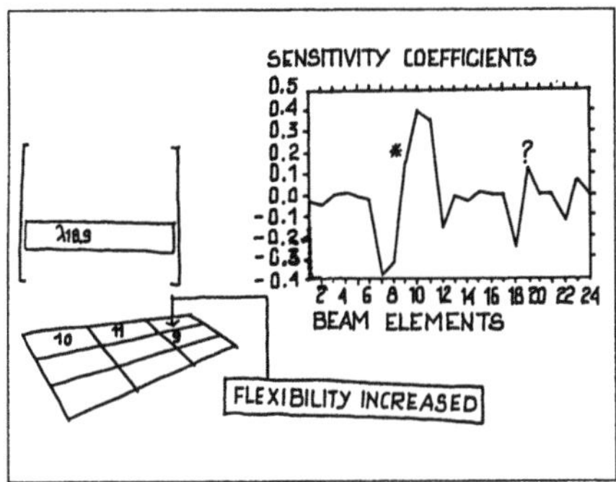

fig.8 - Sensitivity coefficients -3rd run

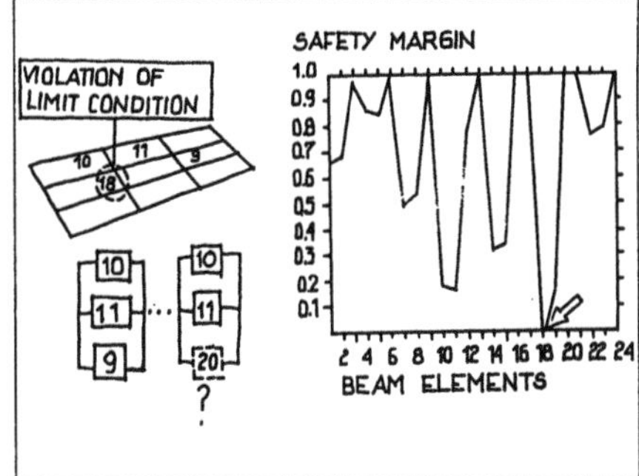

fig.9 - Safety margins - 4th run

6. ITALIAN BRIDGE DECKS

The whole procedure has been applied to assess the probability of failure of the different bridges of the Italian highway system, and intervention priorities are determined on the basis of these results. Table 2 shows the results of a Markov process as the probability of the individual elements arriving at a deteriorated state. In table 3 the same probabilities of failure of individual components are combined to find the failure probability of the cut-sets previously found via the sensitivity analysis. These in turn can be used to calculate the bounds of the system failure probability.

table 2
Failure probabilities as predicted by the Markov model

	time	(years)	
code	**2**	**5**	**10**
170501	0,00016	0,0062	0,0012
170502	0,0269	0,1172	0,2814
171101	0,0054	0,050	0,1817
171102	0,0269	0,1172	0,2814

7. CONCLUSIONS

An attempt has been made to assess the safety of bridges in terms of the probability of arrivng at a deteriorated state, filtered through the structural importance of the different bridge elements. The procedure is used to rank bridges in terms of intervention priorities.

table 3
Sensitivity analysis predictions of failure probabilities

	time	(years)	
code	**2**	**5**	**10**
170501	4,0e-8	2,8e-7	1,0e-6
170502	5,0e-6	5,0e-5	2,4e-4
171101	4,0e-8	2,1e-7	7,2e-7
171102	1,6e-6	2,0e-5	1,3e-4

REFERENCES

1. Ang, A.H.S., and Tang, W. H.(1984). "Probability Concepts in Engineering Planning and Design, vol. 2: Decision, Risk,and Reliability", John Wiley & Sons.
2. Melchers, R.E.(1987). Structural Reliability: Analysis and Prediction, John Wiley & Sons.
3.-MSC/Nastran (1993). User's Guide for Design Sensitivity and Optimization v67.
4. Mancino E., and Pardi L. (1994). "Application of Sensitivity Analysis to Predict the Behaviour of Bridges", Proc. 2nd Int.Conf. on Computational Stochastic Mechanics, Athens (to be published)

23

Interactive Structural Optimization with Quasi-Newton Algorithms

C. Pedersen & P. Thoft-Christensen

Aalborg University, Sohngaardsholmsvej 57, DK-9000 Aalborg, Denmark

Abstract: This paper presents a prototype of an interactive optimization system IROS (Interactive Reliability-Based Optimization System), where the designer during the optimization process is able to adjust simple bounds, fix or relax design variables and include/exclude constraints. The optimization is based on a modified quasi-Newton algorithm, which includes a subproblem that minimizes the condition number of the approximate Hessian matrix in each iteration.

Keywords: Interactive Optimization, Quasi-Newton Algorithms, Reliability-Based Optimization.

1 INTRODUCTION

During the last decades, one of the most important renewals in the fields of structural and especially mechanical design has been the innovation of fully automatic systems and computer based analysis tools. However, fully automated systems are generally not considered feasible or preferable within the field of structural optimization. Therefore, interactive systems are assumed to be used within several engineering fields in the future.

Considering structural constrained optimization problems only, the class of algorithms based on quasi-Newton schemes and Lagrange multipliers have proven to be most efficient due to their relatively low number of function and gradient evaluations.

This paper presents a prototype of an interactive optimization system based on a quasi-Newton algorithm, where the designer during the optimization process is able to adjust simple bounds, fix or relax design variables and include/exclude constraints. The optimization is based on a modified quasi-Newton algorithm, which includes a subproblem that minimizes the condition number of the approximate Hessian matrix in each iteration. The purpose is to minimize the condition number of the approximate Hessian matrix and thereby stabilize the determination of the search direction in each iteration after major interactions.

The interactive optimization is carried out through a shell called IROS (Interactive Reliability-Based Optimization System). IROS is mainly intended to constitute a platform for testing of various algorithms and interactive capabilities rather than being a commercial programme. In order to demonstrate some of the facilities of an interactive optimization system, a simple example of a reliability-based optimization of a plane portal frame of steel with I-shaped cross-sections is shown.

2 RELIABILITY-BASED OPTIMIZATION PROBLEM

In order to introduce the terminology used, the general formulation of a non-linear constrained optimization problem is stated as

$$\min_{\mathbf{z}} \quad C(\mathbf{z}) \tag{1}$$

$$\text{s.t.} \quad c_j(\mathbf{z}) = 0 \qquad j = 1, ..., m_e \tag{2}$$

$$\hat{c}_j(\mathbf{z}) \geq 0 \qquad j = 1, ..., m_i \tag{3}$$

$$z_i^{\min} \leq z_i \leq z_i^{\max} \qquad i = 1, ..., n \tag{4}$$

where $\mathbf{z}^T = (z_1, z_2, ..., z_n)$ denote the design variables. Introducing the Lagrange multipliers λ_i, the Lagrange function $L(\cdot)$ is defined as (5), where the $2n$ simple bounds are included as inequality constraints of the standard form (3).

$$L(\mathbf{z}, \boldsymbol{\lambda}) = C(\mathbf{z}) - \sum_{j=1}^{m_e} \lambda_j c_j(\mathbf{z}) - \sum_{j=1}^{m_i+2n} \lambda_{m_e+j} \hat{c}_j(\mathbf{z}) \tag{5}$$

The reliability-based optimization problem on element level is obtained when a subset of the constraints (3) is of the form

$$\hat{c}_j(\mathbf{z}) = \beta_j(\mathbf{z}) - \beta_j^{\min} \tag{6}$$

Introducing the stochastic variables $\mathbf{X}$, the corresponding standardized independent and normally distributed $\mathbf{U}$-space obtained by the transformation $\mathbf{X} = \mathbf{T}(\mathbf{U})$ and the failure function $g_j(\mathbf{x}, \mathbf{z})$, the probability of failure $P_f = \Phi^{-1}(\beta_j)$ can be approximated by the FORM solution, where the reliability index $\beta_j(\cdot)$ in (6) is obtained from

$$\beta_j = \min_{\mathbf{u}} \quad \sqrt{\mathbf{u}^T \mathbf{u}} \tag{7}$$

$$\text{s.t.} \quad g_j(\mathbf{T}(\mathbf{u}), \mathbf{z}) = 0 \tag{8}$$

A closer description can be found in Madsen et al. (1986) and others.

Using a quasi-Newton algorithm, optimum $\mathbf{z}^*$ is found iteratively as the limit of the sequence $\{\mathbf{z}^{(k)}, k = 1, 2, ...\}$. The iterate $\mathbf{z}^{(k+1)}$ is obtained as $\mathbf{z}^{(k+1)} = \mathbf{z}^{(k)} + \alpha \mathbf{d}$, where $\mathbf{d}$ denotes the search direction while the step length α is obtained from a one-dimensional line search in the direction $\mathbf{d}$ in a merit function $\phi(\cdot)$.

The search direction $\mathbf{d}$ in each iteration is determined as the solution to a sequential quadratic programming (SQP) subproblem of the form

$$\min_{\mathbf{d}} \quad \nabla_z C(\mathbf{z}^{(k)})^T \mathbf{d} + \frac{1}{2} \mathbf{d}^T \mathbf{B}^{(k)} \mathbf{d} \tag{9}$$

$$\text{s.t.} \quad c_j(\mathbf{z}^{(k)}) + \nabla_z c_j(\mathbf{z}^{(k)})^T \mathbf{d} = 0 \qquad j = 1, ..., m_e \tag{10}$$

$$\hat{c}_j(\mathbf{z}^{(k)}) + \nabla_z \hat{c}_j(\mathbf{z}^{(k)})^T \mathbf{d} \geq 0 \qquad j = 1, ..., m_a \tag{11}$$

$$z_i^{(k)} - z_i^{\min} \leq d_i \leq z_i^{\max} - z_i^{(k)} \qquad i = 1, ..., n \tag{12}$$

where only a subset of all m_i inequality constraints in (3) is included in the linearized constraints (11). $\mathbf{B}^{(k)}$ denotes a positive definite matrix which gradually is updated by use of a quasi-Newton scheme (refer to section 4) to approach the Hessian matrix of the Lagrange function $L(\cdot)$.

Introducing the column vector $\mathbf{h}(\mathbf{z})$ in which all equality and active inequality constraints are assembled, the solution of (9)-(12) is obtained from the linear system

$$\begin{bmatrix} \mathbf{B}^{(k)} & -\nabla_z \mathbf{h}(\mathbf{z}^{(k)}) \\ -\nabla_z \mathbf{h}(\mathbf{z}^{(k)})^T & \mathbf{0} \end{bmatrix} \begin{bmatrix} \mathbf{d} \\ \boldsymbol{\lambda}^{(k+1)} \end{bmatrix} = \begin{bmatrix} -\nabla_z f(\mathbf{z}^{(k)}) \\ \mathbf{h}(\mathbf{z}^{(k)}) \end{bmatrix} \tag{13}$$

Standard solution techniques are outlined Ringertz (1993) or Golub & Van Loan (1989), while a modification ensuring that the linearized constraints (10)-(11) on $\mathbf{d}$ are consistent can be found in Powell (1977).

3 INTERACTIVE PROCEDURES

In order to provide the interactive facilities of include/exclude constraints, change simple bounds and the change current value and fix/relax design variables, modifications of standard quasi-Newton algorithms are necessary. A qualitative discussion of various possibilities can be found in Pedersen & Thoft-Christensen (1993).

Considering active set changes with respect to the design variables (i.e. only a subset of all n design variables $\mathbf{z}$ is allowed to vary), a straightforward strategy is to adjust the dimension of the subproblem (9)-(12). Unfortunately, a frequent change of the dimension of the Hessian matrix $\mathbf{B}^{(k)}$ results in loss of already obtained information in cases where temporarily fixed design variables are re-included in the optimization problem. Furthermore, the choice of the additional rows and columns must fulfil certain requirements in order to preserve positive definiteness of $\mathbf{B}^{(k)}$.

An alternative approach is to include an additional equality constraint per fixed design variable in (9)-(12) whereby the full dimension of $\mathbf{B}^{(k)}$ and positive definiteness is preserved. Thus, from the modified solution scheme written below for the search direction subproblem in the case where design variables z_i is fixed, it is easily seen that d_i is equal to zero.

$$\begin{bmatrix} \mathbf{B}^{(k)} & -\nabla_z \mathbf{h}(\mathbf{z}^{(k)}) & \mathbf{e}_i \\ -\nabla_z \mathbf{h}(\mathbf{z}^{(k)})^T & 0 & 0 \\ \mathbf{e}_i{}^T & 0 & 0 \end{bmatrix} \begin{bmatrix} \mathbf{d} \\ \boldsymbol{\lambda}^{(k+1)} \end{bmatrix} = \begin{bmatrix} -\nabla_z f(\mathbf{z}^{(k)}) \\ \mathbf{h}(\mathbf{z}^{(k)}) \\ 0 \end{bmatrix} \tag{14}$$

where the column vector $\mathbf{e}_i = \{0, ...0, 1, 0, ...0\}^T$ contains zeroes except at the ith position. Due to the presence of the additional constraint, the corresponding active constraint set $\mathbf{h}(\mathbf{z}^{(k)})$ is different from the active set that corresponds to the standard problem where all design variables are allowed to vary. Finally, the last element in the vector $\boldsymbol{\lambda}^{(k+1)}$ is seen to correspond to the artificial constraint $d_i = 0$.

With respect to the interactive inclusion/exclusion of constraints, change of current design point and simple bounds, the major difficulties originates from the fact that the Hessian matrix $\mathbf{B}^{(k)}$ and the gradients may be inconsistent. Since the search direction $\mathbf{d}$ obtained from the subproblem (9)-(12) with an ill-conditioned Hessian matrix (ill-conditioned in terms of a large condition number - i.e. the ratio between the largest and smallest eigenvalues) is strongly dependent upon the actual values of the gradients, a minimization of $\text{cond}(\mathbf{B}^{(k+1)})$ is considered in the next section.

4 REVISION OF THE APPROXIMATE HESSIAN MATRIX

The revision of the approximate Hessian matrix $\mathbf{B}^{(k)}$ is required to satisfy the quasi-Newton condition - see e.g. Gill et al. (1981)

$$\mathbf{B}^{(k+1)}\mathbf{p}^{(k)} = \mathbf{q}^{(k)} \tag{15}$$

where $\mathbf{p}^{(k)}$ is the difference between the current and previous design point

$$\mathbf{p}^{(k)} = \mathbf{z}^{(k)} - \mathbf{z}^{(k-1)} = \alpha\mathbf{d}^{(k)} \tag{16}$$

while $\mathbf{q}^{(k)}$ is dependent upon the difference between the current and previous derivatives of the Lagrangian $L(\mathbf{z}, \boldsymbol{\lambda})$. Among the most frequently used definitions of $\mathbf{q}^{(k)}$ is

$$\mathbf{q}^{(k)} = \theta_k(\nabla_z L(\mathbf{z}^{(k)}, \boldsymbol{\lambda}^{(k)}) - \nabla_z L(\mathbf{z}^{(k)}, \boldsymbol{\lambda}^{(k)})) + (1-\theta_k)\mathbf{B}^{(k)}\mathbf{p}^{(k)} \tag{17}$$

proposed by Powell (1977) where only the latest set of Lagrange multipliers is used and the additional parameter θ_k is introduced in order to preserve positive definiteness. In Powell (1977), $\theta_k \in [0,1]$ is chosen in such manner that $\mathbf{q}^{(k)T}\mathbf{p}^{(k)} \geq 0.2\mathbf{p}^{(k)T}\mathbf{B}^{(k)}\mathbf{p}^{(k)}$ which ever since has been widely used in numerous algorithms - e.g. NLPQL by Schittkowski.

With the above definition of $\mathbf{p}^{(k)}$ and $\mathbf{q}^{(k)}$, $\mathbf{B}^{(k)}$ is usually revised using a symmetric rank-two update formula. In accordance with Gill et al. (1981) and Luenberger (1984) the one-parameter Broyden family update formula may be written as

$$\mathbf{B}^{(k+1)} = \mathbf{B}^{(k)} - \frac{\mathbf{B}^{(k)}\mathbf{p}^{(k)}\mathbf{p}^{(k)T}\mathbf{B}^{(k)}}{\mathbf{p}^{(k)T}\mathbf{B}^{(k)}\mathbf{p}^{(k)}} + \frac{\mathbf{q}^{(k)}\mathbf{q}^{(k)T}}{\mathbf{q}^{(k)T}\mathbf{p}^{(k)}} + \phi_k\left(\mathbf{p}^{(k)T}\mathbf{B}^{(k)}\mathbf{p}^{(k)}\right)\mathbf{w}\mathbf{w}^T \tag{18}$$

where $\mathbf{w}$ is defined as

$$\mathbf{w} = \frac{\mathbf{q}^{(k)}}{\mathbf{q}^{(k)T}\mathbf{p}^{(k)}} - \frac{\mathbf{B}^{(k)}\mathbf{p}^{(k)}}{\mathbf{p}^{(k)T}\mathbf{B}^{(k)}\mathbf{p}^{(k)}} \tag{19}$$

Choosing $\phi_k = 1$ the formula (18) is termed the DFP update while $\phi_k = 0$ is resulting in the efficient and even more commonly used BFGS update formula. Furthermore, since both the DFP and BFGS updates satisfy (15), it is easily seen by premultiplying (19) by $\mathbf{p}^{(k)}$ that any choice of ϕ_k also satisfies (15).

In an interactive environment constraints are included/excluded, the current design point changed manually, etc. Therefore, a well-conditioned Hessian matrix is desirable in order to obtain a stable solution in terms of a feasible descent direction $\mathbf{d}$ even though the gradients in (9)-(11) change rapidly due to the above-mentioned interactions.

Since any choice of ϕ_k will satisfy (15), a straightforward choice of $\mathbf{B}^{(k+1)}$ can be obtained from the following additional subproblem which must be solved in each iteration

$$\min_{\phi_k} \quad \text{cond}(\mathbf{B}^{(k+1)}) = f(\mathbf{B}^{(k)}, \mathbf{p}^{(k)}, \mathbf{q}^{(k)}, \phi_k) \tag{20}$$

$$\text{s.t.} \quad \lambda_1(\mathbf{B}^{(k+1)}) = \lambda_{\min} > 0 \tag{21}$$

Another update scheme can be obtained if the parameter ϕ_k in the Broyden formula is kept fixed, while the scalar θ_k in (17) is variable. In the limit $\theta_k = 0$, it is seen from (17)-(18) that the BFGS formula results in the trivial result $\mathbf{B}^{(k+1)} = \mathbf{B}^{(k)}$. Thus, the purpose of the following additional subproblem is to reduce the influence of the relaxation term $\mathbf{B}^{(k)}\mathbf{p}^{(k)}$ in (17) subject to the constraints that $\mathbf{B}^{(k+1)}$ is positive definite and the increase in the condition number is less than a specified factor, say const = max(n, 10), i.e.

$$\min_{\theta_k} \quad (1-\theta_k) \tag{22}$$

$$\text{s.t.} \quad \lambda_1(\mathbf{B}^{(k+1)}) = \lambda_{\min} > 0 \tag{23}$$

$$\text{const} - \text{cond}(\mathbf{B}^{(k+1)})/\text{cond}(\mathbf{B}^{(k)}) > 0 \tag{24}$$

$$0 \geq \theta_k \geq 1 \tag{25}$$

Considering highly non-linear problems, information from the very first iterations may differ significantly from the last iterations - especially after major interactions. Pursuing the idea of discarding obsolete information, $\mathbf{B}^{(k+1)}$ can be constructed from information from the last K iterations only. Utilizing this approach, the Hessian matrix is first initialized to $\mathbf{I}$ after which the quasi-Newton update formula is applied successively K times - see also Pedersen & Thoft-Christensen (1993).

At the present stage, the above-mentioned updating schemes have been tested in various combinations on a few examples only. Compared with the effective algorithm NLPQL, 2-5 extra iterations are generally required in non-interactive tests although quicker convergence has been obtained for some types of problems. Based on these preliminary results, the Hessian matrix update of the form (22)-(25) is chosen for the example in section 6.

5 IROS - INTERACTIVE RELIABILITY-BASED OPTIMIZATION SYSTEM

The interactive optimization is carried out through the a shell called IROS - Interactive Reliability-Based Optimization System. Based on the desirable capabilities of an interactive optimization system outlined in Arora & Tseng (1988), Arora (1989) and others, IROS is being developed for the purpose of testing various interactive facilities and algorithms. IROS is developed for PC/DOS and implemented in the programming language C/C++.

In brief, the core constituents in IROS can be listed as follows:

- *Graphics module*, i.e. a superior controlling module in which the user is given the opportunity to define various graphical displays, e.g. the iteration history of the object function, selected constraints, sensitivities, design variables, Lagrange multipliers, convergence parameters, stchastic variables, etc.
- *Modification module* in which the current value of design variables can be changed, design variables be fixed to given values, simple bounds altered, constraints included/excluded from the active set, convergence parameters or even the optimization algorithm itself can be refined or modified.
- *Communication or control module* in which the information to and from the exterior optimization module is controlled. After a full iteration graphical displays are updated and, in the case where no interaction is detected or no convergence achieved, the iterative process is continued automatically.
- *Exterior optimization module* in which the optimization algorithm is implemented, i.e. any suitable optimization algorithm can be used if a proper interface is provided.

6 EXAMPLE - PLANE PORTAL FRAME

In order to illustrate some of the interactive capabilities described in section 5, this section shows a simple reliability-based optimization of a simple plane frame with I-shaped cross-sections. The overall geometry, applied load and cross-section are shown in figure 1 while the design variables $\mathbf{z}$ and stochastic variables $\mathbf{x}$ are listed in table 1.

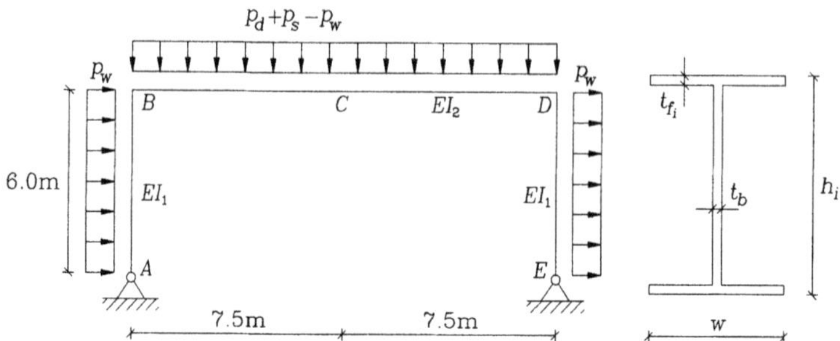

Figure 1: Geometry of frame, applied load and cross-section.

Design Variables					
Variable	Designation	Lower	Initial value		Upper
$z_1 = h_1$	Height of cross-section 1	200.0	250.0	[mm]	400.0
$z_2 = h_2$	Height of cross-section 2	200.0	300.0	[mm]	400.0
$z_3 = w$	Width of flange	75.0	100.0	[mm]	150.0
$z_4 = t_{f1}$	Thickness of flange 1	4.0	8.0	[mm]	20.0
$z_5 = t_{f2}$	Thickness of flange 2	4.0	8.0	[mm]	20.0
$z_6 = t_b$	Thickness of body	4.0	6.0	[mm]	15.0
Stochastic Variables					
Variable	Designation	Distrib.	Exp. value		Var. coeff.
$x_1 = f_y$	Yielding strength	LN	450.0	[N/mm^2]	0.08
$x_2 = E$	Modulus of elasticity	LN	$206 \cdot 10^3$	[N/mm^2]	0.04
$x_3 = p_d$	Dead load	N	3.0	[kN/m]	0.05
$x_4 = p_s$	Applied load, snow	N	4.8	[kN/m]	0.20
$x_5 = p_w$	Applied load, wind	N	3.0	[kN/m]	0.40

Table 1: Design variables $\mathbf{z}$ and stochastic variables $\mathbf{x}$.

In the optimization problem of the standardized form (1)-(4), the volume of the frame is used as the object function $W(\mathbf{z})$ while the reliability-based constraints are related to the bending capacity at 3 critical points, i.e.

$$\hat{c}_j(\mathbf{z}) = \beta_j(\mathbf{z}) - \beta_j^{\min} \geq 0 \quad , \quad j = 1,3 \tag{26}$$

The corresponding jth failure functions can be written as $g_j(\mathbf{x}) = M_j^u(\mathbf{x}) - M_j(\mathbf{x})$, where $M_j^u(\mathbf{x})$ and $M_j(\mathbf{x})$ denote the bending capacity and the moment due to the total applied load, respectively. Due to the direction of the wind, the inequality $|M_B| \geq |M_D|$ always holds. Therefore, the 3 failure functions consist of 2 failure functions at point B (column and girder, respectively) and 1 failure function at point C at the centre of the girder.

In addition, 4 simple deterministic, geometric constraints $\hat{c}_4(\mathbf{z})$ - $\hat{c}_7(\mathbf{z})$ with respect to local instability are included: 2 constraints requiring that the ratio between the height of the cross-section h_j and body thickness t_b does not exceed 50 and 2 constraints requiring that the ratio between the width w and the flange thickness t_{fj} does not exceed 20.

In order to demonstrate some of the interactive capabilities of IROS, the above defined optimization problem is solved according to the following scheme:

1. Optimization where all 6 design variables are allowed to vary and all 7 constraints (3 reliability-based and 4 geometric constraints) are included.
2. Based on the above continuous solution, the 3 design variables that describe thicknesses (t_{f1}, t_{f2} and t_b) are fixed to adjacent discrete values. From this point the optimization is continued with the remaining 3 free design variables.
3. Finally, the 4 geometric constraints $\hat{c}_4(\mathbf{z})$ - $\hat{c}_7(\mathbf{z})$ are excluded and all 6 design variables allowed to vary. A continuous solution prone to local instability is hereby obtained.

The iteration history of design variables, object function and reliability-based constraints are shown at the screen-dump from IROS in figure 2. In addition, the values of the design variables and object function at the start and the end for the 3 optimizations are listed in table 2.

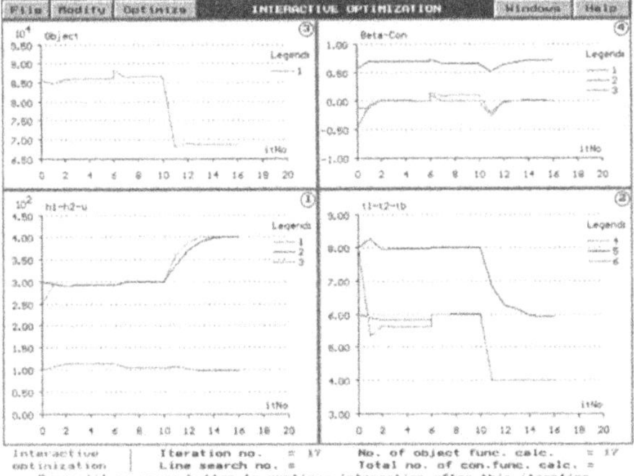

Figure 2: Iteration history of design variables, object function and rel.-based constraints.

From figure 2 and table 2 it is seen, that the continuous solution from the 1st run is obtained after 6 iterations. At this point, the 3 design variables describing thicknesses $\{t_{f1}, t_{f2}, t_b\} = \{5.63, 7.97, 5.84\}$ are changed and fixed interactively to the adjacent standard dimensions $\{6.00, 8.00, 6.00\}$.

After this interactive change the optimization is continued to iteration 10 (2nd run) as shown in figure 2. At this point, a new "discrete" optimum is achieved. A feasible solution with standard dimensions suitable for practical used is hereby obtained with an increase in the object function by less than 1%.

Optimization (iteration no.)		1st run (1-6)		2nd run (7-10)		3rd run (11-17)	
Variable	Designation	Start	End	Start	End	Start	End
$z_1 = h_1$	Height of cross-sec. 1	250.0	291.8	291.8	300.0	300.0	400.0
$z_2 = h_1$	Height of cross-sec. 2	300.0	291.8	291.8	300.0	300.0	400.0
$z_3 = w$	Width of flange	100.0	112.6	112.6	104.6	104.6	96.7
$z_4 = t_{f1}$	Thickness of flange 1	8.00	5.63	6.00	6.00	6.00	4.00
$z_5 = t_{f2}$	Thickness of flange 2	8.00	7.97	8.00	8.00	8.00	5.91
$z_6 = t_b$	Thickness of body	6.00	5.84	6.00	6.00	6.00	4.00
Object function		-	85.9	-	86.5	-	68.5

Table 2: Design variables $\mathbf{z}$ during the 3 stages of the optimization.

Finally, all design variables are relaxed and allowed to vary in the 3rd optimization, where only the 3 reliability-based constraints are included. Thus, based on the information obtained in the first 10 iterations, the optimization is now continued for another 7 iterations as shown in figure 2. A dramatically change in the design and decrease of the object function by approximately 20% is hereby obtained even though 4 of the 6 design variables have reached the simple bounds.

7 ACKNOWLEDGEMENT

This paper is supported by the research project "Marine Structural Design" sponsored by the Danish Technical Research Council.

REFERENCES

[1] Arora, J.S.: *Introduction to Optimum Design*, McGraw-Hill, 1989.

[2] Arora, J.S. & C.H. Tseng: "Interactive Design Optimization", *Engineering Optimization*, Vol. 13, 1988, pp. 173-188.

[3] Gill, P.E., W. Murray & M.H. Wright: *Practical Optimization*, Academic Press, 1981.

[4] Golub, G.H. & C.F. Van Loan: *Matrix Computations*, 2nd edition, Johns Hopkins University Press, 1989.

[5] Luenberger, D.G.: *Linear and Nonlinear Programming*, 2nd edition, Addison-Wesley, Reading, Massachusetts, 1984.

[6] Madsen, H.O., S. Krenk & N.C. Lind (1986): *Methods of Structural Safety*, Prentice-Hall, Englewood Cliffs, New Jersey, 1986.

[7] Pedersen, C. & P. Thoft-Christensen: "Interactive Quasi-Newton Optimization Algorithms", *Structural Reliability Theory*, Paper No. 120, ISSN 0902-7513 R9346, University of Aalborg, Denmark, 1993.

[8] Powell, M.J.D.: "A Fast Algorithm for Nonlinear Constrained Optimization Calculations", *Lecture Notes in Mathematics*, No. 630, Numerical Analysis, Proceedings, Biennial Conference, Dundee 1977, pp. 144-157, Springer-Verlag, 1977.

[9] Ringertz, U.T.: *Numerical Optimization: With Applications in Structural Analysis and Design*, Lecture Notes, Vol. 1, Comett Course: Computer Aided Optimum Design of Strucures, Institute of Mechanical Engineering, Aalborg University, Aug. 1993.

24

Reliability of Laminated Structures by an Energy Based Failure Criterion

S. Plica, R. Rackwitz, Technische Universität München, 80290 München, Germany

Abstract

Studies about the reliability of fibre reinforced laminated structures have mostly concentrated on single layers with simple failure criteria such as the well known by Tsai/Wu. In laminated structures failure frequently is defined as interfibre failure. A realistic criterion taking account of the multi-axial stress states has recently been proposed by Hashin and Puck. This criterion is briefly discussed and related to energy considerations. The criterion then is used in the analysis of a realistic tube structure whose laminate stresses are determined by classical laminate theory. Reliabilities are determined by FORM methods. Correlations among the various failure modes and between the elements of the FEM-model for the tube are taken into account. The size effect is studied with particular reference to the meshing.

1 Introduction

The properties of fibre reinforced plastics (FRP), i. e. high strength combined with low weight, together with the possibility to design materials with different elasticity and strength properties in different directions have led to an increased use, especially in the air- and spacecraft industry. Unfortunately a special problem when designing a laminate is that the fracture behaviour of laminates is not yet fully understood. Theoretical predictions and experimental results still can differ substantially, although much research work has been done on this topic during the last decades.

While the inhomogenity of laminates can be circumvented in a fracture analysis by simply calculating the stresses in the individual layers, the strength of a lamina poses more complex problems. Due to their orthotropy the unidirectional layers which are used to build up a laminate show extremly different strengths in fibre direction and in transverse direction. Two major failure modes can be identified, the failure of the fibres (FF) under tensile or compressive load and the failure of the matrix material between the fibres caused by transverse stresses or shear stresses. Although the failure of the matrix material shows several possible modes like tensile, compressive, longitudinal or transverse shear failure, all these are regarded as one mode, the interfibre failure (IFF).

In case of failure a fracture criterion should indicate whether FF or IFF has taken place. For practical purposes this information is important as a structure may be designed for fibreparallel strength, whence IFF does not affect its usability. However, for some structures IFF does not only cause unserviceability but also leads to large scale disintegration of the structure. A fracture criterion proposed by Hashin [3] and recently further developped to a practicable state by Puck [5] distinguishes between FF and IFF and therefore will be used in this study.

Another severe problem is the variability of loads and strength properties taking into account that such materials are used in highly optimized structures. Due to this optimisation and the fact, that many of the FRP used in practical applications show brittle fracture, laminated structures virtually are nonredundant, i. e. structural failure occurs as soon as local failure occurs.

As the stress distribution over the entire structure in general is not constant, the reliability analysis must be performed in multiple points of the structure. The stresses in laminated structures usually are calculated by Finite-Element-Analysis (FEA), which leads already to a discretization of the structure — the modelling of the structure by a mesh of finite elements. If it is then assumed that every finite element can be regarded as one failure element with a more or less constant stress state, the whole structure can be modelled as a series system of failure elements for the reliability analysis. However the entire structure's probability of failure must be independent from the number of elements in the FEA-mesh.

In the following a method will be presented that allows to calculate the probability of failure for structures showing series-system behaviour. The method will be applied to an example to test its efficiency.

2 A multi-modal fracture criterion for unidirectionally fibre-reinforced plastics

Unidirectionally fibre-reinforced plastics show two failure modes, fibre failure and matrix failure. Hashin proposed to use two independent criteria for FF and IFF [3]. He assumed that in the case of FF the fracture always occurs on a plane normal to the fibres. In case of IFF he postulated that a crack occurs parallel to the plane of maximum stresses in the matrix material and normal to the plane of layer isotropy (see figure 1). The stress σ_{nn} acting normal to this plane provoques a mode-I crack of the matrix material. The longitudinal shear stress τ_{n1} and the transverse shear stress τ_{nt} acting on this plane provoque mixed mode-II-mode-III failure.

He then proposed a simple fracture criterion for IFF where the stresses σ_{nn}, τ_{n1} and τ_{nt} depend on the angle θ of the possible crackplane

$$f\left(\sigma_{nn},\tau_{n1},\tau_{nt}\right)=\left(\frac{\sigma_{nn}}{R_{nn}}\right)^2+\left(\frac{\tau_{n1}}{R_{n1}}\right)^2+\left(\frac{\tau_{nt}}{R_{nt}}\right)^2\leq 1 \tag{1}$$

$$\begin{aligned}
{\sigma_{nn}}^2 &= \left[\frac{1}{2}\left(\sigma_{22}+\sigma_{33}\right)+\frac{1}{2}\left(\sigma_{22}-\sigma_{33}\right)\cos 2\theta+\tau_{23}\cdot\sin 2\theta\right]^2\\
{\tau_{nt}}^2 &= \left[-\frac{1}{2}\left(\sigma_{22}-\sigma_{33}\right)\sin 2\theta+\tau_{23}\cdot\cos 2\theta\right]^2\\
{\tau_{n1}}^2 &= \frac{1}{2}\left({\tau_{12}}^2+{\tau_{13}}^2\right)+\frac{1}{2}\left({\tau_{12}}^2-{\tau_{13}}^2\right)\cos 2\theta+\tau_{12}\cdot\tau_{13}\cdot\sin 2\theta
\end{aligned}$$

The angle θ must be found by maximizing the left side of (1). Hashin considered the maximization of (1) as too expensive. Furthermore he did not see how to determine the compressive strength R^c_{nn} for a mode I crack, which must be used in the first term of (1) for R_{nn} in case of $\sigma_{nn} \leq 0$.

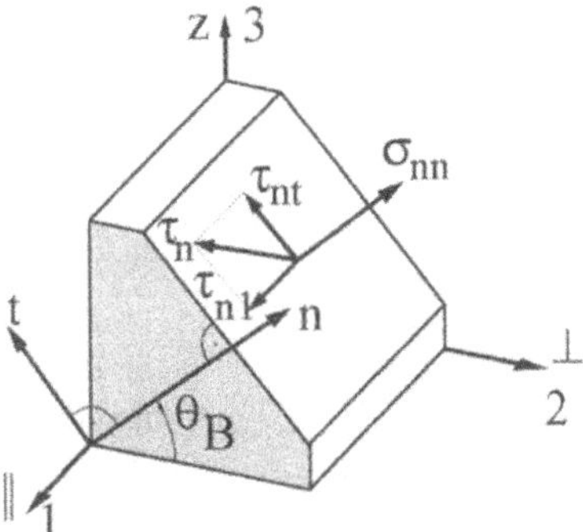

Figure 1: Definition of the coordinate systems for a lamina. The system 1-2-3 points with axis 1 (or ||) in fibre direction, with axis 2 (or ⊥) in transverse direction and with axis 3 (or z) through the thickness of the laminate.

Later Hahn, Tsai and Erikson showed, that the direction of a matrix crack in a unidirectional composite under combined loading can be determined by applying the energy release rate approach [2]. Assuming that the stresses σ_{nn}, τ_{n1} and τ_{nt} cause mode-I, mode-II and mode-III failure, respectively, they generated for the σ_{22}-τ_{12}-interaction a failure surface showing exactly the same shape and characteristics as Hashin's criterion (1), [5]. They also showed that for $\sigma_{nn} \leq 0$ the first term in (1) has to disappear as a mode-I crack does not open due to compression normal to the crack plane. Furthermore under pure compression σ_{22} failure takes place as a mode-III crack under approximately $\pm 45°$.

Finally Puck proposed some rules to determine the strengths required for (1) from standard tests [5]:

$$R^t_{nn} = R^t_{22} \quad R^c_{nn} \to \infty \quad R_{n1} = R_{12} \quad R_{nt} = \frac{R^c_{22}}{2} \tag{2}$$

The index i of a resistance quantity R_{ij} specifies the plane parallel to the crack, the index j is the direction of the stress acting on this plane (see figure 1). A superscript t is used for tensile strength, the superscript c for compressive strength, no superscript for shear strengths. The failure surface for combined loading σ_{22} - τ_{12} is shown in figure 3 as curve A with the associated angles of the crackplane θ.

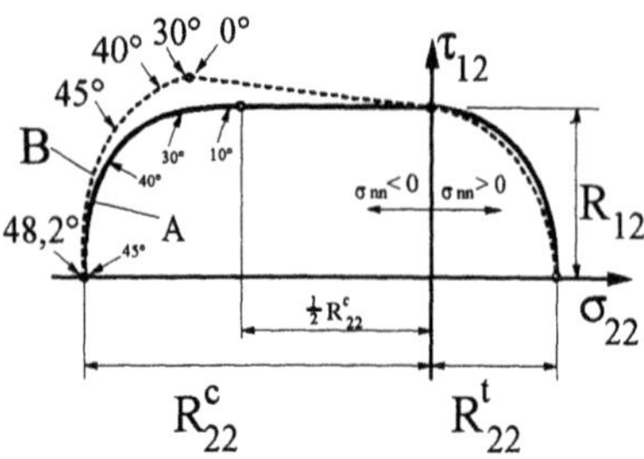

Figure 2: Fracture surfaces given by the Hashin-Puck criterion for the criterion (1) (curve A) and the improvement supposed by Puck (curve B).

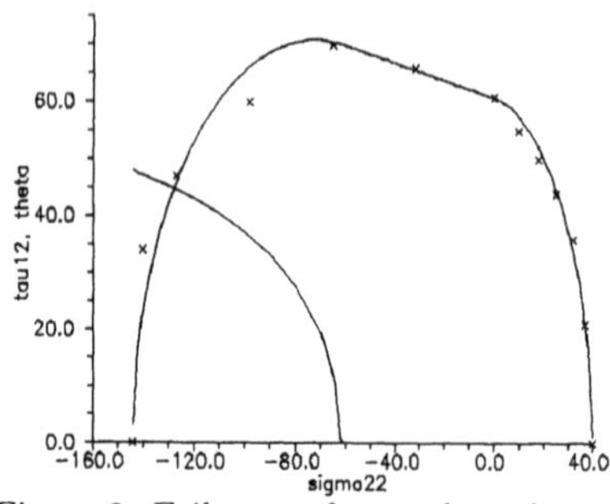

Figure 3: Failure surface and crack angle for criterion (3). Parameters are determined by fitting the curve to experimental data (crosses) with $\mu_{n1} = \mu_{nt} = 0.15$ for a typical GFRP [5].

Puck suggested the introduction of interaction terms into (1), leading to a fracture surface shown as curve B in figure 2 [5]. As compression on the crack surface certainly causes friction and therefore increases the shear strength, a possible modification of (1) is a criterion of the form

$$\left(\frac{\max\{\sigma_{nn},0\}}{R^t_{nn}}\right)^2 + \left(\frac{\tau_{n1}}{R_{n1} - \mu_{n1}\cdot\sigma_{nn}}\right)^2 + \left(\frac{\tau_{nt}}{R_{nt} - \mu_{nt}\cdot\sigma_{nn}}\right)^2 \leq 1 \tag{3}$$

In case of $\sigma_{nn} \leq 0$ the parameters μ_{n1} and μ_{nt} can be considered as coefficients of friction. Thus the parameter μ_{nt} can be determined from the crack angle Ψ of a test specimen under transverse compression in analogy to the fracture hypothesis for soils by Mohr-Coulomb.

$$\mu_{nt} = 2\Psi - \frac{\pi}{2} \qquad R_{nt} = \frac{R^c_{22}}{2\tan\Psi} \tag{4}$$

The parameter μ_{n1} must be determined by fitting (3) to results from tests with combined loading. The fracture surface for a typical glas fibre reinforced plastic (GFRP) under combined loading is shown in figure 3. In contrast to curve B in figure 2 the criterion (3) does not show a somewhat unreasonable kink where θ jumpes from 0° to 30°. In this analysis the criterion (3) will be used for IFF. For FF the criterion proposed by Puck will be applied. A possible influence of shear stresses on FF is ignored.

$$\left(\frac{\sigma_{11}}{R^t_{11}}\right)^2 \leq 1 \quad \text{for } \sigma_{11} \geq 0 \text{ and} \qquad \left(\frac{\sigma_{11}}{R^c_{11}}\right)^2 \leq 1 \quad \text{for } \sigma_{11} \leq 0 \tag{5}$$

3 The size effect in series systems with variable number of components

If the resistance of a brittle material is determined using a test specimen of volume V_0, which is much larger than the size of a single defect, it can be expected that the measured resistance includes the possibility of failure for all defects (with random size and orientation) in the test specimen. Also the local fluctuations of the stress state are incorporated in the specimens strength. Selecting the Weibull distribution as the natural stochastic model for the resistance of the material, the probability of failure of a structural element with volume V_1 under constant stress state σ_0 can be calculated by

$$P_f(\sigma_0) = 1 - exp\left[-\frac{V_1}{V_0}\left(\frac{\sigma_0 - \tau}{w - \tau}\right)^k\right] \quad (6)$$

The stress distribution of complicated structures can be calculated with finite-element-analysis. As one basic assumption of this method is that a structure can be divided into numerous small elements for which a constant stress state can be assumed, formula (6) can be used as a model for the strength of one element. If it is assumed that, firstly, every element of the FEA-model of a structure is a failure element, secondly, the resistances of the individual elements are mutually independent, and, thirdly, the complete structure fails if one element fails, the structure's probability of failure can be calculated using the bounds by Ditlevsen [1] for series systems (F_i are the failure events):

$$P_f = P\left(\bigcup_{i=1}^{n} F_i\right) = \begin{cases} \leq P(F_1) + \sum_{i=2}^{n}\left\{P(F_i) - \max_{j<i}\{P(F_i \cap F_j)\}\right\} \\ \geq P(F_1) + \sum_{i=2}^{n}\max\left\{0, P(F_i) - \sum_{j<i}^{n} P(F_i \cap F_j)\right\} \end{cases} \quad (7)$$

For the reliability analysis of each element FORM methods [4] are used, giving $P_{f,i} \approx \Phi(-\beta_i)$ where $\Phi(x)$ is the standard normal distribution and β the safety index. The safety index is related to the β-point $\underline{u}^*$ by $\underline{u}^* = -\beta \cdot \underline{\alpha}$. As the problem is solved in standard normal space, the correlation coefficient ρ_{ij} of two elements in the structure can be calculated by the scalar product $\rho_{ij} = \underline{\alpha}_i{}^T \cdot \underline{\alpha}_j$. Thus the intersections in (7) take the correlation of the stress states in the individual elements into account.

4 Implementation of the fracture criterion and the size effect into an existing software

The fracture criterion (3) and the size-effect (6) were implemented into an existing software package for componential reliability analysis, COMREL [6]. In order to get the necessary information about the stress state in laminated structures, a three-dimensional anisotropic shell-plate-element — based on classical laminate theory and the Kirchhoff plate theory — was added to the FEA-code NASCOM [7], which serves as preprocessor to the structural reliability software NASREL [8]. The finite elements so far have deterministic elasticity properties. Thus only the variability of the loads and of the local resistances are taken into account. The data flow is shown in figure 4.

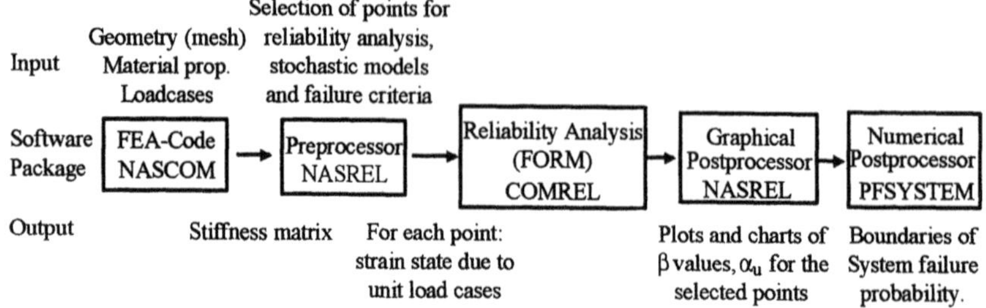

Figure 4: Integration of software packages and data flow for the calculation of the probability of failure of entire structures.

After the reliability analysis for each element of the structure plots of the distribution of β-values and α_u-values allow a visual control of the results. Finally the probability of failure for the entire structure is evaluated using (7).

5 Analysis of a strut under random loading

The method as proposed above is illustrated at a strut. The geometry and the loads are shown in figure 5, the stochastic model and the material properties of the carbon fibre reinforced plastic (CFRP) are given in figure 6. The stacking sequence of the laminate is $(90°/0°/0°/90°)$ with 0° pointing in the longitudinal direction of the strut, the thickness of the layers is $t_i = (0.15/1.35/1.35/0.15)$ (all in [mm]).

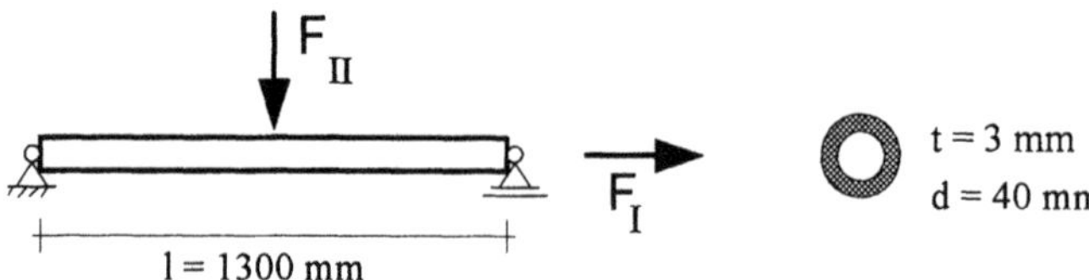

Figure 5: Example: a strut under longitudinal tension F_I, a bending force F_{II} in the middle and constant temperature $\Delta T = -100°C$.

One half of the strut is divided into 16 elements in radial direction and into 13 elements in longitudinal direction (208 in total). As a second case, the strut is divided into 26 elements in longitudinal direction (416 in total). To demonstrate the effect of the Weibull-size-effect the structure is analyzed with the two different meshes. The reliability analysis has to be performed at each boundary of a layer for every element and both FF and IFF must be considered. As an example the influence of the Weibull size effect on the probability of failure is demonstrated for one layer boundary. The results for the outer boundary of the 0°-layer are presented in figure 7. The first column shows the number of elements of the FEA-model for the strut, the second indicates whether the Weibull-scaling

Variable	Mean value	C.o.V.
E_{11}	170000	-
E_{22}	9000	-
ν_{12}	0.28	-
$\alpha_{T,1}$	$-3.0 \cdot 10^{-6}$	-
$\alpha_{T,2}$	$4.0 \cdot 10^{-5}$	-
R_{11}^t	2900	0.04
R_{11}^c	1620	0.08
R_{22}^t	60	0.10
R_{22}^c	165	0.09
R_{12}	86	0.04

Variable	Mean value	C.o.V.	model
Tension	209000	0.1	Lognormal
Bending	3600	0.1	Lognormal
Temperature	-100	0.05	Lognormal
R_{11}^t	2900	0.04	Weibull
R_{11}^c	1620	0.08	Weibull
R_{nn}^t	60	0.10	Weibull
R_{n1}	186	0.04	Weibull
R_{nt}	71	0.1	Weibull
μ_{n1}	0.15	0.1	Lognormal
μ_{nt}	0.15	0.1	Lognormal

Figure 6: Mechanical properties of CFRP used for the example (left table) and set of basic variables for the reliability analysis of the strut (right table).

according to (6) has been taken into account. The third column contains the number of failure components for the series system, the fourth shows the bounds of β, and the last column gives the upper bound of system-P_f. It should be mentioned that the intersection probabilities in (7) of at most the first 800 larger probability events had to be computed in order to obtain the rather narrow reliability bounds.

Meshsize (FEA)	Scaling	Failure components	range of β	P_f
208	yes	3328	$2.81 \leq \beta \leq 2.81$	≤ 0.0025
416	yes	6656	$2.81 \leq \beta \leq 2.81$	≤ 0.0025
208	no	3328	$-0.22 \leq \beta \leq 0.52$	≤ 0.5854
416	no	6656	$-36.62 \leq \beta \leq 0.40$	≤ 1.0000

Figure 7: Probability of failure for one layer boundary, including both FF and IFF.

It can be seen, that ignoring the scaling of resistances produces severe errors. For completeness the probability of failure has been evaluated for the entire strut. For 208 (416) elements having 4 β-points each (due to the geometrical model of the shell-plate-element), 8 layer boundaries and two criteria the number of failure components in the series system is 19968 (39936). The results are given in figure 8.

6 Conclusions

It has been shown that modern FORM methods can be coupled with finite element codes for the analysis of laminated structures involving a large number of failure modes. The strength criteria proposed by Hashin/Puck for both fibre and interfibre failure have been used. It can be demonstrated that the use of some other classical criterion can lead to

Meshsize (FEA)	208	416
Failure components	19998	39936
Componential β	$4.34 \leq \beta \leq 10.96$	$4.40 \leq \beta \leq 11.13$
System bounds (all comp.)	$2.33 \leq \beta \leq 2.34$	$2.34 \leq \beta \leq 2.34$
System prob. of failure	$0.0096 \leq P_f \leq 0.0098$	$0.0096 \leq P_f \leq 0.0096$

Figure 8: Results for the entire strut, including both FF and IFF.

inappropriately designed structures. The outspoken size effect must be taken into account correctly both within the finite element and across the elements. The systems reliability can be narrowly bounded even for a large number of failure modes.

References

1. O. Ditlevsen, "Narrow Reliability Bounds for Structural Systems", *Journal of Structural Mechanics*, Vol. 7, 435 - 451, (1979)

2. H. T. Hahn, J. B. Erikson and S. W. Tsai, "Characterization of Matrix/Interface-Controlled Strength of Unidirectional Composites", *Fracture of Composite Materials*, Martius Nijhoff Publ., Den Haag, 197 - 214, (1982)

3. Z. Hashin, "Failure Criteria for Unidirectional Fibre Composites", *Journal of Applied Mechanics*, Vol. 47, 329 - 334, (1980)

4. M. Hohenbichler, S. Gollwitzer, W. Kruse and R. Rackwitz, "New Light on First- and Second-Order Reliability Methods", *Structural Safety*, Vol. 4, Elsevier Science Publishers B. V., 267 -284, (1987)

5. A. Puck, "Ein Bruchkriterium gibt die Richtung an", *Kunststoffe*, 82. Jahrgang, 607 - 610, (1992)

6. COMREL, Software for componential reliability analysis, RCP GmbH, München, Germany,(1994)

7. NASCOM, Software for finite element analysis, RCP GmbH, München, Germany, (1994)

8. NASREL, Software for reliability analysis of structures modelled by finite elements, RCP GmbH, München, Germany, (1994)

25

Optimization of absorbers in the highway bridges due to traffic flow

P. Śniady, R. Sieniawska, S. Żukowski

Institute of Civil Engineering, Technical University of Wrocław
Wybrzeże Wyspiańskiego 27, 50-370 Wrocław, Poland

1. INTRODUCTION

The problem of reducing the level of vibrations in various constructions is considered for many years and numerous ways and means of preventing unacceptable vibrations are known [1]. One of this ways are various vibration absorbers. Their application play a special role because they can be used during the construction design and also when unsatisfactory dynamic properties appear in a construction during operation. Absorbers are applied in various engineering structures such as shipbuildings, steel chimneys, TV towers, bridges [2, 3], etc. Vibrations of some of this structures are excited by load of random nature. In this cases damped vibration absorbers (DVA) are used to mitigate the random excitation [3] and their parametres are optimized on the condition of the main mass's. The minimum variance of the displacement corresponds to the maximum reliability of the system. The traffic flow is such type of load which causes complicated stress and displacement states in bridges and also causes the material fatigue that ultimately can damage the structure. The application of absorbers can greatly extend its lifetime.

In the paper the problem of optimal choosing of parameters for damping absorbers reducing the random vibration of highway bridges subjected to traffic flow is studied. It is assumed that the traffic is a composite of different types of vehicles Each type of vehicle is modelled by one or more concentrated random forces [4,5,6]. In this idealization the vehicles are regarded as being of random weight. The interrarival times of the vehicles are regarded as random variables. As the optimization criterion of absorbers the minization of the first two probabilistic characteristics (cumulants) of bridge displacements which correspond to the maximum of the bridge's reliability is assumed. The connected coefficient of skewness and kurtosis of displacements are also calculated. To solve the optimization problem the simulation approach is used.

2. FORMULATION OF THE PROBLEM

Let us consider a beam with an absorber fitted at point x_0 under a stream of concentrated random forces moving in the same direction along the beam with constant speed v. The forces, which model the moving vehicles in a traffic flow, arrive at the beam at random times t_k. This arrival times constitute a Poisson stochastic process $N(t)$ with parameter λ. The absorber is modelled by a linear oscillator.

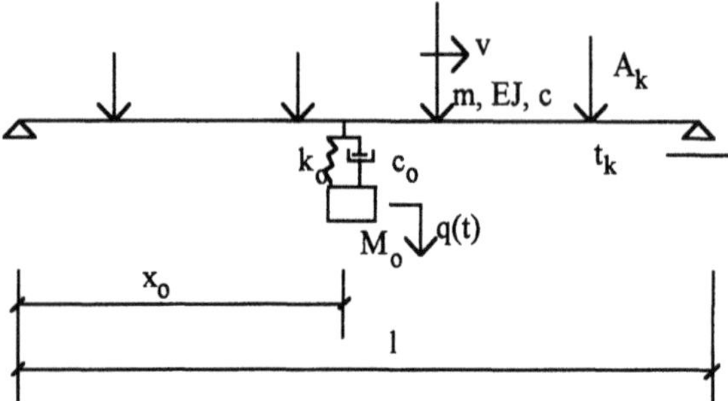

Figure 1. The model of a brigde with an absorber.

The vibrations of the continuous-discrete system consisting of the beam and linear oscillator due to the stream of forces are described by a set of coupled equations

$$EIw^{IV}(x,t)+c\dot{w}(x,t)+m\ddot{w}(x,t)+M_o\ddot{q}(t)\delta(x-x_o)=\sum_{k=1}^{N(t)}A_k\delta(x-v(t-t_k)) \tag{1}$$

and the vibrations of absorber (oscillator) are

$$M_o\ddot{q}(t)+c_o[\dot{q}(t)-\dot{w}(x_o,t)]+k_o[q(t)-w(x_o,t)]=0, \tag{2}$$

where

EI denotes the flexural rigidity of the beam,
c is beam's dampimg coefficient,
m is the mass per unit length,
δ denotes the Dirac delta function,
M_o, c_o, k_o are the mass, the damping coefficient and the spring's stiffness of the model of absorber, respectively,
A_k denotes the random weight of vehicles.

It is assumed, that the random variables A_k are mutually independent and independent of the arrival times t_k and that their probabilistic moments $E[A_k^n]$ are known.
The vertical deflection of the beam can be presented in the modal form as

$$w(x,t)=\sum_{n=1}^{\infty}y_n(t)W_n(x), \tag{3}$$

where

$W_n(x)$ are the normal modes and
y_n are the generalized coordinates.

By such formulation

$$W_n^{IV}(x) - \lambda_n^{IV} W_n(x) = 0 \tag{4}$$

one becomes a set of ordinary differential equations in the domain of generalized coordinates

$$\ddot{y}_n(t) + 2\alpha_n \dot{y}_n(t) + \omega_n^2 y_n(t) + M\ddot{q}(t) W_n(x_o) = f_n(t) \tag{5}$$

$$\ddot{q}(t) + 2\alpha_o \dot{q}(t) + \omega_o^2 q(t) - 2\alpha_o \sum_{n=1}^{\infty} \dot{y}_n(t) W_n(x_o) - \omega_o^2 \sum_{n=1}^{\infty} y_n(t) W_n(x_o) = 0 \tag{6}$$

where

$$2\alpha_n = \frac{c}{m}, \qquad \omega_n^2 = \lambda_n^4 \frac{EJ}{m}, \qquad M = \frac{M_o}{m\gamma_n^2}, \quad \gamma_n^2 = \int_o^l W_n^2(x)dx,$$

$$f_n(t) = \frac{1}{m\gamma_n^2} \sum_{k=1}^{N(t)} A_k W_n(v(t - t_k)), \qquad 2\alpha_o = \frac{c_o}{M_o}, \qquad \omega_o^2 = \frac{k_o}{M_o}$$

The set of equations (5) and (6) can be presented in the matrix form as follows:

$$\boldsymbol{B}\ddot{\boldsymbol{z}} + \boldsymbol{C}\dot{\boldsymbol{z}} + \boldsymbol{K}\boldsymbol{z} = \boldsymbol{F} \tag{7}$$

where

$$\ddot{\boldsymbol{z}} = \begin{bmatrix} \ddot{q} \\ \ddot{y}_1 \\ \ddot{y}_2 \\ \vdots \\ \ddot{y}_n \end{bmatrix}, \quad \dot{\boldsymbol{z}} = \begin{bmatrix} \dot{q} \\ \dot{y}_1 \\ \dot{y}_2 \\ \vdots \\ \dot{y}_n \end{bmatrix}, \quad \ddot{\boldsymbol{z}} = \begin{bmatrix} q \\ y_1 \\ y_2 \\ \vdots \\ y_n \end{bmatrix}, \quad \boldsymbol{F} = \begin{bmatrix} 0 \\ f_1(t) \\ f_2(t) \\ \vdots \\ f_n(t)) \end{bmatrix},$$

$$\boldsymbol{B} = \begin{bmatrix} 1 & 0 & 0 & \cdots & 0 \\ MW_1(x_o) & 1 & 0 & \cdots & 0 \\ MW_2(x_o) & 0 & 1 & \cdots & 0 \\ \vdots & \vdots & \vdots & \vdots & \vdots \\ MW_n(x_o) & 0 & 0 & \cdots & 1 \end{bmatrix},$$

$$\boldsymbol{C} = \begin{bmatrix} 2\alpha_o & -2\alpha_o W_1(x_o) & -2\alpha_o W_2(x_o) & \cdots & -2\alpha_o W_n(x_o) \\ 0 & 2\alpha & 0 & \cdots & 0 \\ 0 & 0 & 2\alpha & \cdots & 0 \\ \vdots & \vdots & \vdots & \vdots & \vdots \\ 0 & 0 & 0 & \cdots & 2\alpha \end{bmatrix},$$

$$\boldsymbol{K} = \begin{bmatrix} \omega_o^2 & -\omega_o^2 W_1(x_o) & -\omega_o^2 W_2(x_o) & \cdots & -\omega_o^2 W_n(x_o) \\ 0 & \omega_1^2 & 0 & \cdots & 0 \\ 0 & 0 & \omega_2^2 & \cdots & 0 \\ \vdots & \vdots & \vdots & \vdots & \vdots \\ 0 & 0 & 0 & \cdots & \omega_n^2 \end{bmatrix}.$$

The solutions of the above equations are stochastic processes and can be calculated by similar approach as presented in [4, 5]. The probabilistic characteristics of the bridge's response are

$$\kappa_w^r(x,t) = E[A^r]\lambda \int_0^t H^r(x,t-\tau)d\tau \tag{8}$$

where the function $H(x,t-\tau)$ is the dynamical influence function which describes the response of the structure under simple load impulse with quantity $A_k = 1$ which is moving along the beam with constant speed v. This function is calculated from the equation (7) in which the right side has the form

$$\boldsymbol{F} = \begin{bmatrix} 0 \\ \frac{1}{m\gamma_1^2} W_1(v(t-\tau)) \\ \frac{1}{m\gamma_2^2} W_2(v(t-\tau)) \\ \vdots \\ \frac{1}{m\gamma_n^2} W_1(v(t-\tau)) \end{bmatrix}.$$

The impulse influence function has two different shapes: the first one describes the structure vibrations when the force is on its ($t-\tau < l/v$) and the second one describes the free vibrations ($t-\tau > l/v$).

Now we formulate the problem of optimal design of absorber.

Find: $\kappa_w^r(x,t) = E[A^r]\lambda \int_0^t H^r(x,t-\tau)d\tau$

which satisfies following conditions

$$0.01 \le \frac{M_o}{ml} \le 0.04, \qquad c_o > 0 \text{ kN}\cdot\text{m/s}, \qquad k_o > 0.$$

3. NUMERICAL RESULTS

The problem of reducing random vibrations in bridge under traffic flow by applying a damping absorber with optimal parameters described above will be illustrated by an example. The bridge is modelled by a simply supported beam with a damping absorber located in the midspan. It has been assumed that the parameters of the beam are as follows: span l =30 m, mass per unit length m = 3500 kg/m, the first natural frequency $\omega_1 = 4s^{-1}$, the damping ratio $\alpha_n = \alpha = 0.04s^{-1}$. The probabilistic characteristics of the vehicle weight are as described in [6] i.e. $E[A]$=31 kN, the variance σ_A^2=320 kN*kN, the minimum weight is 8 kN and the maximum weight is 80 kN. The constant speed of the vehicles varies from $v = 10\,\mathrm{m/s}$ to $35\,\mathrm{m/s}$. The steady state response of the bridge is considered. In order to choose the optimum absorber's parameters the expected value of the bridge's midspan deflection, its standard deviation, coefficient of skewness and kurtosis are calculated for different absorber's parameters. The values of this variables as functions of the damping coefficient are shown in Figures 2, 3, 4, 5, and as functions of spring's stiffness in the Figures 6, 7, 8, 9, respectively. The calculations have been made for M_o = 2 t = 2000 kg, v = 25 m/s, λ = 1. In all Figures the curve No 1 is the curve plotted with continuous line, the curve No 2 is plotted with dashed line, the curve No 3 is plotted with dotted line and the curve No 4 is plotted with dashed-dotted line. In the first four Figures four curves are presented. The curve No 1 has been obtained for the spring's mass k_o=10 kN/m, the curve No 2 for k_o= 15 kN/m, the curve No 3 for k_o= 20 kN/m and the curve No 4 for k_o = 30 kN/m.

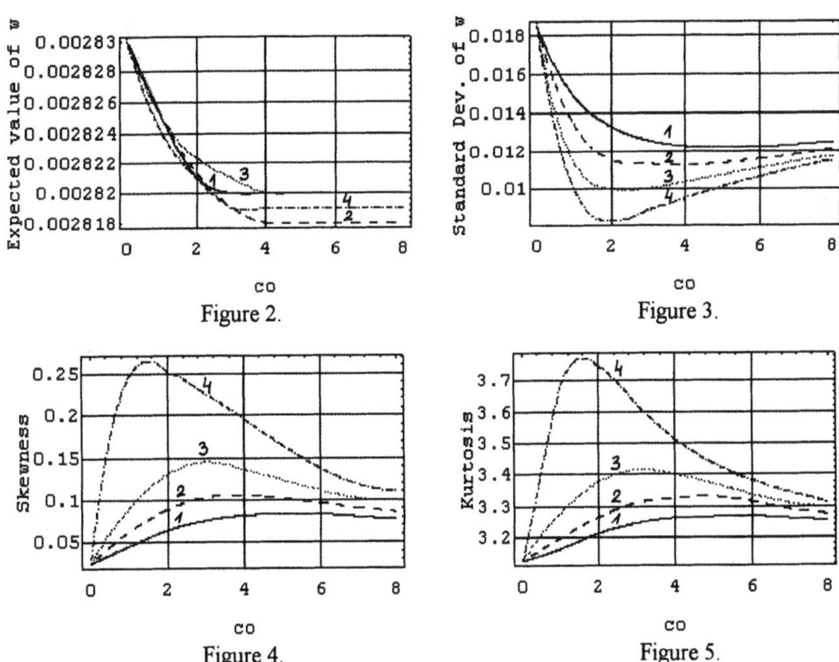

Figure 2.

Figure 3.

Figure 4.

Figure 5.

The curves presented in Figures 6, 7, 8, 9 have been calculated for c_o = 2 kN s/m (the curve No 1), c_o = 3 kN s/m (the curve No 2), c_o = 4 kN s/m (the curve No 3) and c_o = 5 kN s/m (the curve No 4).

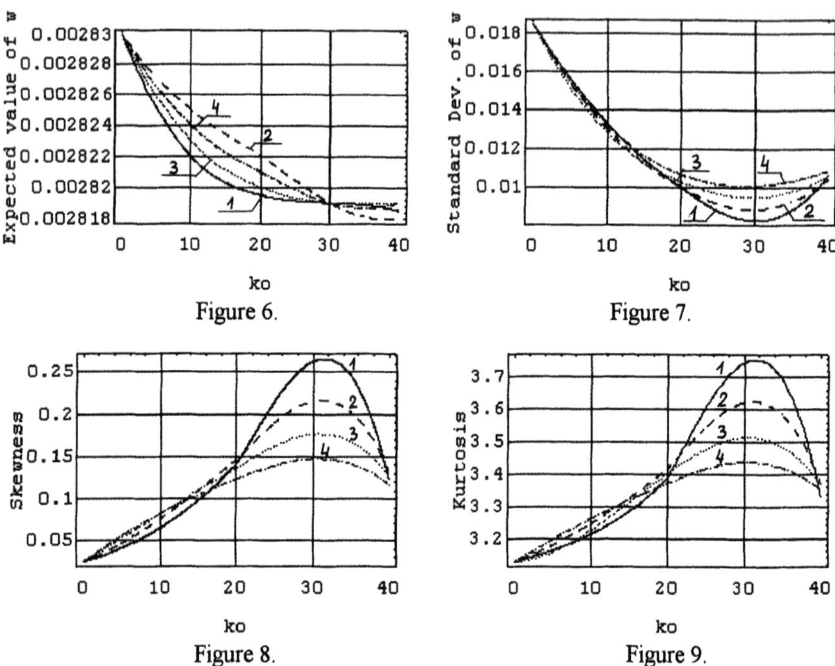

Figure 6.

Figure 7.

Figure 8.

Figure 9.

In Figures 10,11,12,13 the influences of the change of absorber's mass on the bridge's probabilistic characteristics are shown. This calculations have been made for c_o = 2 kNs/m, k_o = 25 kN/m (the curve No 1),.c_o = 4 kNs/m, k_o = 25 kN/m (the curve No 2), c_o = 4 kNs/m, k_o = 15 kN/m (the curve No 3), c_o = 2 kNs/m, k_o = 15 kN/m (the curve No 4).

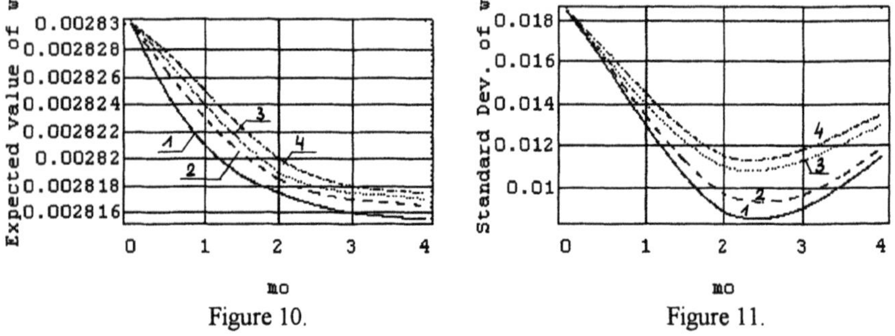

Figure 10.

Figure 11.

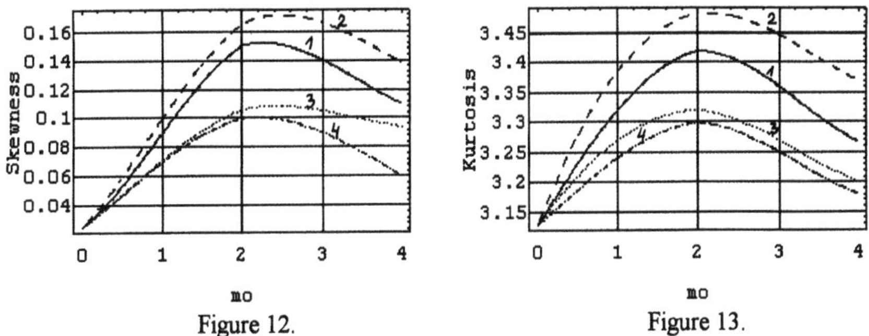

Figure 12.

Figure 13.

To evaluate the level of vibration reduction in the bridge due to applying an absorber the expected value of midspan deflection, its standard deviation, coefficient of skewness and kurtosis calculated for different vehicle speed are compared with the results obtained for the case without absorber and presented in Figures 14, 15, 16, 17. The curve No 1 has been obtained for the bridge without absorber, the curve No 2 for the bridge with an absorber with parameters $c_o = 2$ kNs/m, $k_o = 25$ kN/m, $M_o = 2$ t, the curve No 3 for the bridge with an absorber with parameters $c_o = 2$ kNs/m, $k_o = 25$ kN/m, $M_o = 4$ t, the curve No 4 for the bridge with an absorber with parameters $c_o = 2$ kNs/m, $k_o = 10$ kN/m, $M_o = 2$ t.

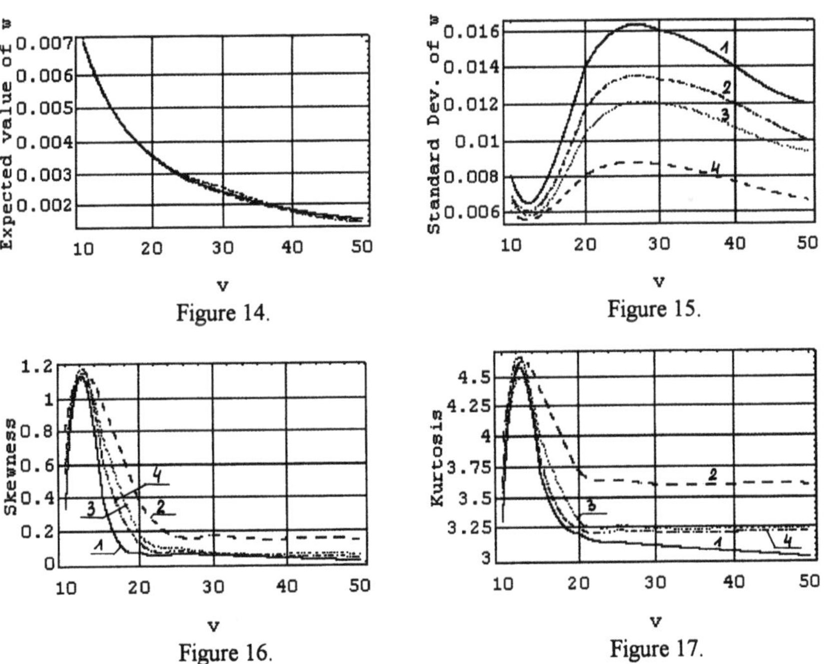

Figure 14.

Figure 15.

Figure 16.

Figure 17.

For this numerical example following conclusions can be made:
- for a given absorber's mass the change of its damping coefficient or spring's stiffness have a little influence on the expected value of the displacement and a significant influence on the standard deviation,
- if for some parameters the standard deviation has the minimum value, the corresponding coefficients of skewness and kurtosis have both the maximum values,
- if the change of the absorber's damping coefficient or spring's stiffness causes increasing of the standard deviation, the corresponding coefficients of skewness and kurtosis decrease.

Acknowledgement

This research was partially supported by the Scientific Research Committe in Warsaw under grant number 772309203.

REFERENCES

1. B. G. Korenev, L.M. Reznikov, Dynamic Vibration Absorbers - Theory and Technical Applications, J.Wiley & Sons, Chichester, 1993.
2. S. El-Bagalaty, M. Klasztorny, Elimination of Excessive Forced Vibrations of Structures Using Dynamic Absorbers, XXXVI Conf. KILiW PAN & KN PZiTB, Krynica, 1990, v.1, 5-10.
3. A. K. Das, S.S. Dey, Effects of Tuned Mass Dampers on Random Response of Bridges, Comp. & Struct., 1992, Vol. 43, No. 4, pp. 745-750.
4. C. C. Tung, Random response of highway bridges to vehicle loads, Proceeding of the American Society of Civil Engineers, Journal of the Engineering Mechanics Division (1967) 93, 73-74
5. R. Sieniawska, P. Śniady, First passage problem of the beam under a random stream of moving forces, Journal of Sound and Vibration (1990) 136(2), 177-185.
6. R. Sieniawska, P. Śniady, Life expectancy of highway bridges due to traffic load, Journal of Sound and Vibration (1990) 140(1), 31-38.
7. P. N. Takaoka, Reliability Analysis in Highway Bridges, PAN, KILiW, Jabłonna 1982.

26

INTERACTIVE RELIABILITY-BASED OPTIMAL DESIGN

J.D. Sørensen*, P. Thoft-Christensen*, A. Siemaszko, J.M.B. Cardoso** & J.L.T. Santos****
*** : CSR, Aalborg, Denmark** **** : IST, IDMEC, Lisboa, Portugal**

Abstract

Interactive design/optimization of large, complex structural systems is considered. The objective function is assumed to model the expected costs. The constraints are reliability-based and/or related to deterministic code requirements. Solution of this optimization problem is divided in four main tasks, namely finite element analyses, sensitivity analyses, reliability analyses and application of an optimization algorithm. In the paper it is shown how these four tasks can be linked effectively and how existing information on design variables, Lagrange multipliers and the Hessian matrix can be used in interactive optimization.

1. Introduction

Reliability-based optimal design of large, complex structural and mechanical systems, e.g. aerospace structures is considered. Uncertain quantities are modelled by stochastic variables and the reliability is estimated using First Order Reliability Methods, see Madsen et. al [1]. Sensitivity information is obtained using the adjoint continuum method of design sensitivity analysis, see Haug et. al [2]. An optimization problem is considered where the design (optimization) variables are assumed to be related to size variables defining the geometry of the structural elements and to shape variables defining the overall geometry of the structure.

Reliability-based optimization of structural systems has been considered in a number of papers, see e.g. Murotsu et. al [3], Frangopol [4], Sørensen & Thoft-Christensen [5] and Enevoldsen & Sørensen [6]. A reliability-based optimization problem can be formulated as minimization of e.g. the structural weight with constraints on element reliability indices and/or on the system reliability. Alternatively the total expected costs in the design lifetime can be minimized or a multiobjective optimization problem can be formulated.

The objective function can in this paper model the expected costs related to the structure in its planned lifetime or it can simply be the volume of the structure. Deterministic constraints can be imposed to ensure that certain response characteristics such as displacements and stresses do not exceed codified critical values. The deterministic constraints can also include general design requirements for the design variables. Further constraints are included to ensure that the reliability of the structure is satisfactory, see section 2.

Special emphasis is given to techniques to perform the optimization interactively, see section 3. Basic interactive operations are described and it is shown how existing information on design variables, Lagrange multipliers and the Hessian matrix can be used in interactive optimization.

Solution of this optimization problem can be divided in four main tasks, namely finite element analyses, sensitivity analyses, reliability analyses and application of an optimization algorithm. For realistic structures solution of a reliability-based optimization problem can be very computertime demanding. It is shown how these four tasks performed by individual computer programs can be linked effectively, see section 4. An example with size and shape variables is shown, see section 5. Special effort is given to demonstrate the effect of taking into account the available information in interactive optimization.

2. General formulation of optimization (design) problem

The following general reliability-based structural optimization problem is considered

$$\min \quad C(\mathbf{b}) = \sum_{i=1}^{m_D} C_{I_i} V_i(\mathbf{b}) + \sum_{i=1}^{m_P} C_{f_i} \Phi(-\beta_i(\mathbf{b})) \tag{1}$$

$$s.t. \quad \beta_i(\mathbf{b}) \geq \beta_i^{min} \quad , i = 1, \ldots, M \tag{2}$$

$$B_i(\mathbf{b}) \geq 0 \quad , i = 1, \ldots, m \tag{3}$$

$$b_i^l \leq b_i \leq b_i^u \quad , i = 1, \ldots, N \tag{4}$$

where $\mathbf{b} = (b_1, \ldots, b_N)$ are the design (or optimization) variables. The optimization variables are assumed to be related to parameters defining the geometry of the structure (for example diameters and thicknesses of tubular elements) and coordinates (or related parameters) defining the geometry (shape) of the structural system.

The objective function C consists of a deterministic and a probabilistic part with m_D and m_P terms, respectively. V_i is e.g. a volume in the ith deterministic term and C_{I_i} is the cost per volume of the ith term modelling the construction costs. V_i is assumed to be deterministic. If stochastic variables influence V_i then design values, see below, are assumed to be used to calculate V_i. In the probabilistic part C_{f_i} is the cost due to failure of failure mode i. Φ is the standard Normal distribution function and $\beta_i, i = 1, ..., m_P$ are reliability indices for m_P failure modes. The general formulation of (1) allows the objective function to model both the structural weight and the total expected costs of construction and failure.

The constraints in (2) are based on the reliability indices $\beta_i, i = 1, ..., M$ for M failure modes. $\beta_i^{min}, i = 1, ..., M$ are the corresponding lower limits on the reliabilities. $B_i, i = 1, ..., m$ define the deterministic inequality constraints in (3) which can ensure that response characteristics such as displacements and stresses do not exceed codified critical values. Determination of the inequality constraints usually includes finite element analyses of the structural system. The inequality constraints can also include general design requirements for the design variables. The constraints in (4) are so-called simple bounds.

The variables (parameters) used to model the structure to be analysed are characterized as *stochastic or deterministic* if the variable can be modelled as stochastic or deterministic and *design or fixed* if the variable can be a design (optimization) variable or a fixed constant.

The stochastic variables are denoted $\mathbf{X}$ and deterministic variables are denoted $\mathbf{x}^D$. The design variables $\mathbf{b}$ can be expected values of stochastic variables or deterministic variables. The reliability indices in (1) and (2) are determined on basis of limit state functions

$$g_i(\mathbf{x}(\mathbf{b}), \mathbf{x}^D(\mathbf{b}), \mathbf{Y}(\mathbf{x}(\mathbf{b}), \mathbf{x}^D(\mathbf{b}))) = 0 \qquad i = 1, ..., M \tag{5}$$

where $\mathbf{x}$ are realisations of the stochastic variables $\mathbf{X}$. $\mathbf{Y}$ are performances such as displacements and stresses calculated by structural analysis, see section 4. The deterministic constraints in (3) are assumed to be related to the limit state functions g_j by

$$B_j = g_j(\mathbf{x}^d(\mathbf{b}), \mathbf{x}^D(\mathbf{b}), \mathbf{Y}(\mathbf{x}^d(\mathbf{b}), \mathbf{x}^D(\mathbf{b}))) \qquad j = 1, ..., m \tag{6}$$

x_i^d is a design value calculated from $x_i^d = \gamma_i(\mu_{X_i} + k_i \sigma_{X_i})$ where k_i is a factor defining the characteristic value and γ_i is a partial safety factor. μ_{X_i} and σ_{X_i} are expected value and standard deviation of X_i, respectively.

3. Interactive optimization

The basic types of interactive optimization which influences the formulation of the optimization problems are, see Arora [7]: • include (delete) a design (optimization) variable, • include (delete) a constraint, • modify a constraint or • modify (change) the objective function.

In order to investigate the effect of interactive optimization on the optimality criteria (1) - (4) is restated as an optimization problem with objective function $C(\mathbf{b})$, equality constraints $c_i(\mathbf{b}) = 0, i = 1, \ldots, m_e$ and inequality constraints $c_i(\mathbf{b}) \geq 0, i = m_e + 1, \ldots, m$. First order **necessary** conditions that have to be satisfied at a (local) optimum point $\mathbf{b}^*$ are given by the Kuhn-Tucker conditions. If the optimization process has almost converged, a good guess on the optimal design is available. A modification of the optimization problem is then specified by the user. The influence on the optimality criteria is described in the following, see Sørensen [8] for details.

Include/delete an optimization variable : Let one of the fixed parameters p in the model of the structural system be transferred from the list of fixed parameters to the list of optimization variables. Let the new optimization variable be denoted b_{N+1}. The new optimal point is then denoted $\mathbf{b}^+ = (b_1^+, ..., b_N^+, b_{N+1}^+)$ and the Kuhn-Tucker conditions are modified to

$$\frac{dC(\mathbf{b}^+)}{db_j^+} - \sum_{i=1}^{m} \lambda_i^+ \frac{dc_i(\mathbf{b}^+)}{db_j^+} = 0 \quad , j = 1, 2, ..., N+1 \tag{7}$$

$$\lambda_i^+ c_i(\mathbf{b}^+) = 0 \qquad , \lambda_i^+ \geq 0 \quad , i = m_e + 1, \ldots, m \tag{8}$$

$$c_i(\mathbf{b}^+) = 0 \quad , i = 1, \ldots, m_e \tag{9}$$

$$c_i(\mathbf{b}^+) \geq 0 \quad , i = m_e + 1, \ldots, m \tag{10}$$

where $\lambda_i^+, i = 1, ..., m$ are the multipliers corresponding to the optimum point. Compared with the unmodified case an extra equation is introduced in (7). The variables in the modified optimum point $\mathbf{b}^+$ can be expected to be very close to the point $(\mathbf{b}^*, p)$ because all the constraints in (8)-(10) are satisfied with $\mathbf{b}^*$ and the parameter p. Also the first N equations in (7) are satisfied in the point $(\mathbf{b}^*, p)$. If optimization variable b_N is deleted from the list of optimization variables and included as a parameter p then the new optimal point $(\mathbf{b}^-, p)$ can also be expected to be very close to $\mathbf{b}^*$.

The number of Lagrangian multipliers $\boldsymbol{\lambda}^+$ is unchanged and the numerical values can be expected to be almost unchanged. If an optimization variable is included then the size of the Hessian matrix $\mathbf{H}$ of the Lagrangian function (which is used in some optimization algorithms) is increased with one compared with the old Hessain matrix $\mathbf{H}^{\text{old}}$. A first estimate of the updated Hessian matrix is $\mathbf{H} = \begin{bmatrix} \mathbf{H}^{old} & 0 \\ 0^T & 1 \end{bmatrix}$ where $0^T = (0, 0, ..., 0)$. If an optimization variable is deleted, the size of the Hessian matrix is decreased with one. A first estmate of the updated Hessian matrix is simply to delete the row and column corresponding to the deleted optimization variable. The updated Hessian matrix can be used as input to restarted optimization algorithms.

Include/delete a constraint : Let a new inequality constraint be introduced as constraint no. $m+1$. The new optimal point is denoted $\mathbf{b}^+ = (b_1^+, ..., b_N^+)$. The number of equations in (8) and (10) is increased by one and a new multiplier λ_{m+1}^+ is introduced. The new optimum $\mathbf{b}^+$ can be quite different from $\mathbf{b}^*$, depending on the new constraint c_{m+1}. The size of the Hessian matrix is not changed and a first estimate of $\mathbf{H}$ can simply be the old Hessian matrix. If a constraint is deleted then the same considerations as for inclusion of a constraint can be stated. If the new constraint or the deleted constraint is not active then the optimality criteria is satisfied with $\mathbf{b}^*$.

Modify a constraint : Let the constraint no. m be modified to $c_m'(\mathbf{b})$. The new optimal point is denoted $\mathbf{b}^+ = (b_1^+, ..., b_N^+)$. In this case the number of equations is unchanged but the new optimum $\mathbf{b}^+$ can be quite different from $\mathbf{b}^*$, depending on the modified constraint c_m. The size of the Hessian matrix is also unchanged and a first estimate of $\mathbf{H}$ can simply be the old Hessian matrix.

Modify the objective function : The objective function in (1) can be modified by changing the performances in the deterministic part, the failure modes in the probabilistic part or by changing the cost factors C_{I_i} and C_{f_i}. Modification of the objective function in (1) has almost the same effect on the Kuhn-Tucker conditions as modification of a constraint.

4. Integrated optimization techniques

The main modules in the intergrated optimization system CARBOS, [10] are *OPT* : Reliability-based interactive optimization by RELOPT, [11], *DSA* : Design sensitivity analysis by the continuum method, [12], *FEA* : Finite element calculations by ASASNL, [13], *REL* : Reliability analysis by RELIAB, [14] or COMREL, [15] and *UI* : User interface module which generates ASCII files with interactive input to OPT, REL and FEA. The data flow is shown in figure 1.

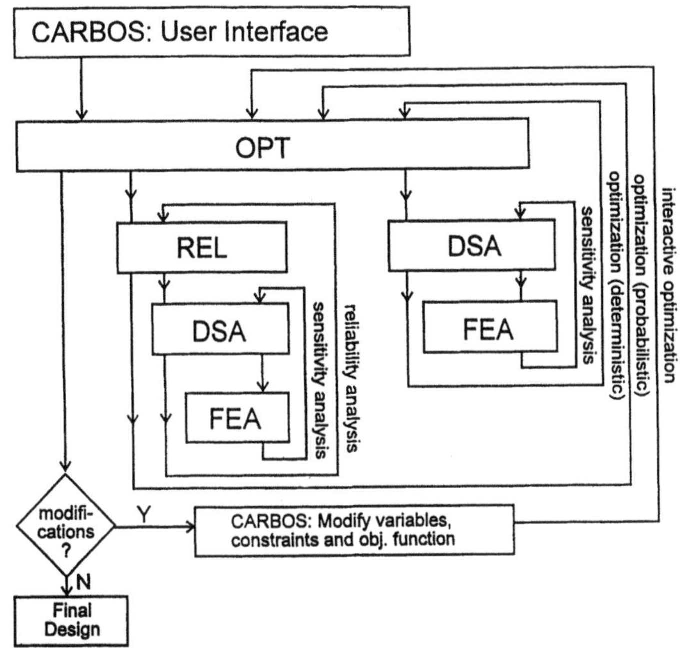

Figure 1. Data flow in interactive optimization by CARBOS.

5. Example

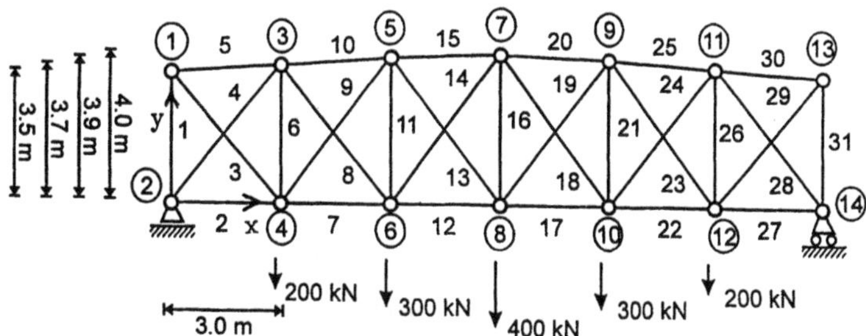

Figure 2. Bridge consisting of 31 elements and 14 nodes.

A simple 2D truss model of a bridge structure is considered, see figure 2. The truss elements are all tubular elements and are divided into three groups with identical diameters. The bottom truss elements (2,7,12,17,22 and 27) are group 1, the top truss elements (5,10,15,20,25 and 30) and the two outer diagonals (3 and 29) are group 2 and the remaining are belonging to the 3rd material group.

Reliability model : The stochastic variables used are shown in table 1, where SIGMAX models the maximun acceptable stress σ_{max} and ADM models the maximun acceptable displacement u_{max}.

Table 1. Stochastic variables. μ: expected value, σ: standard deviation. N: normal, LN: lognormal, G: Gumbel. $b_1 ... b_{10}$ are design variables.

	distribution	μ	σ
Load multiplier	G	1	0.40
Diameter group 1	N	b_1	0.1 b_1
Diameter group 2	N	b_2	0.1 b_2
Diameter group 3	N	b_3	0.1 b_3
y-coor. node 1,3,5,7,9,11 and 13	N	b_4, ..., b_{10}	50 mm
SIGMAX	LN	300 N/mm^2	30 N/mm^2
ADM	N	45 mm	4.5 mm

Optimization problem : To specify the optimization problem 19 limit states are defined. The limit state functions are

$$g_1 = \text{ADM} - \text{PER2}$$
$$g_j = \text{SIGMAX} - \text{PER}(j+1) \qquad ,j = 2, ..., 19$$

The optimization variables are $\mathbf{b} = (d_1, d_2, d_3, y_1, y_3, y_5, y_7, y_9, y_{11}, y_{13})$, see table 1. The initial values are $\mathbf{b}$=(150,150,150,3500,3700,3900,4000,3900,3700,3500) [mm]. The objective function is $C(\mathbf{b})$ = Volume and the constraints are $\beta_1 \geq 3.29$ and $B_j = g_j \geq 0, j = 2, 3, ..., 19$. β_1 is calculated using g_1. In the deterministic constraints the design values are calculated using k = -2 and $\gamma = 0.8$ for SIGMAX, $k = 2$ and $\gamma = 1.2$ for the load and $k = 0$ and $\gamma = 1.0$ for all other parameters. The reliability analyses are performed using RELIAB [14]. The optimization problem is solved using RELOPT [11] and the NLPQL optimizer [9] with convergence limit $\epsilon = 10^{-7}$.

Interactive optimization results : The optimization problem is solved interactively in three main steps where interaction is possible for each 4 iterations:

1. The optimization problem is solved with only the reliability constraint included. First 4 iterations are performed with the intitial values of the optimization variables as start point. Next restart is done where the 'old' Hessian matrix and Lagrange multipliers are used. 4 iterations is performed. This restart procedure is repeated until convergence.
2. Deterministic constraints are included and optimization is continued with 2a or 2b:
   2a. All 'old' values of design variables, Hessian matrix and Lagrange multipliers are used as start conditions. This restart procedure is repeated in blocks of 4 iterations until convergence.

2b. First 4 iterations are performed where only the current values of the optimization variables are used as starting info. Next restarts are performed as in 2a until convergence.

3. The diameters of the struss elements are fixed on reasonable values and unimportant stress constraints are deleted. The optimization is continued with 3a or 3b :

   3a. All 'old' values of design variables, Hessian matrix and Lagrange multipliers are used as start conditions. This restart procedure is repeated in blocks of 4 iterations until convergence.

   3b. First 4 iterations are performed where only the current values of the optimization variables are used as starting info. Next restarts are performed as in 2a until convergence.

The initial volume is $C(\mathbf{b}) = 21.2\ 10^8$. The results of the optimization are:
Step 1: $\mathbf{b} = (d_1, d_2, d_3, y_1, y_3, y_5, y_7, y_9, y_{11}, y_{13})$ = (102.1, 108.4, 51.5, 2000, 3357, 4759, 5000, 4759, 3357, 2000), $C(\mathbf{b})$=5.66 10^8.
Step 2: $\mathbf{b} = (d_1, d_2, d_3, y_1, y_3, y_5, y_7, y_9, y_{11}, y_{13})$ = (103.3, 110.0, 51.7, 3102, 3389, 4806, 5000, 4806, 3389, 3102), $C(\mathbf{b})$=5.82 10^8.
Step 3: $\mathbf{b} = (d_1, d_2, d_3, y_1, y_3, y_5, y_7, y_9, y_{11}, y_{13})$ = (110 (fixed), 110 (fixed), 60 (fixed), 2000, 3234, 4835, 5000, 4835, 3234, 2000), $C(\mathbf{b})$= 6.52 10^8.

To measure the computation time the number of finite element analyses of the structure is counted. This number can be considered as a good indication of the computational effort. The results are shown in table 2. It is seen that in this example the amount of computations is decreased if information in the form of Lagrange multipliers and Hessian matrices are used when the optimization problem is restartet after a modification. The optimized structure is shown in figure 3.

Table 2. Total number of finite element analyses using the different four strategies.

	1 + 2a + 3a	1 + 2a + 3b	1 + 2b + 3a	1 + 2b + 3b
step 1	59	59	59	59
step 2	76	76	111	111
step 3	96	109	125	144

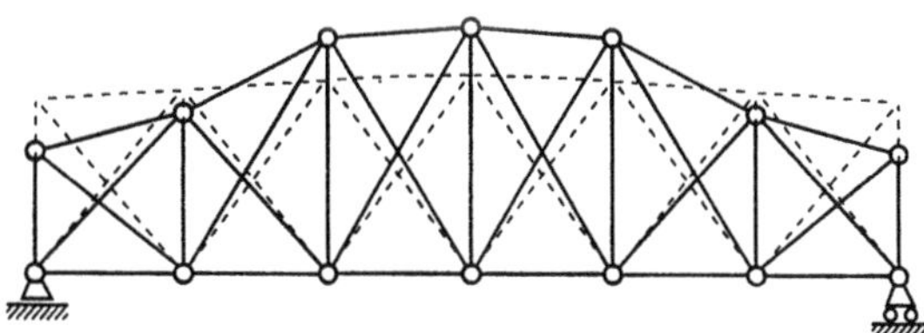

Figure 3. Initial (dashed lines) and optimized structure.

6. Conclusions

A general reliability-based optimization problem for interactive design/optimization of large, complex structural systems is described. Techniques for interactive solution of this optimization problem are discussed, especially how to use information from Lagrange multipliers and from the Hessian matrix. Integration of the main tasks in solution of this optimization problem, finite element analyses, sensitivity analyses, reliability analyses and application of an optimization algorithm is discussed briefly. An example illustrates the interactive optimization.

7. Acknowledgements

The paper presents parts of results of the BRITE/EURAM P4194 Project "Reliability Based Optimal Design with Applications to Advanced Aerospace Structures". This project, which started in September 1991, is funded through a 50 % contribution from the Directorate General for Science, Research and Development of the Commision of the European Communities and another 50 % contribute from the partners in the project: • WS Atkins Science and Technology, Epsom, UK, • Computational Safety and Reliability (CSR), Aalborg, Denmark, • Instituto Superior Tecnico, Lisbon, Portugal, • MAN Technologie, Munich, Germany and • Reliability Consulting and Programs (RCP), Munich, Germany.

8. References

[1] Madsen, H.O., S. Krenk & N.C. Lind: Methods of Structural Safety. Prentice-Hall, 1986.

[2] Haug, E.J., K.K. Choi & V. Komkov: Design Sensitivity Analysis of Structural Systems. Academic press, 1986.

[3] Murotsu, Y., M. Kishi, H. Okada, M. Yonezawa & K. Taguchi: Probabilistically Optimum Design of Frame Structures. Proc. 11*th* IFIP Conf. on "System Modelling and Optimization". Springer Verlag, 1984, pp. 545-554.

[4] Frangopol, D.: Sensitivity of Reliability-Based Optimum Design. ASCE, Journal of Structural Engineering, Vol. 111, No. 8, 1985, pp. 1703-1721.

[5] Sørensen, J.D. & P. Thoft-Christensen: Structural Optimization with Reliability Constraints. Proc. 12*th* IFIP Conf. on "System Modelling and Optimization". Springer Verlag, 1985, pp. 876-885.

[6] Enevoldsen, I. & J.D. Sørensen: Reliability-Based Optimization in Structural Engineering. Structural Safety, Vol. 15, 1994, pp. 169-196.

[7] Arora, J.S.: Introduction to Optimum Design. McGraw-Hill, 1989.

[8] Sørensen, J.D.: Optimization methods and interactive optimization. BRITE/EURAM P4194 report B(V.1) - 4.2, CSR, March 1994.

[9] Schittkowski, K.: NLPQL: A FORTRAN Subroutine Solving Constrained Non-Linear Programming Problems. Annals of Operations Research, 1986.

[10] CARBOS: BRITE/EURAM P4194 report, WS Atkins, July 1994.

[11] RELOPT: RELiabilty-based OPTimization. CSR-Software, CSR, Aalborg, Denmark, August 1994.

[12] IST: Interactive Sensitivity Tools. User's guide, Version 2.0B, CEMUL, IST, Lisbon, 1993.

[13] ASASNL: User's Guide, Version 18, WS Atkins, UK, 1990.

[14] RELIAB: Reliabilty analysis. CSR-Software, CSR, Aalborg, Denmark, February 1994.

[15] COMREL-TI: User's manual. RCP GmbH, Munich, Gernany, 1992.

27

RELIABILITY ANALYSIS AND OPTIMAL DESIGN OF MONOLITHIC VERTICAL WALL BREAKWATERS

J.D. Sørensen, H.F. Burcharth & E. Christiani
Aalborg University, Sohngaardsholmsvej 57, DK-9000 Aalborg, Denmark

Abstract

Reliability analysis and reliability-based design of monolithic vertical wall breakwaters are considered. Probabilistic models of the most important failure modes, sliding failure, failure of the foundation and overturning failure are described. Relevant design variables are identified and reliability-based optimization problems are formulated. Results from an illustrative example are given.

1. Introduction

Coastal structures are normally designed on a deterministic basis using level 1 codes. However, the design wave which defines the dominating load on the structures is estimated on a probabilistic basis, for example as the characteristic significant wave height having a return period of 100 years. All other uncertainties are taken into account using safety factors or partial safety factors.

Breakwater structures are used under quite different conditions. The expected lifetime can be from 5 years (interim structure) to 100 years (permanent structure) and the accepted level of probability of failure in the expected lifetime can vary from a very small number, e.g. 10^{-4} if failure of the breakwater results in significant damage to large probabilities, e.g. 0.5 if the consequences are insignificant, see Burcharth [1]. Further a number of serious failures of breakwaters have been reported during the last 20 years. In order to obtain more rational and consistent estimates of the reliability of breakwater structures and in order to be able to perform a reliability-based design optimization, the paper describes a probabilistic model of the failure modes of typical vertical wall breakwaters and of the uncertainties related to these failure modes.

As an example monolithic vertical wall breakwaters built of reinforced concrete caissons filled with sand are considered. Important failure modes for this type of breakwaters are identified, namely sliding failure, failure of the foundation and overturning failure. Limit state functions are formulated and stochastic models for the uncertain variables are identified in section 2. Next reliability-based design optimization of breakwaters are considered in section 3. Results of a reliability analysis and a reliability-based design of a monolithic vertical wall breakwater are presented in section 4.

2. Failure modes for monolithic vertical wall breakwaters

Vertical wall breakwaters can be constructed in a number of different ways. The most common types are shown in figure 1a and 1b, namely one type where the caisson is placed on a thin bedding layer and one type where the caisson is placed on a rubble mound.

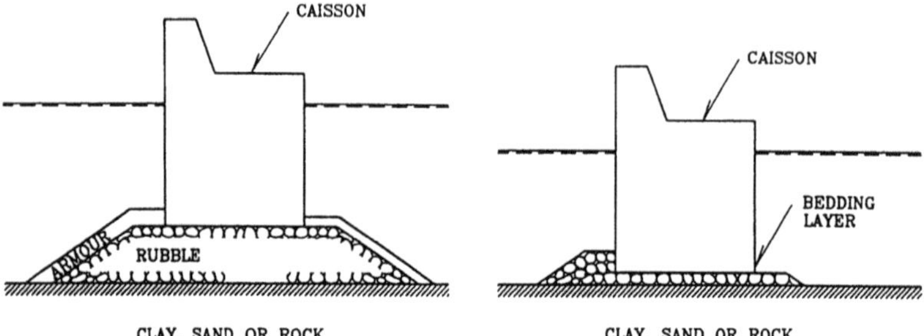

Figure 1a and 1b. Main types of vertical wall breakwaters.

Modelling of wave load by the Goda model

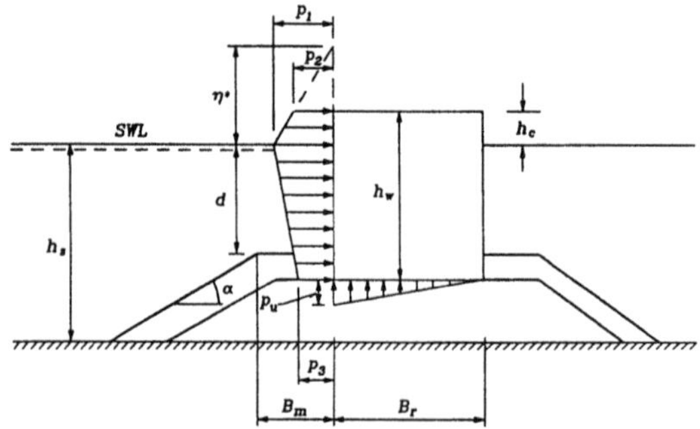

Figure 2. Wave induced pressure under a wave crest modelled by the Goda model.

Models for estimating the wave load on vertical wall breakwaters have been developed by Goda [2] and Goda et al. [3]. The wave load is used to formulate safety margins for the most important failure modes for vertical wall breakwaters, namely sliding, foundation failure and overturning. Failure of the caisson structure and hydraulic stability failure of the armour layer and mound are not considered in this paper.

Figure 2 shows the wave pressure model suggested by Goda for the case of a conventional vertical front where wave breaking is not enhanced by a steep sea bed or structural configurations. The model is based on model tests and includes the effect of breaking waves to the extent of normal accidental (non-provoked) wave breaking. A detailed description is given in Goda [4] and Burcharth et al. [5]. p_1, p_2, p_3 and p_u in figure 2 are

$$p_1 = 0.5(1+\cos\beta)(\alpha_1 + \alpha_2 \cos^2\beta)\rho_w g H_d \quad (1)$$

$$p_3 = \alpha_3 p_1 \quad (2)$$

$$p_2 = \begin{cases} (1 - \frac{h_c}{\eta^*})p_1 & \text{for } \eta^* > h_c \\ 0 & \text{for } \eta^* \leq h_c \end{cases} \quad (3)$$

$$p_u = 0.5(1 + \cos\beta)\alpha_1\alpha_3\rho_w g H_d \quad (4)$$

where $\eta^* = 0.75(1+\cos\beta)H_d$, β is the angle of incidence of waves, ρ_w is the mass density of sea water, H_S is the significant wave height, H_d is the wave height to be used in the Goda formula defined as the highest wave in the design sea state just in front of the breakwater. H_d is taken as 1.8 H_S, see Goda [4]. However, H_d cannot be larger than the wave breaking height, see [4]. $\alpha_1 = \alpha_1(H_S, d, h_s, L)$, $\alpha_2 = \alpha_2(H_S, d, h_s, L)$, $\alpha_3 = \alpha_3(H_S, d, h_s, h_c, L)$, see [4] and L is the wave length corresponding to that of the significant wave.

On the basis of the wave pressures defined by p_1, p_2, p_3 and p_u and the tidal elevation ζ the resultant horisontal force $F_H = F_H(H_S, \zeta, d, h_s, h_c, B_r)$ and the vertical (uplift) force $F_U = F_U(H_S, \zeta, d, h_s, h_c, B_r)$ can be determined. Also the corresponding resultant moments about the heel of the caisson can be determined : $M_H = M_H(H_S, \zeta, d, h_s, h_c, B_r)$ and $M_U = M_U(H_S, \zeta, d, h_s, h_c, B_r)$. It should be noted that the Goda model can be modified to take into account impulsive forces, see Takahashi et al. [6].

Sliding failure

Stability failure by sliding can be modelled by the limit state function

$$g = (F_G - ZF_U)\mu - ZF_H \quad (5)$$

where μ is the coefficient of friction between the bedding layer and the caisson, $F_G = F_G(\zeta, h_w, B_r)$ is the weight of the caisson reduced for boyancy and Z is a stochastic variable modelling the model uncertainty connected with the Goda model, see van der Meer et al. [7].

Overturning failure

Failure by overturning can be modelled by the limit state function

$$g = M_G - ZM_U - ZM_H \quad (6)$$

where $M_G = M_G(\zeta, h_w, h_c, B_r)$ is the moment of the weight of the caisson about the heel.

Foundation failure

If it is assumed that the soil consists of clay only, figure 3 shows some of the most important kinematic admissible foundation failure modes corresponding to the two types of structures shown in figure 1. F_R is the resultant wave load from F_G and F_H. As an example a limit state function is described in the following for foundation failure by rotation of the caisson placed on a bedding layer. Figure 4 describes the rotation mechanism. The centre of rotation is $O(x_0, z_0)$. The failure mechanism is based on the assumption of plane strain conditions in the soil, i.e. a two-dimesional solution for the equilibrium of the considered slip surface in the soil. Note that it is assumed that tensile stresses cannot occur under the caisson base plate.

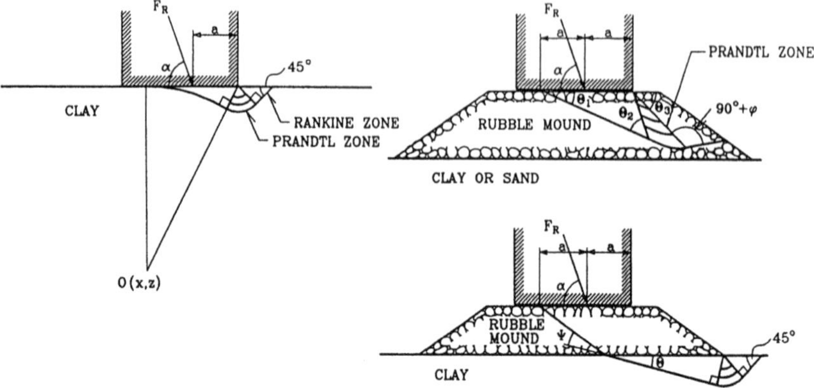

Figure 3. Foundation failure mechanisms.

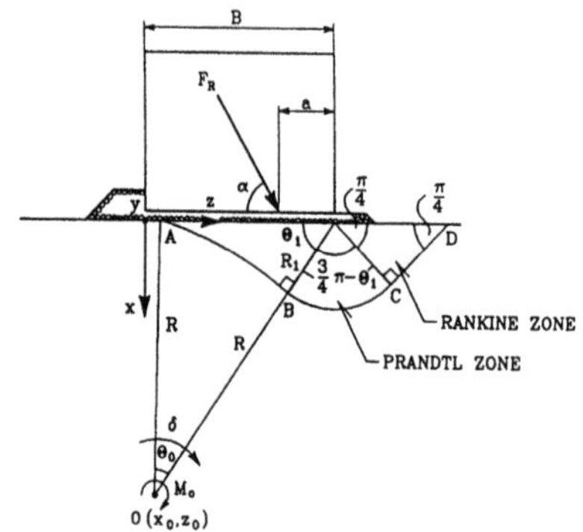

Figure 4. Rotation mechanism.

The increment of the external work W_E is calculated as the moment from the pressures defined in figure 2 around the centre of rotation $O(x_0, z_0)$ multiplied by a unit increment of the rotation angle δ.

The total increment of the internal work W_I covers the rim of all fracture zones and the contribution from the Prandtl and Rankine fracture zones, i.e. $W_I = W_1 + W_2 + W_3 + W_4 + W_5$ where W_1, W_2 and W_3 are the works along the rim of the fracture zones AB, BC and CD. W_4 and W_5 are the internal works in the Prandtl zone and in the Rankine zone.

Internal work along the rim of the fracture zones can be calculated from the point of rotation $O(x_0, z_0)$:

$$W_1 = \delta R^2 \int_0^{\theta_o} c_u(\theta) d\theta \quad W_2 = \delta R R_1 \int_0^{\frac{3\pi}{4}-\theta_1} c_u(\theta) d\theta \quad W_3 = \delta R \int_0^{R_1} c_u(a) da \qquad (7)$$

where c_u is the undrained shear strength of the clay. R, R_1, θ_0 and θ_1 are defined in figure 4. The internal work from the Prandtl and Rankine fracture zones are

$$W_4 = \delta R_1 \int_0^{R_1} \int_0^{\frac{3\pi}{4}-\theta_1} c_u(a,\theta) d\theta \, da \qquad W_5 = \delta \int_0^{R_1} \int_0^{R_1 - a} c_u(a,l) dl \, da \tag{8}$$

The rotation point is determined as the point where the ratio between total internal work W_I and external work W_E is a minimum. In the reliability analysis it is taken into account that the rotation point $O(x_0, z_0)$ is dependent on the stochastic variables modelling the loading and the strength of the soil.

The undrained shear strength of the clay is modelled as a log-Gaussian stochastic field $\{c_u(x,z)\}$, see e.g. Keaveny et al. [8] and Andersen et al. [9]. If only uncertainty related to c_u is considered and the correlation lengths for $\{c_u(x,z)\}$ are small compared to the integration intervals it follows from the central limit theorem that the total internal work can be approximated by a normal distributed stochastic variable W_I with mean value μ_{W_I} and standard deviation σ_{W_I}. The limit state function is written

$$g = \mu_{W_I} + u_W \sigma_{W_I} - W_E = \sum_{i=1}^{5} E[W_{Ri}] + u_W \cdot \left(\sum_{i=1}^{5} \sum_{j=1}^{5} Cov[W_{Ri}, W_{Rj}] \right)^{\frac{1}{2}} - W_E \tag{9}$$

where u_W is a realization of a normal distributed stochastic variable U_W with mean 0 and unit standard deviation.

Further e.g. the mean value of W_4 is

$$E[W_4] = \delta R_1 \int_0^{R_1} \int_0^{\frac{3\pi}{4}-\theta_1} E[c_u(a,\theta)] d\theta \, da \tag{10}$$

where $E[c_u(a,\theta)]$ is the expected value of c_u at the position described by (a,θ) and the variance of W_5 is

$$Var[W_5] = \delta \int_0^{R_1} \int_0^{R_1 - a_1} \int_0^{R_1} \int_0^{R_1 - a_2} Cov[c_u(a_1,l_1), c_u(a_2,l_2)] dl_2 \, da_2 \, dl_1 \, da_1 \tag{11}$$

where $Cov[c_u(a_1,l_1), c_u(a_2,l_2)]$ is the covariancefunction of c_u at the positions corresponding to (a_1,l_1) and (a_2,l_2).

3. General formulation of optimization (design) problem

In design of breakwaters the main objectives are usually that the breakwaters are as inexpensive as possible and have a satisfactory reliability. These two contradictory requirements can be treated by formulating reliability-based optimization problems on a decision-theoretical basis, see Enevoldsen & Sørensen [10]. The design (decision) variables are denoted $\mathbf{b} = (b_1, \ldots, b_N)$, i.e. the number of design variables is N. The design variables can e.g. be the height and the width of the breakwater, see below. If the objective function is chosen as the total expected costs C_T of the structure during the lifetime the optimal design can be found as the solution to the optimization problem

$$\min_{\mathbf{b}} \quad C_T(\mathbf{b}) = C_I(\mathbf{b}) + C_F P_F(\mathbf{b}) \tag{12}$$

$$b_i^l \leq b_i \leq b_i^u \quad , i = 1, \ldots, N \tag{13}$$

where b_i^l and b_i^u are lower and upper bounds to b_i. C_I is the initial/construction costs, C_F is the costs of failure and P_F is the probability of failure during the expected lifetime of the breakwater. If more than one failure mode has to be considered then failure of the structure has to be modelled by a system for which P_F can be estimated using First Order Reliability Methods (FORM), see Madsen et al. [11]. If only one failure mode is considered then $P_F = \Phi(-\beta)$ where the reliability index β is evaluated by FORM on the basis of a limit state function $g(\mathbf{x}) = 0$ where $\mathbf{x}$ is a realisation of the stochastic variables $\mathbf{X} = (X_1, ..., X_n)$ Φ is the standard normal distribution function.

In the optimization problem (12)-(13) design (decision) variables and costs related to inspection, maintenance and repair can be included, see [10]. The optimization problem (12)-(13) can alternatively be formulated as an optimization problem with reliability-based constraints.

Generally the reliability requirements for different failure modes are not the same. Therefore it is of interest to consider the element reliability-index based optimization problem

$$\min_{\mathbf{b}} \quad C_I(\mathbf{b}) \tag{14}$$

$$s.t. \quad \beta_i(\mathbf{b}) \geq \beta_i^{min} \quad , i = 1, \ldots, M \tag{15}$$

$$B_i(\mathbf{b}) \geq 0 \quad , i = 1, \ldots, m \tag{16}$$

$$b_i^l \leq b_i \leq b_i^u \quad , i = 1, \ldots, N \tag{17}$$

where β_i is the reliability index for failure mode i and β_i^{min} is the corresponding lower bound on the reliability index. The above optimization problems are usually non-linear and non-convex. The optimization problems can be solved effectively using non-linear optimization algorithms and FORM, see [10]. The reliability indices in (25) are determined on the basis of limit state functions written as $g_i(\mathbf{x}(\mathbf{b}), \mathbf{b}) = 0, i = 1, ..., M$.

In a traditional deterministic design the design (optimization) problem the constraint (15) is exchanged by the deterministic constraint

$$B_i(\mathbf{b}) = g_i(\mathbf{x}^D(\mathbf{b}, \gamma), \mathbf{b}) \geq 0 \quad , i = 1, \ldots, M \tag{18}$$

where $\mathbf{x}^D$ are design values calculated using the statistical parameters for the stochastic variables $\mathbf{X}$ and e.g. partial safety factors γ. Typically the design variables $\mathbf{b}$ will be related to the expected value of the stochastic variables.

Examples of design variables for vertical wall breakwaters are the main geometrical quantities defining the structures shown in figure 1.

4. Example

A vertical wall breakwaters placed on a bedding layer is considered, see figure 1. The following data are used: $h_s = 13$ m, $d = 9.5$ m, $h_c = 5.5$ m, $h_w = 17$ and $\rho_w = 1$. The tidal elevation ζ is modelled as a stochastic variable with distribution function $F_\zeta(\zeta) = -\frac{1}{\pi}\arccos\left(\frac{\zeta}{\zeta_0} - 1\right) - 1$ where $\zeta_0 = 0.75$ m. The average mass density of the caisson ρ_c including sand ballast is assumed to be normal distributed with mean value 2.15 t/m^3 and a coefficient of variation of 5 %. The friction coefficient μ is assumed to be normal distributed with mean 0.636 and a coefficient of variation of 15 %, see Takayama [12].

The wave direction is assumed head-on to the breakwater. The maximum significant wave height $H_S^{T_L}$ in the design lifetime T_L is modelled as a stochastic variable on the basis of measurements with the distribution function $F_{H_S^{T_L}}(h) = \left[1 - \exp\left(-\left(\frac{h-B}{A}\right)^k\right)\right]^{\lambda T_L}$ where λ is the average number of H_S data values available per year. In this example $B = 1.55$. To model the statistical uncertainty k and A are assumed to be stochastic variables. k is in this example normal distributed with mean 2 and standard deviation 0.18. A is normal distributed with mean 2.5 and standard deviation 0.3 (corresponding to a data sample of size 30), see Burcharth et al. [5]. Based on model tests the model uncertainty variable Z in (5) and (6) is assumed normal distributed with mean 1 and standard deviation 0.2, see van der Meer et al. [7].

The undrained shear strength of the clay c_u is modelled by a log-Gaussian stochastic field $\{c_u(x,z)\}$. The mean value function and covariance function are in this paper assumed to be (with c_u is in kPa and (x,z) is in meters)

$$E[c_u(x,z)] = 150 + 3x$$

$$Cov[c_u(x_1,z_1), c_u(x_2,z_2)] = 41^2 \exp\left(-|\frac{x_1 - x_2}{3.0}|\right) \exp\left(-(\frac{z_1 - z_2}{30.0})^2\right)$$

Reliability analysis

With a design lifetime of $T_L = 50$ years and a width of the caisson of $B_r = 16.4$ m a reliability analysis gives the results: $\beta_{\text{sliding}} = 0.89$, $\beta_{\text{overturning}} = 2.43$ and $\beta_{\text{foundation}} = 0.91$. The reliability levels are seen to be rather low compared with e.g. structural systems. But compared with observed failure rates for breakwaters the reliability levels are realistic. As expected sliding and foundation failures are the critical failure modes. For sliding failure the uncertainties connected with the wave climate are dominating whereas for foundation failure the uncertainty in the shear strength of the clay is dominating.

Reliability-based design

A reliability-based optimization problem is formulated on the basis of (14)-(17). The objective function models the cost of the caisson which is assumed to be proportional to the weight of the caisson. As design variable only the width of the caisson is reasonable here. For caissons placed on rubble mounds the design variables indicated in figure 5 can be used. The lower limit on the reliability indices for the three failure modes are taken as 1.28 ($\approx$ a probability of failure equal to 0.1). The result is an optimal $b^* = B_r = 18.5$ m and the corresponding weight is $C^* = 4544$ t/m.

5. Conclusions

It is shown how reliability analysis and reliability-based design of monolithic vertical wall breakwaters can be formulated. Probabilistic models for sliding, foundation failure and overturning are described. Relevant design variables are identified and reliability-based optimization problems are formulated. As an illustrative example is considered a vertical wall breakwater with the caisson placed on a bedding layer and with clay in the foundation.

6. Acknowledgements

This work is partly supported by the Danish research program "Marin Teknik" sponsored by the Danish Technical Research Council and partly by the MAST II program "Rubble Mound Breakwater Failure Modes" sponsored by the Directorate General for Science, Research and Development of the Commision of the European Communities.

7. References

[1] Burcharth, H.F.: Development of a Partial Coefficient System in the Design of Rubble Mound Breakwaters. Report Subgroup F, PIANC PTC II Working group 12 on Rubble Mound Breakwaters, 1991.

[2] Goda, Y.: A New Method of Wave Pressure Calculation for the Design of Composite Breakwater. Proc. 14th Int. Conf. Coastal Eng., Copenhagen, Denmark, 1974.

[3] Goda, Y. & T. Fukumori: Laboratory Investigation of Wave Pressures exerted upon Vertical and Composite walls. Coastal Engineering in Japan, Vol. 15, 1972, pp. 81-90.

[4] Goda, Y.: Random Seas and Design of Maritime Structures. University of Tokyo Press, 1985.

[5] Burcharth, H.F. & J.D. Sørensen & E. Christiani: On the Evaluation of Failure Probability of Monolithic Vertical Wall Breakwaters. Proc. "Wave Barriers in Deepwater", Port and Harbour Research Institute, Yokosuka, Japan, 1994, pp. 458-468.

[6] Takahashi, S. & K. Tanimoto & K. Shimosako: Dynamic Response and Sliding of Breakwater Caisson against Impulsive Breaking Wave Forces. Report of Port and Harbour Research Inst., 1994.

[7] van der Meer, J.W. & J. Juhl & G. Driel: Probabilistic Calculations of Wave Forces on Vertical Structures. Proc. Final Workshop, MAST G6-S Coastal Structures, Lisbon, 1992.

[8] Keaveny, J.M. & F. Nadim & S. Lacasse: Autocorrelation Function for Offshore Geotechnical Data. Proc. ICOSSAR89, 1989, pp. 263-270.

[9] Andersen, E.Y. & B.S. Andreasen & P. Ostenfeld-Rosenthat: Foundation Reliability of Anchor Block for Suspension Bridge. Proc. IFIP WG7.5, Lecture notes in Eng. Vol. 76, Springer Verlag, 1992 pp. 131-140.

[10] Enevoldsen, I. & J.D. Sørensen: Reliability-Based Optimization in Structural Engineering. Accepted for publication in Structural Safety, 1994.

[11] Madsen, H.O., S. Krenk & N.C. Lind: Methods of Structural Safety. Prentice-Hall, 1986.

[12] Takayama, T.: Estimation of sliding Failure Probability of present Breakwaters for Probabilistic Design. Report of Port and Harbour Research Inst., Vol. 31, No. 5, 1992.

28

Assessment of performance and optimal strategies for inspection and maintenance of concrete structures using reliability based expert systems

P. Thoft-Christensen
CSR, P.O.Box 218, DK-9100 Aalborg, Denmark

SUMMARY

Expert systems for optimal reliability-based inspection and maintenance of reinforced concrete bridges are described. The deterioration is corrosion of the reinforcement due to carbonation and chloride. Expert system module BRIDGE1 is used at the bridge site to assist during the inspection. BRIDGE2 is used after an inspection during the detailed analysis of the bridge when testing in the laboratory has taken place.

1. INTRODUCTION

The main objective of the project is to optimise strategies for inspection, maintenance and repair of reinforced concrete bridges by developing improved methods for modelling the deterioration of existing as well as future structures using reliability based methods and expert systems are presented with special emphasis on reliability assessment, updating, and the software modules. Further, the functionalities and the structure of the expert systems and the optimal strategy is briefly discussed. The paper is based on Thoft-Christensen [8].

2. MODELLING OF THE DETERIORATION

Only one type of deterioration is included namely corrosion of the reinforcement due to carbonation and chloride. The area of the reinforcement A(t) as function of time t, the depth of the carbonation front d_c, and the chloride concentration C are estimated on basis of well known techniques, see Thoft-Christensen [7]. It is assumed that the area A(t) of the n reinforcement bars in a cross section as function of time t can be determined by

$$A(t) = \begin{cases} nD_i^2 \pi / 4 & for\ t \le T_i \\ n(D(t))^2 \pi / 4 & for\ T_i < t \le T_i \\ 0 & for\ t > T_i \end{cases} \tag{1}$$

where $T = T_i + D_i / (0.023 i_{corr})$. The diameter D(t) of a single bar (in mm) at time t is modelled by D(t) = D_i - 0.023 (t-T_i) i_{corr}. D_i is the initial diameter of a single bar, and i_{corr} is the rate of corrosion in terms of a mean corrosion current density in $\mu A / cm^2$. It is assumed that a corrosion current of 1 A / cm^2 is equivalent to an average oxidation or dissolution of approximately 1.6 mm/year from the surface of steel.The corrosion initiation time T_i (in years) , due to carbonation (as function of the carbonation front d_c in mm) and due to chloride (as function of the chloride diffusion coefficient D_c in $cm^2/\sec$) is estimated by:

carbonation: $T_i = (\frac{d}{K})^2$ chloride: $T_i = \frac{d^2}{4D_c}(erf^{-1}(\frac{C_{cr} - C_0}{C_i - C_0}))^{-2}$ (2)

where d is the concrete cover in cm, K is a constant depending of the concrete cover depth d, concrete mix proportions, water/cement ratio, cement content etc. C_0 is the equilibrium chloride concentration (in % by weight of cement) on the concrete surface and C_{cr} is the critical chloride concentration at the reinforcement level. Using Fick's law of diffusion the chloride concentration C(x,t) at depth x from the surface and at time t can be determined by

$$C(x,t) = C_0(1 - erf(\frac{x}{2\sqrt{D_c t}}) \quad (3)$$

Table 1
Definition of stochastic variables

	Stochastic variable	Distribution type
X_1	Concrete cover	Normal
X_2	Height of beam	Normal
X_3	Height of deck	Normal
X_4	Initial diameter of reinforcement	Normal
X_5	With of column	Normal
X_6	Dept of column	Normal
X_7	Compression yield stress, concrete	Normal
X_8	Yield stress of reinforcement	Normal
X_9	Uniformly distributed dead load	Normal
X_{10}	Uniformly distributed traffic load	Gumbel
X_{11}	Point traffic load	Gumbel
X_{12}	Chloride concentration on concrete surface	Normal
X_{13}	Chloride diffusion coefficient	Lognormal
X_{14}	Coefficient rate of carbonation	Normal
X_{15}	Rate of corrosion	Normal
X_{16}	Measurement uncertainties	Normal

3. RELIABILITY ASSESSMENT

The reliability of the bridge is measured using the reliability index β for a single failure element or for the structural system (the bridge) (Thoft-Christensen & Baker [5], Thoft-Christensen & Morutsu [6]). The overall requirement is that the expected reliability index should never be smaller than some minimum reliability index. In the present version of the expert systems 3 failure modes are implemented namely: "positive" and "negative" bending failure of a reinforced T-beam (see figure 1a) and compression failure of a rectangular reinforced column (two models, namely with and without bending moments, see figure 1b). The probability of failure is estimated using the software package RELIAB.

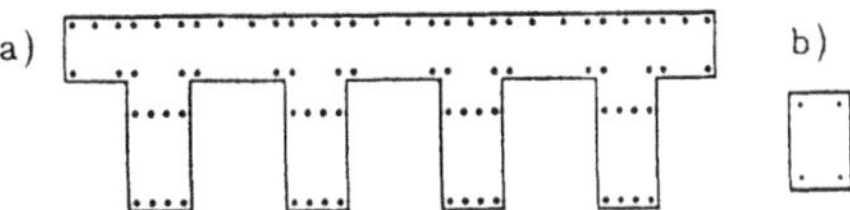

Figure 1. a) reinforced T-beam; b) rectangular reinforced column.

4. FAILURE PROBABILITY UPDATING

Two main types of updating of the probability of failure estimates are considered:

- Updating of stochastic variables based on measured samples of the stochastic variables, e.g. measurements of the yield strength of the reinforcement.
- Updating based on general information, e.g. the observation that the structure is not failed or that a corrosion degree is smaller than a certain value is measured.

Basic variable updating is performed within the framework of Bayesian statistical theory (Lindley [2], Aitchison & Dunsmore [1]). The updating based on general information is mainly based on the Bayesian methods suggested by Madsen [3] and Rackwitz & Schrupp [4].

Let the density function of a stochastic variable X be given by $f_X(x,\Theta)$, where Θ are parameters defining the distribution of X The parameters Θ are treated as uncertain parameters (stochastic variables). $f_X(x,\Theta)$ is therefore a conditional density function $f_X(x|\Theta)$. The initial (or prior) density function for Θ is called $g'_\Theta(\theta)$.

When an inspection is performed n realisations $\overline{x}^* = (x_1,\ldots,x_n)$ of the stochastic variable X are obtained. The inspection results are assumed to be independent. An updated density function Θ taking into account the inspection results is then defined by

$$g''_\Theta(\theta|\overline{x}^*) = \frac{f(\overline{x}^*|\theta)g'_\Theta(\theta)}{\int f_n(\overline{x}^*|\theta)g'_\Theta(\theta)d\theta} \quad , \quad f_X(x|\overline{x}^*) = \prod_{i=1}^{n} f_X(x_i|\theta) \tag{4}$$

The updated density function of X taking into account the realisations $\overline{x}^*$ is then obtained by

$$f_X(x|\overline{x}^*) = \int f_X(x|\theta)g''_\Theta(\theta|\overline{x}^*)d(\theta) \tag{5}$$

In the expert systems the functions $g'_\Theta(\theta)$, $g''_\Theta(\theta)$, and $f_X(x|\overline{x}*)$ are implemented for normal distributions with unknown mean and standard deviations, for the Gumbel distribution with unknown location parameter, for the Weibull distribution with unknown location parameter, for the Frechet distribution with unknown location parameter, and for the exponential distribution with unknown location parameter.

5. THE EXPERT SYSTEM FUNCTIONALITIES

The expert system is divided into two expert system modules which are used in two different situations, namely by the inspector of the bridge during the inspection at the site of the bridge and after the inspector has returned to his office. During the inspection the expert system will supply information on: the causes of observed defects, appropriate diagnosis methods, and related defects. Further the inspector will be asked to record the inspection results so that they can be used later for e.g. assessment of the reliability of the bridge and in the decision whether a detailed structural assessment is needed.

A detailed analysis of the state of the bridge after an inspection takes place when the inspector has returned to his office and after testing in the laboratory has taken place. The output of the analysis includes an updated estimation of the reliability of the bridge, decision whether a structural assessment should take place, decision whether repair should take place, relevant repair procedures, and the time for repair. Expert knowledge is used to improve the quality of the decisions.

6. ARCHITECTURE OF THE EXPERT SYSTEMS

The architecture of the expert system is shown in figure 2. It consists of two main modules called BRIDGE1 and BRIDGE2. The BRIDGE1 expert system is used at the site of the bridge. BRIDGE2 is used by the engineer in the office.

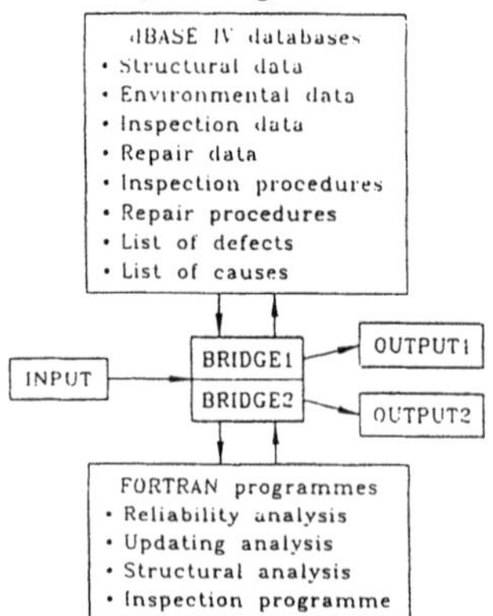

Figure 2. Architecture of the expert systems.

7. APPLICATION OF THE EXPERT SYSTEMS

The general inspection, maintenance, and repair model from inspection no. i at time t_i to inspection no. i+1 at time $t_{i+1} = t_i + \Delta t$ is indicated in figure 3 , where also the application of the modules BRIDGE1 and BRIDGE2 is shown. C, D and A are *current* , *detailed inspections* and *structural assessments* respectively. M is *maintenance* and repair of minor defects and R is Structural *repair*. B1 indicates application of *BRIDGE1*. B2(M), B2(I) and B2(R) are the *maintenance subsystem*, the *inspection module and* the *repair subsystem* in *BRIDGE2* respectively.

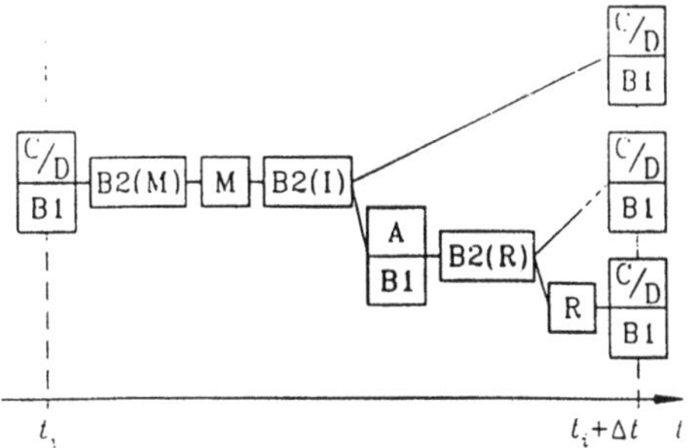

Figure 3. The inspection, maintenance, and repair model.

After a current or a detailed inspection BRIDGE2 is used to rate the maintenance and minor repair work needed and to decide if a structural assessment has to be performed. The decision is based partly on estimates of the reliability of the bridge and partly on expert knowledge. The decision does not include economic considerations.

After a structural assessment BRIDGE2 is used to decide if a repair has to be performed and also to give the optimal point of time for the repair. Expert knowledge as well as numerical algorithms are used. The decisions are partly based on a cost-based optimisation where different repair possibilities (selected by expert knowledge) and no repair are compared. The total expected costs are minimised using the FORTRAN inspection module.

8. DECISION MODEL WITH REGARD TO STRUCTURAL ASSESSMENT

Let t_i be the time of a periodic inspection and let the updated reliability index at time t be $\beta(t,t_i)$. The *general decision model* with regard to the structural assessment can then be formulated as:

- If $\beta(t_{i+1},t_i) > \beta^{min}$ then the inspection at time t_{i+1} should be a current or detailed inspection unless the damage is so serious that a structural assessment is needed. This decision is based on expert knowledge. β^{min} is the minimum acceptable reliability index (e.g. 3.72).
- If $\beta(t_{i+1},t_i) \leq \beta^{min}$ then a structural assessment should be performed before the next periodic inspection

9. MODELLING OF REPAIR

After a structural assessment it must be decided whether the bridge should be repaired and if so, how the repair is performed. In order to decide which repair type is optimal after a structural assessment, the following optimisation problem is considered for each repair technique:

$$\max_{T_R, N_R} W(T_R, N_R) = B(T_R, N_R) - C_R(T_R, N_R) - C_F(T_R, N_R) \quad (6)$$

$$s.t. \quad \beta^U(T_L, T_R, N_R) \geq \beta^{min} \quad (7)$$

where the optimisation variables are the expected number of repair N_R in the remaining lifetime and the time T_R of the first repair. W is the total expected benefits minus costs in the remaining lifetime of the bridge. C_R is the repair cost capitalised to time t=0 in the remaining lifetime of the bridge. C_F is the expected failure costs capitalised to time t=0 in the remaining lifetime of the bridge. T_L is the expected lifetime of the bridge. β^U is the updated reliability index. β^{min} is the minimum reliability index for the bridge (related to a critical element or to the total system). The repair decision is then based on the results of solving this optimisation problem but also on expert knowledge.

10. BRIDGE1

The expert system module BRIDGE1 contains useful information concerning the bridge being inspected and the defects being observed. The information includes: general information about the bridge, appropriate diagnosis methods for each defect, probable causes for each defect, and other defects related to a defect. It is also possible to create a provisional defect report. The general information about the bridge stored in the database for the selected bridge can be reviewed. The database contains information about: bridge site, design, budget, traffic, strength, load, deterioration, factors that models costs, and the cross-sections entered for the bridge. New cross-sections can be entered for the selected bridge. The information stored in the database for each cross-section contains: cross-section identification, geometry of cross-section (detailed description of the reinforcement layers for cross-sections in the deck), failure mode, and load data. Technical support can be provided for a defect. The technical support include:

- List of diagnosis methods that can be used to observe a selected defect.
- List of probable causes of a selected defect.
- List of defects associated with the selected defect.

11. BRIDGE2

The expert system module BRIDGE2 is used to make a detailed analysis of the bridge after an inspection when testing in the laboratory has taken place. New bridges and cross-sections can be entered in the database and existing bridges and cross-sections can be edited. For the bridges in the database the following options are available: review provisional defect reports, enter inspection results, estimate the reliability index, plan maintenance work and estimate costs, plan structural repair work and estimate costs, and review the agenda of inspection for one bridge or all bridges. Further, the database can be updated after repair.

New bridges can be entered and existing bridges can be edited. The general information about the bridges stored in the database contains information about: bridge site, design, budget, traffic, strength, load, deterioration, factors that model the costs, and the cross-sections entered for each bridge.

The reliability index for the bridge can be estimated by the integrated FORTRAN program RELIAB. Both the reliability index when no inspection results are taken into account and the updated reliability index when all inspections performed for the bridge are taken into account can be estimated.

The following submodules are integrated in BRIDGE2:

- BRIDGE2(M) is the maintenance/small repair submodule. This submodule is always used after a current or detailed inspection. It assists in selecting the maintenance work and repair of minor structural defects to be performed and estimate the maintenance costs. The defects are rated based on the defect classification in terms of rehabilitation urgency, importance of the structure's stability, and affected traffic recorded during the inspection .
- BRIDGE2(I) is the inspection strategy submodule. This submodule is always used after a current or detailed inspection. It assists in the decision whether a structural assessment is needed before the next periodic inspection. The decision taken in BRIDGE2(I) is mainly based on the updated reliability index for the bridge calculated by RELIAB (see figure 8). If the value of the updated reliability index for the bridge is acceptable then each of the defects detected at the latest periodic inspection and the combination of defects are investigated. Based on expert knowledge it is investigated whether from a structural point of view a defect or combinations of defects require a structural assessment.

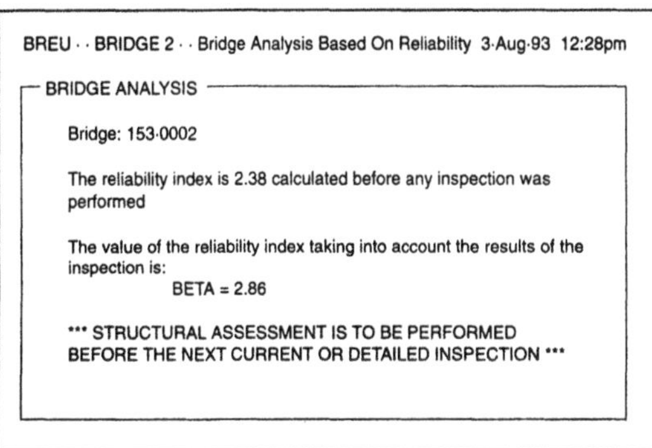

Figure 4. Decision tool related to structural assessment.

- BRIDGE2(R) is the repair submodule. This submodule is always used after a structural assessment. It assists in selecting the optimal structural repair technique (including no repair) to be performed, when the repair should be performed, and the number of repairs in the remaining lifetime of the bridge. Further, the expected benefits minus costs are estimated. The repair plan is optimised based on a cost-benefit analysis by the FORTRAN program INSPEC (see figure 5).

The FORTRAN program RELIAB can be used to estimate the reliability of a reinforced concrete bridge. Two different failure modes are considered, namely bending failure of the main beam of a bridge and compression failure of a column.

The FORTRAN program INSPEC can be used to estimate the optimal repair time and number of repairs for a given repair method. The estimation is based on a cost-benefit analysis for the bridge. The total expected benefits minus expected repair and failure costs in the remaining lifetime of the bridge is optimised.

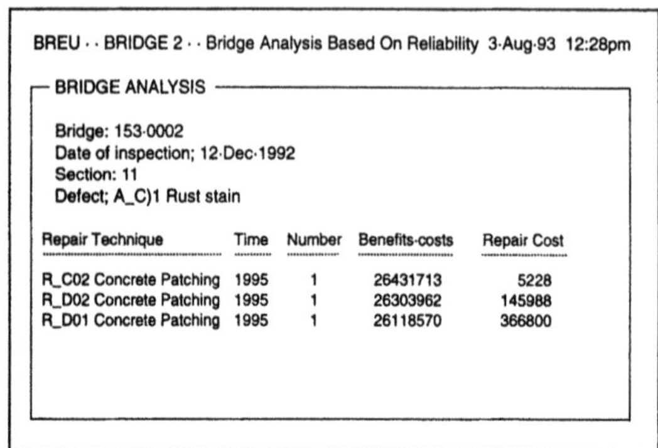
BREU - - BRIDGE 2 - - Bridge Analysis Based On Reliability 3-Aug-93 12:28pm

BRIDGE ANALYSIS

Bridge: 153-0002
Date of inspection; 12-Dec-1992
Section: 11
Defect; A_C)1 Rust stain

Repair Technique	Time	Number	Benefits-costs	Repair Cost
R_C02 Concrete Patching	1995	1	26431713	5228
R_D02 Concrete Patching	1995	1	26303962	145988
R_D01 Concrete Patching	1995	1	26118570	366800

Figure 5. Optimised repair plan for the defect 'rust stain'

ACKNOWLEDGEMENT

This paper presents results of the CEC supported research project BREU P3091 "Assessment of Performance and Optimal Strategies for Inspection and Maintenance of Concrete Structures Using Reliability Based Expert Systems". The partners in the project are CSR, Aalborg, Denmark; University of Aberdeen, Aberdeen, UK/Sheffield Hallam University, Sheffield, UK; Jahn Ingenieurbureau, Hellevoetsluis, Holland; Instituto Superior Técnico, Lisboa, Portugal , and LABEIN, Bilbao, Spain

REFERENCES

1. Aitchison, J. & I.R. Dunsmore. *Statistical Prediction Analysis* , Cambridge University Press, Cambridge, 1975.
2. Lindley, D.V. *Introduction to Probability and Statistics from a Bayesian Viewpoint* , Vol. 1+2, Cambridge University Press, Cambridge, 1976.
3. Madsen, H.O. *Model Updating In Reliability Theory* , Proc. ICASP5, pp. 564--577, 1987.
4. Rackwitz, R. & K. Schrupp . *Quality Control, Proof Testing and Structural Reliability*, Structural Safety, Vol. 2, pp. 239--244, 1985.
5. Thoft-Christensen, P. & M.J. Baker. *Structural Reliability Theory and Its Applications* , Springer Verlag, 1982.
6. Thoft-Christensen, P. & Y. Murotsu . *Application of Structural Systems Reliability Theory* , Springer Verlag, 1986.
7. Thoft-Christensen, P. & H.I. Hansen. Optimal Strategy for Maintenance of Concrete Bridges Using Expert Systems. Proc. ICOSSAR '93, Innsbruck, August 1993.
8. Thoft-Christensen, P. *Assessment of Performance and Optimal Strategies for Inspection and Maintenance of Concrete Structures Using Reliability Based Expert Systems* . Proc. ESReDA seminar on "Maintenance and System Effectiveness", Charmonix, France, April 1994, pp. 97-112.

29

Seismic reliability of electric power transmission systems

I. Vanzi [b], R. Giannini [a], P.E. Pinto [b]

[a] Department of Structural Engineering (DISAT), Università de L'Aquila, Italy
[b] Department of Structural and Geotechnical Engineering (DISG), Università di Roma "La Sapienza", Italy

Electric power transmission systems are networks of nodes (stations) connected by lines (power transmission lines), dedicated to the production, transformation and distribution of electric power to users. The network contains many seismically fragile elements, the stations being composed of slender steel-ceramic members, and is a complex system because it is capacitive and capable of insulating short-circuits. In the present work the functional logic of a generic electric network is analyzed and a model suitable for reliability assessment is estabilished. The model is then applied to evaluate the seismic reliability of a realistic network comprising the central part of Italy, using simulation methods. Various measures can be adopted to assess the performance of a network: the one selected in this study is a global index related to the quantity and the quality of the power delivered.

1. ELEMENTS OF THE NETWORK: STATIONS AND POWER TRANSMISSION LINES

In fig.1 an example high-tension electric network (Rome compartment, central Italy) is shown.The two basic components of the network, i.e. stations (nodes) and power transmission lines (lines), are recognizable. Notice that in fig.1 each line may also represent two or more coincident lines. Stations are of three different types: 1)power plants: where electric energy is produced; different types of power plants exist, e.g. thermic, hydroelectric and nuclear power plants. 2)transformation stations: where electric voltage is transformed.; high voltage is used for long-distance transmission; voltage is then transformed to lower values when approaching consumers; for reference, in Italy the following tensions are used : 380, 220, 150, 132, 66, 45, 30, 20, 15, 10, 6, 3, 0.380, 0.220 KV. 3)distribution stations: these deliver electricity to consumers; they very often are at the same location as transformation stations. Power transmission lines are generally grouped into three categories, depending on the voltage of the transported electricity : 1)high voltage, 380 to 220 KV (AT); for long-distance transmission; there is little redundancy in AT transmission networks. 2)Medium voltage, 150 to 30 KV (MT); for regional transmission; more redundant than AT 3)Low voltage, below 30 KV (BT);

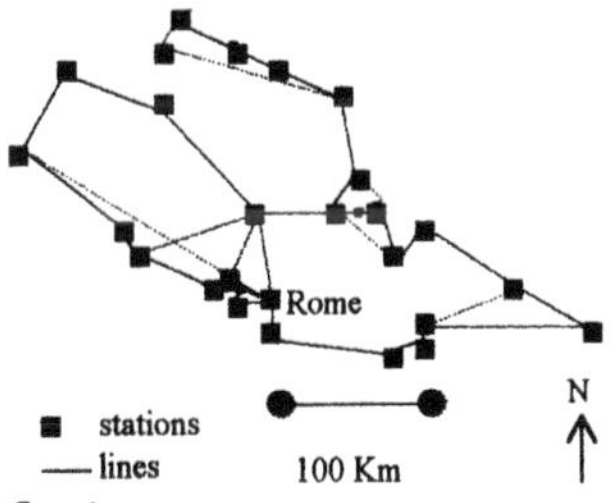

fig. 1
Example high-tension electric network (Rome compartment, central Italy)

to final consumers; very redundant networks. Transmission lines are composed of steel towers (or wood poles for lower tension), carrying ceramic suspension insulators and copper or aluminium cables. They are much less fragile than the stations' components versus seismic actions, which allows to consider them as not vulnerable. Electric power transmission systems are complex systems since they are redundant, capacitive, with time-varying configuration (the latter is auto-adaptive to the demand of electric power) and *intelligent* because protected by mechanisms which isolate breakdowns, driven by electronic logic.The present work only deals with AT network and assumes that only transformation and distribution stations contain seismically fragile elements. The reason to restrain the study to the AT network is that its failures are much more important than lower tension networks' failures. Any major failure in the AT network implies, in fact, consequences on a national or even international scale. For example the italian and french AT electric networks have several links and are pretty sensitive to each other's problems. On the other hand, any failure in the MT or BT network would only have regional or local consequences.

2. FUNCTIONING OF THE NETWORK; PERFORMANCE INDEX

The components of an electric network are designed to produce, transform and distribute electric energy. Electric power originates from production stations usually at medium voltage; after transformation to higher voltage (320 or 220 KV) electric energy flows into the transportation network. The network (lines+stations) is a capacitive system, i.e. is capable of carrying a finite amount of power to the demand nodes which must be fed with electric power at prescribed tension. A global index of the quality of the functioning of the network in a specified condition has to account for the amount and quality (in terms of shift from the target voltage at the nodes) of the electric power delivered at the demand nodes, compared with the values of voltage and power in an optimal condition. The following form is proposed:

$$q(c)=\frac{\sum_{d}\left(\mathrm{P}_{r}(c)\right)_{d}\cdot\left(1-\frac{\left\|V_{d}(c)\right|-\left|V_{d}(oc)\right\|}{\left|V_{d}(oc)\right|}\right)\cdot step\left[\left(\frac{\left\|V_{d}(c)\right|-\left|V_{d}(oc)\right\|}{\left|V_{d}(oc)\right|}\right)-\alpha\right]}{\sum_{d}\left(\mathrm{P}_{r}(oc)\right)_{d}} \qquad (1)$$

with: q=global quality index; $\mathrm{P}_r(*)$=real part of the electric power in condition * (only the real part of P actually gives power to consumers); $|V_d(*)|$=modulus of the tension in condition *; α=maximum allowed shift from the target voltage (we assume 10%); d=demand node; c=current condition; oc=optimal (reference) condition. In Equation (1) the delivered power in condition c is weighted using as weight the shift from the target voltage.

3. STATIONS

3.1 Description and logical model

The seismically fragile components are concentrated in the nodes (stations) of the network.
The typical layout of a distribution-transformation station is shown in fig. 3; microcomponents are shown in fig. 2. The components may be grouped as follows: (i) measuring

instruments:TA,TV (ii) short-circuit and lightning protection instruments: switches, dischargers (iii) various supports and mechanical devices: horizontal and vertical sectionalising switches, coil support, bar support (iv) other components: line boxes, power supply to protection system, transformers. (i), (ii) and (iii) are slender, vertical and steel-ceramic members. The following aspects may be noted to help understanding and modeling the functioning of a distribution-transformation station: 1)two sides: high and low tension (380 and 132 KV) 2)symmetrical with respect to transformers 3)some macro-components may be identified: (1)*lines in-out* (montante di linea) : TV+ coil support+ sectionalising switch+ TA+ protection switch+ supports; (2)*bars-connecting line* (montante parallelo) : TA+ protection switch+ sectionalising switch+ supports; (3)*bars*: TV+ supports; (4)*line boxes*: containing measurement and electronic equipment; (5)*power supply to protection system*: small transformers+ building+ diesel engine+ batteries; (6)*autotransformer line* (montante ATR): transformer+ dischargers+ TA's+ protection switches+ supports; 4)power generally flows from high-voltage to low-voltage side but flow inversions and/or insulation of low or high tension side (hence no more transformation, only distribution) are possible;5)only active microcomponent: protection switches, driven by the corresponding line box *and* the station's power supply to protection system (chain system), which switch open to insulate short-circuits originating from collapses.

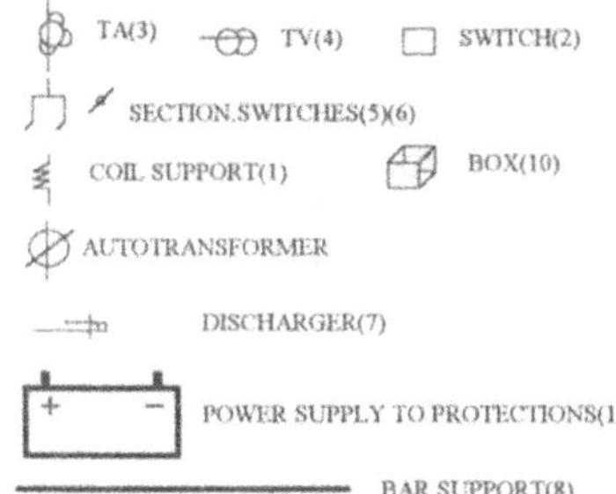

fig.2 macrocomponents

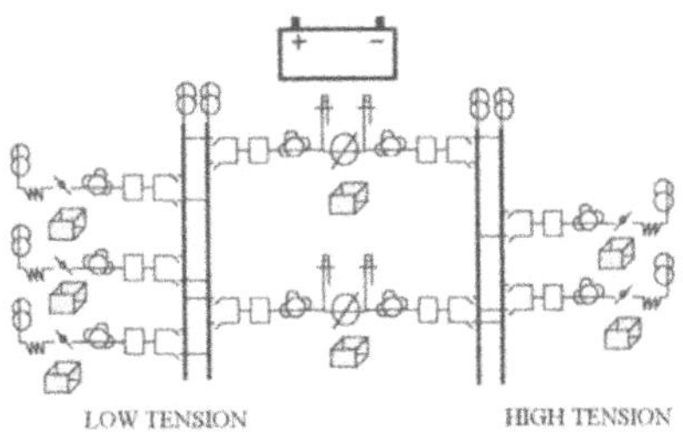

fig. 3 scheme of a trans.-distr. station

Macrocomponents are recognized by searching the longest and simple enough chain (i.e. series elements) of microcomponents.

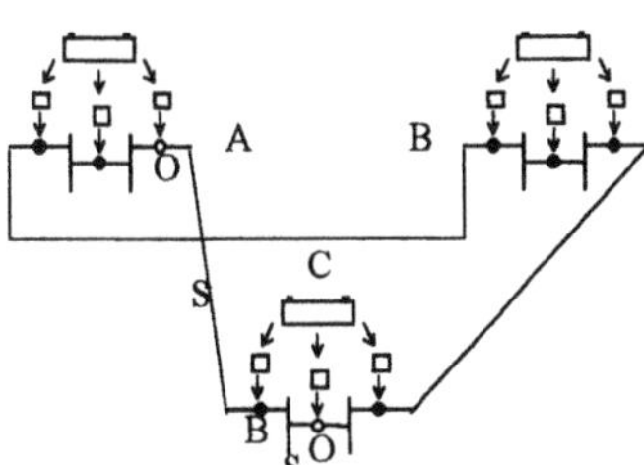

fig.4 example of the functioning of a station

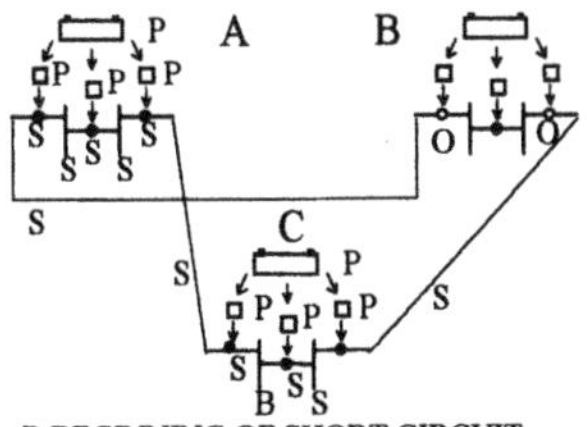

B:BEGINNING OF SHORT CIRCUIT
S:SPREAD OF SHORT-CIRCUIT
O:INSULATED MACROELEMENT AND OPEN SWITCH
P:PASSIVE ELEMENTS

fig.5 example of the functioning of a station

The functioning of the protections and the interdependencies of the short-circuits and insulation of the stations in a network may be visualized considering the two examples of fig. 4 and 5 using a simplified 3 stations network. In the first one (fig.4) all of the macrocomponents in the stations are properly working apart from a short-circuit on a line of station C. The protection switches open and insulate the remaining parts of the stations. Notice that, while station B is fully functional, stations A and C are only partially working. Fig. 5 clarifies the central role of power supply to protections. In fact, with non-working protection switches, a single short-circuit similar to that of the previous case, makes the whole system to collapse (station C is directly involved in the short-circuit while stations A and B are insulated).

2.2 Fragilities of the macrocomponents

Most of the microelements of a reference distribution-trasformation station have been recently subjected to shaking table tests(/1/) . These tests allowed the corresponding mechanical model to be calibrated , from which the fragilities have been evaluated using the standard FORM/ SORM methods. The intensity factor for the seismic motion is the peak ground acceleration. The generic macrocomponent (lines, boxes, bars, power supply to protection) is a series system composed of microelements. Assuming their failures are independent it is straightforward that :

$$\mathrm{P_s}(Macro) = \mathrm{P}\left(\bigcap_m S_m\right) = \prod_m \mathrm{P}(S_m) = \prod_m [1-\mathrm{P}(F_m)];\ \mathrm{P_f}(Macro) = 1 - \prod_m [1-\mathrm{P}(F_m)] \quad (2)$$

m = index for microcomponent in the Macro

Fig. 6 to 9 show the obtained fragilities for micro and, from (2), for macrocomponents.

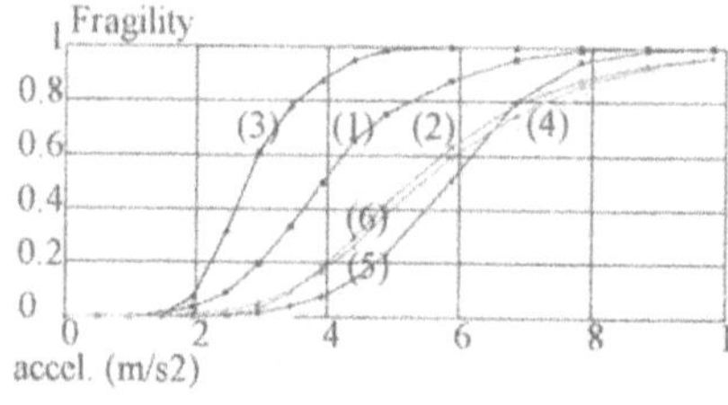

fig.6 Fragilities of microcomponents

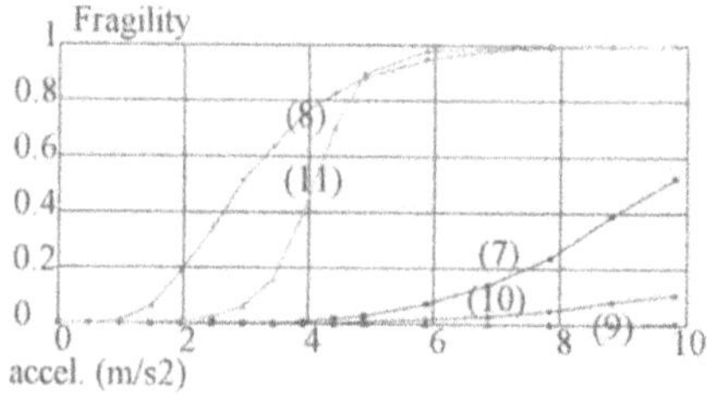

fig.7 Fragilities of microcomponents

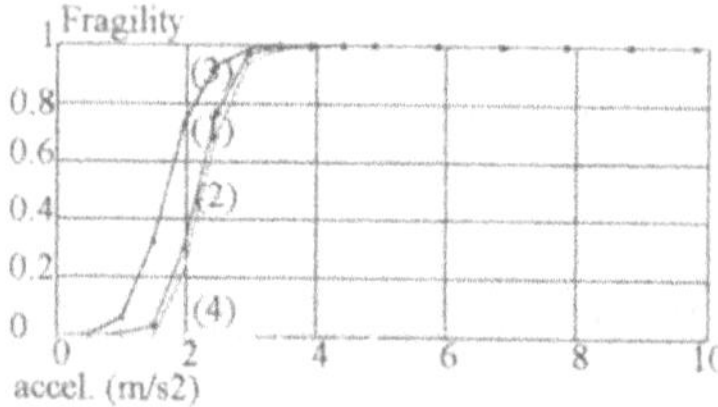

fig.8 Fragilities of macrocomponents

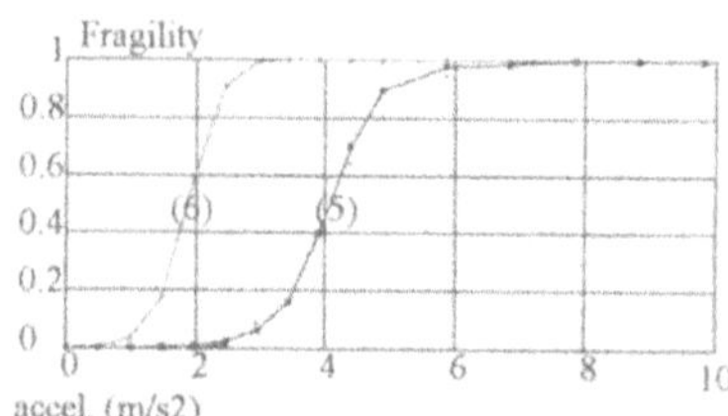

fig.9 Fragilities of macrocomponents

In short, the reliability model for the stations includes a number of fragile macrocomponents and the actual intelligent response of the stations both to element failures and to collapses and short-circuits.Clearly the station can be described as a multi-state component.

3. LOAD-FLOW ANALYSIS

The system is modeled as a capacitive network, i.e. is capable of transporting a finite amount of power from production to users, via the transportation lines. The elements of the network are the nodes (N stations, divided into C load, G supply and 1 balance stations) and the lines (L transportation lines). Boundary conditions are specified in terms of power at the C nodes (2xC conditions), real power and tension modulus at the G nodes (2xG conditions) and tension modulus and phase at the balance node (2 conditions). The electrical characteristics of the lines from node i to node j (i.e. longitudinal and transversal impedances, conceptually the power required by the line to let through a flow of electric energy) influence both the value of the elements (i,i), (j,j), and (i,j) of the *admittance matrix* Y (NxN complex matrix), which accounts for the state of the lines of the network, and the power at the nodes. In fact the relationships:

$$Y_{p,p} = \sum_q y^L{}_{p,q} + \sum_q \frac{y^T{}_{p,q}}{2}; \quad Y_{p,q} = -y^L{}_{p,q}; \quad P_n = E_n \cdot \overline{[Y \cdot E]}_n \tag{3}$$

with :

$y^L{}_{p,q}$ = longitudinal admittance of line p,q; $y^T{}_{p,q}$ = transversal admittance of line p,q

$Y_{p,q}$ = p,q element of Y matrix; E_n = complex tension at node n; P_n = complex power at node

hold. Unknowns of the load-flow problem are the 2xN real and complex tension values at the nodes. The problem is mathematically formulated as a system of 2N algebraic non linear equations in E_n , solved by trial and error procedure (which may not converge in the case of unstable conditions). Collapses of the macrocomponents of the station i imply insulation of one or more (i,j) lines; matrix Y and power flow along the network sides change and hence the network is in a variated condition.

fig 10

4. AN EXAMPLE APPLICATION : CENTRAL ITALY

4.1 Electric network and earthquake model

As an example application the electric network of central Italy has been chosen. The network consists of 60 nodes (8 supply and 51 demand stations, plus one balance node) and 82 lines. This network is about one third of the italian AT network. It is located in a rather active seismogenetic region.Thirteen seismogenetic areas have been considered, which are shown in fig. 10. The Cornell model for the seismic hazard has been chosen.

In each seismogenetic area, the coordinates of the epicenter are indipendent random variables with constant distribution between the minimum and maximum values of the coordinates of the points of the area. The activity of the seismogenetic area is expressed via the doubly truncated Gutemberg- Richter expression :

$$\lambda(i) = \lambda(i_{MIN}) \cdot [1 - F_I(i)]; F_I(i) = 1 - \frac{\exp(-\beta \cdot i) - \exp(-\beta \cdot i_{MAX})}{\exp(-\beta \cdot i_{MIN}) - \exp(-\beta \cdot i_{MAX})} \quad (4)$$

β = severity parameter ; $i_{MIN}[i_{MAX}]$ = lower[upper] bound for the earthquake's intensity

The parameters of the Gutemberg- Richter law are evaluated on the basis of the historical seismicity and are shown in table 1.The adopted attenuation law is (/2/): $\Delta I_s(R) = a + b \cdot (R + R_o) + c \cdot \log(R + R_o) + \varepsilon$ with parameters : R_0 = 3 Km; a = −0.5; b=4.43; c=0.056 , ε =N(0; σ_ε = 1.037).

The law (/3/) $a[cm/s^2] = 10^{MM \cdot 0.237 - 0.594} + \varepsilon$ with ε =N(0; σ_ε = 0.35) has been adopted to convert the earthquake intensity from MM to cm/s2 , since the fragilities of the macrocomponents are expressed as a function of the soil acceleration.

region	β	λ	i_{min}	i_{max}
1	1.043	0.886	4.0	11.0
2	1.246	1.257	4.0	11.0
3	0.819	0.827	4.0	10.0
4	1.167	4.348	4.0	11.0
5	1.29	1.561	4.0	10.0
6	1.574	1.074	4.0	9.0
7	1.145	1.092	4.0	11.0
8	0.868	0.368	4.0	12.0
9	0.991	0.719	4.0	11.0
10	1.236	1.494	4.0	11.0
11	0.891	0.653	4.0	12.0
12	1.113	0.944	4.0	12.0
13	0.471	0.176	4.0	9.0

table 1 seism. areas

4.2 Montecarlo procedure

A Montecarlo procedure has been set up to assess the seismic reliability of the electric network. The steps may be listed as follows:

Cicle on all the seismogenetic areas
  Generate (from G.R. law) a vaue for the intensity of the earthquake
  Cicle on the N sites of the stations
  Attenuate the intensity
  Generate (from N(0;σε)) a value for the error due to the attenuation
  Convert to acceleration
  Generate (from N(0;σε)) a value for the error due to the conversion
  Cicle on the M macroelements of the station
    Generate (from the fragility/acc.) the state of M (surviv./failed)
  Spread short-circuits and/or insulation of subareas of the station
Spread short-circuits and/or isolation of subareas of the network among the N stations
Load-Flow analysis---->q
Estimate P[q belong to q1;q2] and the c.o.v. of the estimator
If c.o.v.>maximum allowed error generate one more value for the intensity of the earthquake and repeat the analysis.

5. RESULTS

The results shown in figs.11 and 12 have been obtained. Fig.11 shows the cdf of the random variable q, as defined in paragraph 2, given an earthquake has occurred in the region being studied. The probability that q is equal to 100% (optimal functioning) is around 90%. The remaining probability mass (10%) is mainly concentrated in the range q=[80;99]%. If an assumption is made about the repair time before optimal functioning after an earthquake, it is then possible to compute the moments of the random variable energy not distributed in 1 year. Fig. 12 shows the mean and standard deviation of this r.v. obtained with the assumption of repair time equal to 4 hours. Seismogenetic areas 3 and 4 have proven to be the most influential with respect to the other ones. The large values of standard deviations as compared to the mean values are due to the fact that the underlying pmf shows a clear bimodal shape with maximum at the nul value and second larger value of pmf at the largest value of the random variable.

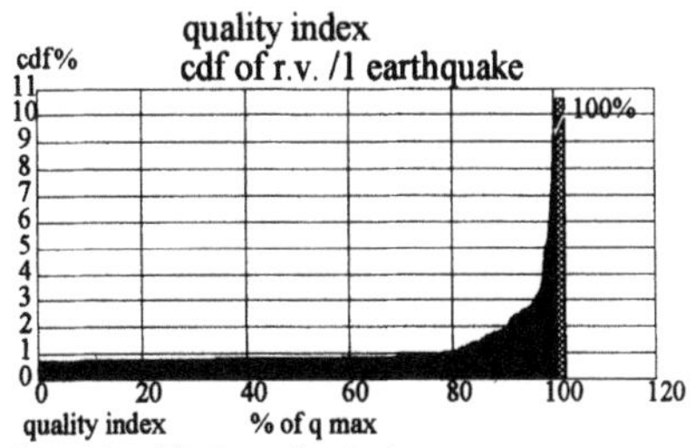

fig. 11 cdf of quality index

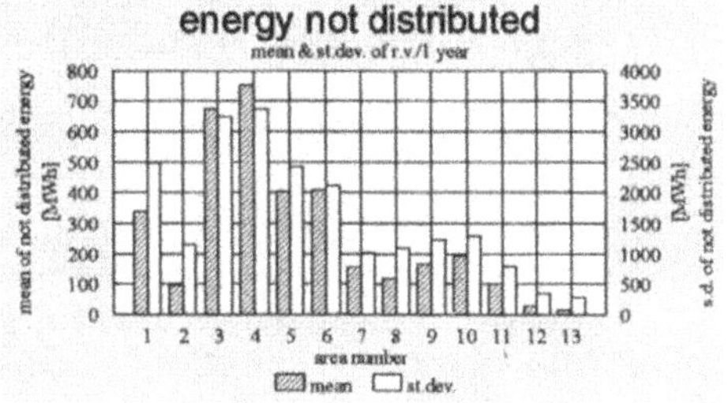

fig. 12 means of energy not distributed

6. CONCLUSIONS

The method developed allows to determine the global reliability of electric networks.

As a general remark on the model and results, one should notice the following points:

- important interactions exist between adjacent AT networks; on the other hand, when limiting the space extension of the electric network, neighboroughing systems are modeled via a few load or supply nodes with prescribed power.
  The extension in space of the studied network is likely to be too small and the inclusion of the whole italian AT network is next programmed.
- the results are to a great extent dependent on the macrocomponents' fragilities. As for boxes and power supply system a more accurate analysis is desirable.
- the performance index should also take into account quantities like the number of broken pieces of equipment and the overload of some of them, the repair times, the economical and social difference in lack of power supply to big or small centers.
- the results obtained here only consider the AT network, wich is the one containing the smaller number of fragile components with respect to MT and BT networks, and are dependent on the activity and severity of the seismogenetic areas. Central Italy seismogenetic areas are pretty active but only moderately severe with respect to the other italian ones (for instance the ones in Friuli, Irpinia and Calabria). An extension of the study is under progress.

ACKNOWLEDGEMENTS

The authors wish to thank E.N.E.L. (the italian state electricity board) and in particular Dr. Camensig, Ing. Giannì, Ing. Salvati and Mr. Dall'Antonia for the support given in the course of the present study.

REFERENCES

/1/ Ismes S.p.A., **Attività numeriche svolte sulle apparecchiature AT della sezione 380 KV della stazione elettrica ENEL di Selargius (CA),** 1993

/2/ Giannini, R., Giuffrè, A., Nuti, C., Ortolani, F., Pinto, P.E. : **Valutazione quantitativa del rischio sismico. Metodologie ed applicazione alla città di Subiaco,** Ingegneria Sismica, Anno I, n. 1, Novembre 1984

/3/ Decanini,L.D., Gavarini, C., Mollaioli, F. : **Algunas consideraciones sobre la correlaciones entre intensidad macrosismica y parametros del movimento del suelo**, IX international seminar on earthquake prognostics, San Josè (Costa Rica), 1994

/4/ Elgerd, O.I. : **Electric energy systems theory : an introduction,** Mc-Graw Hill Book Co., 1977

/5/ Ciampoli, M., Giannini, R., Nuti, C., Pagnoni, T., Pinto, P.E.: **Valutazione del rischio sismico delle reti di trasmissione dell'energia elettrica,** V convegno nazionale l'Ingegneria Sismica in Italia, Palermo, Settembre 1991

/6/ Ang, A.H.S., Pires, J., Viallverde, R., Schinzinger, R. : **Seismic probabilistic assessment of electric power transmission systems,** X W.C.E.E., Madrid (Spain), 1992

/7/ Silvestri, A. : **Impostazione e metodi di risoluzione dei problemi di load-flow,** l'Elettrotecnica, vol. LXIII, n.8, August 1976

/8/ Ciampoli, M., Pagnoni, T., Giannini, R. : **Seismic reliability assessment of power transmission networks by simulation technique,** X W.C.E.E., Madrid (Spain), 1992

30

Reliability Analysis of Load Path-Dependent Structures

Wei Wang*, Ross B. Corotis†, and Martin R. Ramirez‡

*Research Associate, Wind Engineering Research Center, Texas Tech Univ., Lubbock, Texas 79409, USA

†Dean, College of Engineering and Applied Science, Univ. of Colorado, Boulder, Colorado 80309, USA

‡Dean, Greenfield Virtual Univ., UDM, 4001 West McNichols Rd., Detroit, Michigan 48219, USA

1. INTRODUCTION

The state of a structure can be described in terms of the complete set of state variables at any given time in its lifetime [1,2,4,5,7]. If the state of the structure is independent of the loading history, the structure is referred to as load path-independent. For such a case, the state of the structure can be expressed in terms of the current values of the loads and common response statistics. In these cases, there exists a unique surface in the basic variable space which divides the space into two regions, representing safety and failure of the structure. This limit state can be written in the form $g(\mathbf{x}) = 0$. As will be discussed, the concept of such a function is not valid for load path-dependent structures. Instead, the concept of survival and failure paths must be used, and incipient failure directions determined for critical points.

As a first step toward the goal of developing general path-dependent methods which are more efficient than Monte Carlo methods and can provide insight into the nature of the problem, some fundamental concepts in time-varying reliability analysis of path-independent problems will be examined in the context of load path-dependency. New concepts that can help in the understanding of path-dependent reliability will be introduced.

2. LIMIT STATE REGIONS IN THE BASIC VARIABLE SPACE

Let $\mathbf{X}(t)$ be a vector of random processes, and define the *basic variable space*, $\mathcal{X}$, as a space whose axes are the sample space of each component of $\mathbf{X}(t)$. Then, for structures whose response is not load path-dependent, there is, in general, a unique limit state surface that divides $\mathcal{X}$ into a safe region and a failure region [10]. When the structural response is load path-dependent, however, there is no such division between a failure region and safe region [8,12]. At some point

$\mathbf{x} \in \mathcal{X}$, the structure may or may not fail, depending on the path which the vector process $\mathbf{X}(t)$ has taken to reach $\mathbf{x}$. To see how the points in $\mathcal{X}$ are related to the state of the structure under investigation, the following definitions are helpful.

A structure is **critical** if an infinitesimal change of $\mathbf{X}(t)$ from its current value in the next moment will bring it to failure. A point $\mathbf{x} \in \mathcal{X}$ is a **critical point** if the structure with basic variables $\mathbf{X}(t) = \mathbf{x}$ is critical. A **path to x** is a sample function, $\mathbf{x}(\tau)$, $\tau \leq t$, of the vector process $\mathbf{X}(t)$ with $\mathbf{x}(t) = \mathbf{x}$. A **survival path to x** is a path to $\mathbf{x}$, under which the structure has not failed before moment t, and is safe or critical at t. A **final survival path to x** is a path to $\mathbf{x}$ for which the structure may or may not have failed before moment t, but is safe or critical at t. A structure may have "failed" before moment t, but be safe at t in the case, for example, of a serviceability limit state. A survival path to $\mathbf{x}$ is also a final survival path to $\mathbf{x}$. A final survival path may not be defined if some of the basic variable processes, such as material properties, have no definition after the failure of the structure. An **immediate survival direction at x** is a direction in which an *infinitesimal* step away from $\mathbf{x}$ will not cause failure to the structure. An **immediate failure direction at x** is a direction in which an *infinitesimal* step away from $\mathbf{x}$ will cause failure to the structure.

It appears that there are only three types of points in the basic variable space:

Type f: Points which have no survival path reaching them.

Type t: Points which for some survival paths reaching them, have both immediate survival and failure directions.

Type s: Points which have no immediate failure directions for any survival path that ends at them.

Regions in $\mathcal{X}$ can be defined correspondingly,

$$\begin{array}{lll} \textbf{failure region:} & D_f = & \{\text{All type f points}\} \\ \textbf{transition region:} & D_t = & \{\text{All type t points}\} \\ \textbf{safe region:} & D_s = & \{\text{All type s points}\} \end{array} \tag{1}$$

These three regions are mutually exclusive and collectively exhaustive.

For any point $\mathbf{x} \in D_t$, there will, in general, be paths to $\mathbf{x}$ which do not have failure directions (i.e., failure is not imminent). By the definition of a type t point, however, there exists at least one path to $\mathbf{x}$ which has failure directions (i.e., $\mathbf{x}$ is critical). Define $q(\mathbf{x};t)$ to be the likelihood that $\mathbf{x}$ is critical. Then

$$q(\mathbf{x};t) = q_1(\mathbf{x};t)q_2(\mathbf{x};t) \tag{2}$$

where

$$q_1(\mathbf{x};t) \equiv \text{the likelihood that there is a survival path to } \mathbf{x} \tag{3}$$

and

$$q_2(\mathbf{x};t) \equiv \text{the likelihood that there is an immediate failure direction at } \mathbf{x} \text{ for any survival path to } \mathbf{x} \tag{4}$$

q_1 can be viewed as the traditional reliability function L_T over space, and q_2 the traditional hazard function $h(t)$ over space. $q_2(\mathbf{x},t)$ can be defined to equal unity over region D_f without conflict with Eq. 4.

Regions D_s, D_t and D_f, and quantities q and q_2 are illustrated in Fig. 1. It is conceivable that one or even two of the three regions may not exist, depending on the specific problem.

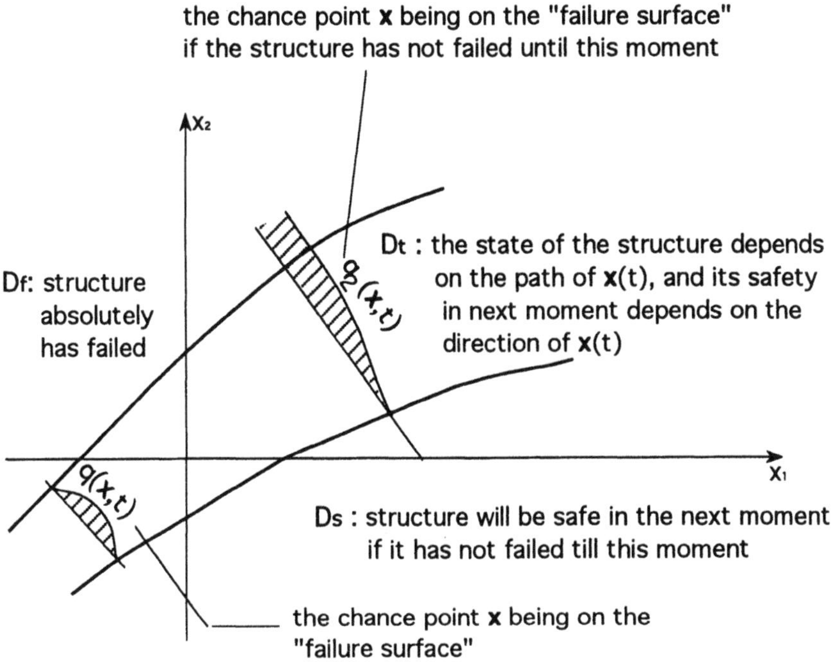

Fig. 1 Regions in the Basic Variable Space

Consider an arbitrary point $\mathbf{x} \in D_t$, and an arbitrary survival path to $\mathbf{x}$ which has failure directions at $\mathbf{x}$. If there exists *a segment of surface which passes through* $\mathbf{x}$ *and separates the failure directions from the survival directions*, this segment of surface can then be defined as the limit state surface of the structure *at point* $\mathbf{x}$ *of path* $\mathbf{x}(t)$. It is similar to the limit state surface in the path-independent problems, except that it must be defined pointwise and pathwise.

If the segment of the failure surface at $\mathbf{x}$ of path $\mathbf{x}(t)$ is smooth, it is necessarily a hyperplane. A failure direction, $\mathbf{n}$, of unit length perpendicular to the hyperplane can be defined as the unit outnormal vector to the failure surface at $\mathbf{x}$ by path $\mathbf{x}(t)$. It is dependent on path $\mathbf{x}(t)$ and time t:

$$\mathbf{N} = \mathbf{n}(\mathbf{X}(t); t) \tag{5}$$

The upper case $\mathbf{N}$ is used in Eq. 5 to reflect the fact that the outnormal vector is a random quantity. The probability density function of $\mathbf{N}$ over the basic variable space $\mathcal{X}$, if it exists, should depend on time t and location $\mathbf{x}$, and therefore has the form,

$$f_{\mathbf{N}}(\mathbf{n}; \mathbf{x}, t) \tag{6}$$

When the segment of surface is not smooth, a normal vector does not exist. The density function in Eq. 6 is nevertheless useful if it is redefined as the likelihood that an arbitrary direction $\mathbf{n}$ is an immediate failure direction.

With the basic definitions of regions and failure directions, it is possible to address the reliability problem in terms of vector outcrossing theory. A general approach is introduced next.

3. OUTCROSSING RATE ANALYSIS

In path-independent problems, the rate at which the basic variable processes cross the limit state surface into the failure zone can be related directly to the probability of failure of the structure if outcrossing is a rare event [3, 13]. This should hold for path-dependent problems also, if the limit state surface is taken as defined in the previous section. The formulas for outcrossing rate computation are derived for path-dependent problems for the one- and multi- dimensional cases and for discrete $\mathbf{X}(t)$ in this section.

3.1 upcrossing rate – one-dimensional case

Consider a problem involving two random processes, the load process $X(t)$, and the barrier process $B(t)$. $B(t)$ depend on the entire history of X up to time t. Failure is defined as $X \geq B$.

For an upcrossing to take place during $(t, t+\Delta t)$, the following conditions must be met [9]:

$$X(t) < B(t) \quad \text{and} \quad X(t+\Delta t) > B(t+\Delta t) \tag{7}$$

That is, $X(t)$ must be below the barrier at time t and above it at time $t+\Delta t$.

When Δt is small, the approach of Rice [1944] can be used, and the second inequality in Eq. 7 can be written as

$$X(t) + \dot{X}(t)\Delta t \geq B(t) + \dot{B}(t)\Delta t \tag{8}$$

or equivalently,

$$X(t) \geq B(t) - [\dot{X}(t) - \dot{B}(t)]\Delta t \tag{9}$$

The conditions for an upcrossing to occur is thus,

$$B(t) \geq X(t) \geq B(t) - [\dot{X}(t) - \dot{B}(t)]\Delta t \tag{10}$$

Note that in Eq. 10, $b(t)$ and $\dot{b}(t)$, realizations of $B(t)$ and $\dot{B}(t)$ respectively, are different for different sample functions, $x(\tau)$, of $X(\tau)$ $(0 \leq \tau \leq t)$. This is the critical difference from the path-independent problems, where $b(t)$ and $\dot{b}(t)$ are constants for fixed t, x, and $\dot{x}$. $B(t)$ and $\dot{B}(t)$ now are also random variables whose distributions depend on the nature of the process $X(t)$.

Assume that the joint distribution of $X(t), \dot{X}(t), B(t)$ and $\dot{B}(t)$ exists. The probability associated with Eq. 10 for given $b(t)$ and $\dot{b}(t)$ is the integration of the joint pdf over the shaded area in Fig. 2. Then the probability that an upcrossing (U.C.) takes place during $(t + \Delta t)$, is,

$$P[U.C.] = \int_{-\infty}^{\infty} db \int_{-\infty}^{\infty} d\dot{b} \int_{\dot{b}(t)}^{\infty} d\dot{x} \int_{b(t)-[\dot{x}(t)-\dot{b}(t)]\Delta t}^{b(t)} f_{B(t),\dot{B}(t),X(t),\dot{X}(t)}(b, \dot{b}, x, \dot{x}; t) dx \quad (11)$$

Equation 11 can be simplified using the mean value theorem for integration,

$$P[U.C.] = \Delta t \int_{-\infty}^{\infty} db \int_{-\infty}^{\infty} d\dot{b} \int_{\dot{b}(t)}^{\infty} [\dot{x}(t) - \dot{b}(t)] f_{B(t),\dot{B}(t),X(t),\dot{X}(t)}(b, \dot{b}, b, \dot{x}; t) d\dot{x} \quad (12)$$

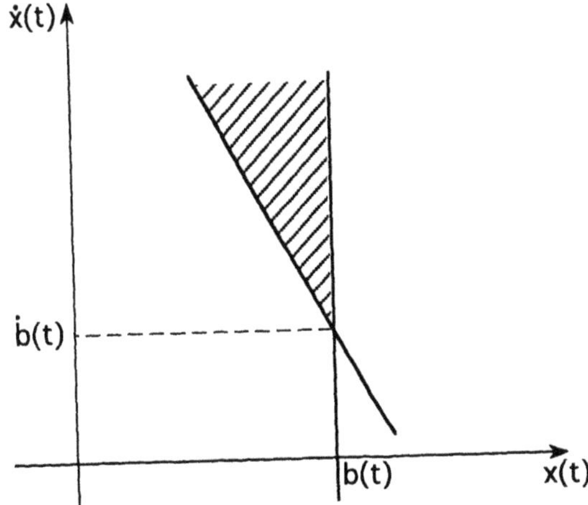

Fig. 2 Likelihood of an Upcrossing

Assume that Δt is so small that the occurrence of more than one upcrossing during $(t, t + \Delta t)$ is negligible, and define N^+ to be the number of upcrossings during this period. The expected number of outcrossings in $(t, t + \Delta t)$ is then,

$$E[N^+] = 1 \cdot \text{Prob}[U.C.] + 0 \cdot \text{Prob}[\text{no } U.C.] = \text{Prob}[U.C.] \quad (13)$$

Hence, the mean rate of upcrossing at time t is given by,

$$\begin{aligned} \nu^+(t) &= \frac{E[N^+]}{\Delta t} \\ &= \int_{-\infty}^{\infty} db \int_{-\infty}^{\infty} d\dot{b} \int_{\dot{b}(t)}^{\infty} [\dot{x}(t) - \dot{b}(t)] f_{B(t),\dot{B}(t),X(t),\dot{X}(t)}(b, \dot{b}, b, \dot{x}; t) d\dot{x} \end{aligned} \quad (14)$$

or, equivalently,

$$\nu^{+}(t)=\int_{-\infty}^{\infty} f_{X(t)}(b)db\int_{-\infty}^{\infty} d\dot{b}\int_{\dot{b}(t)}^{\infty}[\dot{x}(t)-\dot{b}(t)]f_{(B(t),\dot{B}(t),\dot{X}(t))|X(t)}(b,\dot{b},\dot{x};t)d\dot{x} \quad (15)$$

3.2 Outcrossing Rate – Multi-Dimensional Case

In outcrossing analysis, the attention is focused on the time period $(t, t+\Delta t)$ [6,11]. What happened before t is not important as long as the structure is not in a failure state at t. Accordingly, for the convenience of outcrossing analysis, points in $\mathcal{X}$ can be re-classified using the concept of final survival path described previously.

Define,

Type fo: Points which have no *final* survival path entering them.

Type to: Points which for some *final* survival paths entering them, have both immediate survival and failure directions.

Type so: Points which have no failure directions for *any final* survival path that ends at them.

and, correspondingly,

$$\begin{array}{lll} \textbf{failure region:} & D_{fo} = & \{\text{All type fo points}\} \\ \textbf{transition region:} & D_{to} = & \{\text{All type to points}\} \\ \textbf{safe region:} & D_{so} = & \{\text{All type so points}\} \end{array} \quad (16)$$

D_{so}, D_{to} and D_{fo} are also mutually exclusive and collectively exhaustive.

With these definitions, outcrossings can only take place in region D_{to}. A failure surface at point $\mathbf{x}$ for a path $\mathbf{x}(t)$ is now a segment of surface that divides the immediate failure directions from the immediate survival directions of a *final* survival path $\mathbf{x}(t)$. The chance that an arbitrary point $\mathbf{x} \in D_t$ is on a failure surface of a *final* survival path will be denoted as $q^o(\mathbf{x};t)$, which can be decomposed into two components,

$$q^o(\mathbf{x};t) = q_1^o(\mathbf{x};t)q_2^o(\mathbf{x};t) \quad (17)$$

where

$$q_1^o(\mathbf{x};t) \equiv \text{the likelihood that there is a } \textit{final} \text{ survival path to } \mathbf{x} \quad (18)$$

and

$$q_2^o(\mathbf{x};t) \equiv \text{the likelihood that there is a failure direction at } \mathbf{x} \text{ for any } \textit{final} \text{ survival path to } \mathbf{x} \quad (19)$$

Since a survival path is also a *final* survival path, the following relationships hold,

$$D_f \supseteq D_{fo} \quad D_t \subseteq D_{to} \quad D_s \supseteq D_{so} \quad (20)$$

Moreover,

$$q_1^o(\mathbf{x};t) \geq q_1(\mathbf{x};t) \text{ and } q_2^o(\mathbf{x};t) \geq q_2(\mathbf{x};t) \tag{21}$$

$$\longrightarrow \quad q^o(\mathbf{x};t) \geq q(\mathbf{x};t) \tag{22}$$

For any path reaching $\mathbf{x}$ at time t, only those velocity vectors, $\dot{\mathbf{X}}(t)$ that satisfy

$$\dot{x}_n \equiv \dot{\mathbf{x}} \cdot \mathbf{n} > 0 \tag{23}$$

will cause failure at the next moment, where $\mathbf{n}$ is the outnormal vector to a failure surface, or an immediate failure direction if the surface is not smooth.

Assume that the joint pdf of $\mathbf{N}(t), \mathbf{X}(t), \dot{\mathbf{X}}(t)$ at time t exists. Emulating the extension of Rice's upcrossing formula to the multi-dimensional case [Melchers, 1987], the likelihood of an outcrossing at time t, which equals the mean rate of outcrossings at t when outcrossing is a rare event, is

$$\nu^+(t) = \int_{\mathcal{X}} d\mathbf{x} \int\int_{\dot{x}_n \geq 0} (\dot{\mathbf{x}} \cdot \mathbf{n}) f_{(\mathbf{N},\dot{\mathbf{X}},\mathbf{X})}(\mathbf{n}, \dot{\mathbf{x}}; \mathbf{x}, t) q^o(\mathbf{x};t) d\mathbf{n} d\dot{\mathbf{x}} \tag{24}$$

In Eq. 24, q^o is the likelihood that $\mathbf{x}$ is on a limit state surface; $f_{\mathbf{N},\dot{\mathbf{X}},\mathbf{X}}$ is the likelihood that a path $\mathbf{x}(t)$ reaches point $\mathbf{x}$, having velocity vector $\dot{\mathbf{x}}$ and outnormal vector $\mathbf{n}$; and $\dot{\mathbf{x}} \cdot \mathbf{n}$ is analogous to $\dot{x} - \dot{b}$ in Eq. 15. $f_{\mathbf{N},\dot{\mathbf{X}},\mathbf{X}}$ vanishes in regions D_{fo} and D_{so} by the definitions of these two regions.

Define,

$$E^+[X_n|\mathbf{X} = \mathbf{x}] = \int\int_{\dot{x}_n > 0} (\dot{\mathbf{x}} \cdot \mathbf{n}) f_{(\mathbf{N},\dot{\mathbf{X}})|\mathbf{X}}(\mathbf{n}, \dot{\mathbf{x}}; \mathbf{x}, t) q^o(\mathbf{x};t) d\mathbf{n} d\dot{\mathbf{x}} \tag{25}$$

which determines the mean rate of outcrossing at a given point $\mathbf{x} \in \mathcal{X}$. Equation 25 then takes the form,

$$\nu^+(t) = \int_{\mathcal{X}} E^+[X_n|\mathbf{X} = \mathbf{x}] f_{\mathbf{X}}(\mathbf{x};t) d\mathbf{x} \tag{26}$$

where $f_{\mathbf{X}}$ is the joint pdf of $\mathbf{X}$. Equation 26 is very similar to its counterpart for the path-independent problems. However, the integrand and the integration domain here have quite different interpretations.

4. SUMMARY AND CONCLUSION

The concept of limit state surface in the basic variable space is discussed in this paper. The traditional concept of limit state surface for the load path-independent problems is not applicable to load path-dependent problems. Instead, there is a transition zone in which each point may be on a limit state surface, which is a segment of surface that separates the immediate failure directions from the immediate survival directions. This piece of failure surface is in general different for different points in the transition zone, and is different for different paths to the same point.

Formulas computing the outcrossing rates are derived for load path-dependent problems. It can be observed that the major difference between these formulas and their counterparts in the path-independent problems is that an extra term, which represents the chance that the current point is on a limit state surface, is present in the path-dependent formulation.

ACKNOWLEDGMENT

The support of the Department of Civil Engineering, The Johns Hopkins University, and the National Science Foundation, grant MSS-9016018, is gratefully acknowledged.

REFERENCES

[1] Bogdanoff, J.L., Kozin, F. (1985). **Probabilistic Models of Cumulative Damage**. John Wiley & Sons, New York.

[2] Cordts, D., Kollmann, F. G. (1986). "An Implicit Time Integration Scheme for Inelastic Constitutive Equations With Internal State Variables." *Int J for Num Meth in Eng*, v 23 p 533-554.

[3] Ditlevsen, O. (1983). "Gaussian Outcrossings From Safe Convex Polyhedrons". *J of Eng Mech Div, ASCE*, v 109, n 1, p 127-148.

[4] Ditlevsen, O. (1990). " Asymptotic First-Passage Time Distributions in Compound Poisson Processes." *Structural Safety*, v 8, p 327-336.

[5] Ditlevsen, O. (1991). " Gaussian Excited Elasto-Plastic Oscillator With Rare Visits to the Plastic Domain." *J of Sound & Vib*, v 145, n 3, p 443-456.

[6] Hagen, O. and Tvedt, L. (1991). "Vector Process Out- Crossing as Parallel System Sensitivity Measure." *J Eng Mech, ASCE*, v 117, n 10, p 2201-2220.

[7] Mukherjee, S. (1982). **Boundary Element Methods in Creep and Fracture**. Applied Science Publishers, New York.

[8] Rackwitz, R., and Fiessler, B. (1978). "Structural Reliability Under Combined Random Load Sequences." *Computers and Structures*, v 9, n 2 p 489-494.

[9] Rice, S. O. (1944). "Mathematical Analysis of Random Noise." *Bell System Technical Journal*, v 23, n 282.

[10] Thoft-Christensen, P., Murotsu, Y. (1986). **Application of Structural Systems Reliability Theory**. Springer-Verlag, New York.

[11] Veneziano, D., Grigoriu, M., Cornell, C. A. (1977). "Vector-Process Models for System Reliability." *J Engng Mech Div*, ASCE, v 103, n EM3, p 441-460.

[12] Wang, W., Corotis, R. B., Ramirez, M. R. (1993). "Modal Failure Probability and System Performance." *Structural Engineering in Natural Hazards Mitigation, Proc Structural Congress'93*. v 2, p 1312-1317.

[13] Wen, Y. K. and Chen, H.-C. (1989). "System Reliability Under Time Varying Loads: I." *J Eng Mech, ASCE*, v 115, n 4, p 808-823.

31

RANDOM VIBRATIONS OF STRUCTURES UNDER PROPAGATING EXCITATIONS

by Zbigniew ZEMBATY
Technical University of Opole, ul.Mikołajczyka 5, 45-233 Opole, POLAND

1.Introduction

The problem of vibrations of structures under incoherent or, in particular, propagating excitations is important for large, extended civil engineering structures like bridges, lifelines, dams, offshore structures or for aircraft structures. In seismic engineering spatial ground motion models have been studied for more than a decade but credible, stochastic characteristics are available only since SMART-1 accelerograph array is in operation at Lotung in Taiwan.

Based on the spatial ground motion models the structural response can be analyzed in form of spatial response spectra [1-4] or for various specific types of structures, e.g. [5-11].

In present paper the problem is analyzed again and illustrated by an example of random vibrations of a bridge structure under kinematic wave excitations.

2.Equations of motion and mean square response

Consider equation of motion of discrete systems under kinematic excitations:

$$[M]\{\ddot{q}^t\} + [C]\{\dot{q}^t\} + [K]\{q^t\} = \{0\} \tag{1}$$

where [M], [C], [K] are mass, damping and stiffness matrices, vector $\{q^t\} = \{q_1^t, q_2^t, \dots, q_n^t\}^T$ represents total displacements (fixed reference) and symbol ^{T} stands for transposition. These n degrees of freedom can be divided onto n_s structural degrees and n_g degrees associated with ground motion ($n=n_s+n_g$). Then eq.1 takes form

$$\begin{bmatrix} [M_{ss}] & [M_{sg}] \\ [M_{gs}] & [M_{gg}] \end{bmatrix} \begin{Bmatrix} \{\ddot{q}_s^t\} \\ \{\ddot{u}\} \end{Bmatrix} + \begin{bmatrix} [C_{ss}] & [C_{sg}] \\ [C_{gs}] & [C_{gg}] \end{bmatrix} \begin{Bmatrix} \{\dot{q}_s^t\} \\ \{\dot{u}\} \end{Bmatrix} + \begin{bmatrix} [K_{ss}] & [K_{sg}] \\ [K_{gs}] & [K_{gg}] \end{bmatrix} \begin{Bmatrix} \{q_s^t\} \\ \{u\} \end{Bmatrix} = \begin{Bmatrix} \{0\} \\ \{0\} \end{Bmatrix} \tag{2}$$

where vector $\{u\} = \{u_1, u_2, ..., u_{n_g}\}^T$ describes the free field motion (excitations). The total response $\{q^t\}$ can be separated onto dynamic $\{q\}$ and pseudo-static $\{q^p\}$.

$$\left\{q^t\right\} = \left\{\begin{matrix}\{q^p\}\\ \{u\}\end{matrix}\right\} + \left\{\begin{matrix}\{q\}\\ \{0\}\end{matrix}\right\}. \tag{3}$$

Substituting eq.3 into the equation of motion (2) and dropping dynamic terms one obtains the pseudo-static motion

$$\left\{q^p\right\} = -\left[K_{ss}\right]^{-1}\left[K_{sg}\right]\left\{u\right\}. \tag{4}$$

Substituting again eqs. 4 and 3 into 2 and assuming either the stiffness proportional damping or that the damping contribution to effective force is negligible leads to:

$$\left[M_{ss}\right]\left\{\ddot{q}\right\} + \left[C_{ss}\right]\left\{\dot{q}\right\} + \left[K_{ss}\right]\left\{q\right\} = \left(\left[M_{ss}\right]\left[K_{ss}\right]^{-1}\left[K_{sg}\right] - \left[M_{sg}\right]\right)\left\{\ddot{u}\right\}. \tag{5}$$

For all the analyzed degrees of freedom vector of the associated generalized elastic forces is equal to

$$\left\{f\right\} = \left[K\right]\left\{q^t\right\}. \tag{6}$$

Separating the forces onto structural $\{f_s\}$ and support $\{f_g\}$

$$\left\{\begin{matrix}\{f_s\}\\ \{f_g\}\end{matrix}\right\} = \begin{bmatrix}[K_{ss}] & [K_{sg}]\\ [K_{gs}] & [K_{gg}]\end{bmatrix}\left\{\begin{matrix}\{q_s^t\}\\ \{u\}\end{matrix}\right\} \tag{7}$$

where $\{f_s\} = \{f_1, ..., f_{n_s}\}^T$ and $\{f_g\} = \{f_1, ..., f_{n_{fg}}\}^T$. From eq.11

$$\left\{f_s\right\} = \left[K_{ss}\right]\left\{q_s^t\right\} + \left[K_{sg}\right]\left\{u\right\}, \tag{8}$$

$$\left\{f_g\right\} = \left[K_{gs}\right]\left\{q_s^t\right\} + \left[K_{gg}\right]\left\{u\right\} \tag{9}$$

Substituting from eq.3 $\{q_s^t\}=\{q^p\}+\{q\}$ and applying eq.4 one obtains

$$\left\{f_s\right\} = \left[K_{ss}\right]\left\{q\right\} \tag{10}$$

$$\left\{f_g\right\} = \left[K_{gs}\right]\left\{q\right\} + \left(\left[K_{gg}\right] - \left[K_{gs}\right]\left[K_{ss}\right]^{-1}\left[K_{sg}\right]\right)\left\{u\right\}. \tag{11}$$

It is interesting to note from eq.10 that the forces associated with structural degrees of freedom depend only on dynamic displacements of the structure which, in turn, depend on "averaged" excitations of the structure. On the other hand the forces associated with support degrees of freedom (eq.11) depend on two terms: dynamic and pseudo-static. The calculated elastic forces may be applied in calculations of any desired inner forces (shear, axial or moments). In practice however the forces in the structure are are calculated using the classic formulae of Finite Element Method (FEM) which are more effective in numerical calculations than formulae (6-11). Regardless of the applied method the calculated forces are linear combinations of structural displacements and the stiffness properties of the structure. The support elastic forces $\{f_g\}$ better represent the propagation effects on structure than dynamic displacements of the structure as they combine both dynamic and pseudo-static response.

Solving the eigenproblem of the analyzed discrete system leads to eigenmatrix $[W]$ and natural frequencies $\omega_1, \omega_2, \ldots, \omega_{n_s}$. Introducing normal coordinates $\underline{q}_1, \underline{q}_2, \ldots, \underline{q}_{n_s}$ and applying the modal transformation

$$\{q\} = [W]\{\underline{q}\}, \tag{12}$$

leads, after some algebra [15], to following formulas for mean square displacements

$$\sigma^2_{q_k} \cong \sum_{i=1}^{n_w}\sum_{p=1}^{n_w} W_{ki}W_{kp}\int_{-\infty}^{\infty} H_i(\omega)H_p^*(\omega)\sum_{j=1}^{n_g}\sum_{r=1}^{n_g} g_{ij}g_{pr}S_{jr}(\omega)d\omega, \tag{13}$$

and forces

$$\begin{aligned}\sigma^2_{f_k} \cong & \sum_{p=1}^{n_s}\sum_{m=1}^{n_s} K^{gs}_{kp}K^{gs}_{km}\sum_{i=1}^{n_w}\sum_{n=1}^{n_w} W_{pi}W_{mn}\int_{-\infty}^{\infty} H_i(\omega)H_n^*(\omega)\sum_{j=1}^{n_g}\sum_{r=1}^{n_g} g_{ij}g_{nr}S^{\ddot{u}\ddot{u}}_{j\,r}(\omega)d\omega \; + \\ & + \sum_{p=1}^{n_s} K^{gs}_{kp}\sum_{i=1}^{n_w} W_{pi}\sum_{j=1}^{n_g} g_{ij}\sum_{r=1}^{n_g} K^{g}_{kr}\int_{-\infty}^{\infty} Re\left(H_i(\omega)S^{u\ddot{u}}_{j\,r}(\omega)\right)d\omega \; + \\ & + \sum_{j=1}^{n_g}\sum_{r=1}^{n_g} K^{g}_{kj}K^{g}_{kr}\int_{-\infty}^{\infty} S^{uu}_{j\,r}(\omega)d\omega\,,\end{aligned} \tag{14}$$

where

W_{ij} are elements of eigenmatrix $[W]$,

$H_i(\omega) = (\omega_i^2 - \omega^2 + 2i\xi_i\omega_i\omega)^{-1}$ are modal frequency response functions, $i = \sqrt{-1}$

g_{ij} are elements of matrix $[g] = diag\left[\frac{1}{m_i}\right][W]^T\left([M_{ss}][K_{ss}]^{-1}[K_{sg}] - [M_{sg}]\right)$,

$S^{\ddot{u}\ddot{u}}_{j\,r}(\omega) = S_{jr}(\omega)$ are cross spectral densities of excitation accelerations,

$S^{u\ddot{u}}_{j\,r}(\omega) = -S_{jr}(\omega)/\omega^2$ are cross spectral densities displacements-accelerations,

$S^{uu}_{j\,r}(\omega) = S_{jr}(\omega)/\omega^4$ are cross spectral densities of displacements,

K^g_{kr} are elements of matrix $[K_g] = [K_{gg}] - [K_{gs}][K_{ss}]^{-1}[K_{sg}]$,

K^{gs}_{kp} are elements of matrix $[K_{gs}]$,

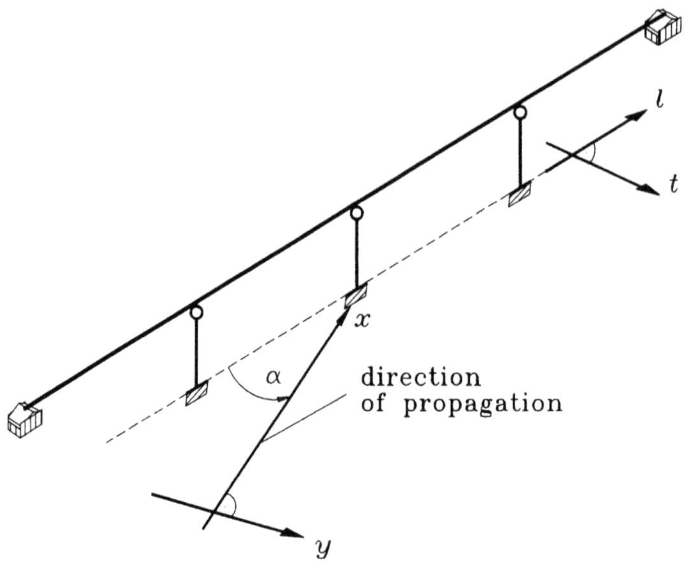

Fig.1 Bridge structure with local (*l-t*), and principal coordinate system (*x-y*)

3.Numerical example

Consider a bridge structure with coordinate system *l,t* arbitrarily situated with respect to a system of principal axes *x,y* in which the axis *x* is directed towards an epicenter and axis *y* is perpendicular to *x*. Axis *x* is inclined with angle α to axis *l* (Fig.1). The horizontal ground motions $u_x(t,x,y)$ and $u_y(t,x,y)$. They form a random field with uncorrelated u_x and u_y [13]. Assuming $S^{\ddot{u}\ddot{u}}_{yy}(\omega) = b^2 S^{\ddot{u}\ddot{u}}_{xx}(\omega)$ gives:

$$\left[S^{\ddot{u}\ddot{u}}_{xy}(\omega)\right] = \begin{bmatrix} S^{\ddot{u}\ddot{u}}_{xx}(\omega) & 0 \\ 0 & S^{\ddot{u}\ddot{u}}_{yy}(\omega) \end{bmatrix} = \begin{bmatrix} S(\omega) & 0 \\ 0 & b^2 S(\omega) \end{bmatrix} = \begin{bmatrix} 1 & 0 \\ 0 & b^2 \end{bmatrix} S(\omega) \tag{15}$$

where $S(\omega)$ is the horizontal spectral density of excitation accelerations along propagation direction (axis x) and $b^2=0.7$ here. The transformation of displacements u_x and u_y to u_l and u_t is given through familiar coordinate transformation

$$\begin{Bmatrix} u_l \\ u_t \end{Bmatrix} = \begin{bmatrix} cos(\alpha) & -sin(\alpha) \\ sin(\alpha) & cos(\alpha) \end{bmatrix} \begin{Bmatrix} u_x \\ u_y \end{Bmatrix} \tag{16}$$

The analyzed structure is a r/c highway viaduct modeled using FEM as a 3D frame. The data of the bridge were based on an example from Ref.14. The structure can vibrate

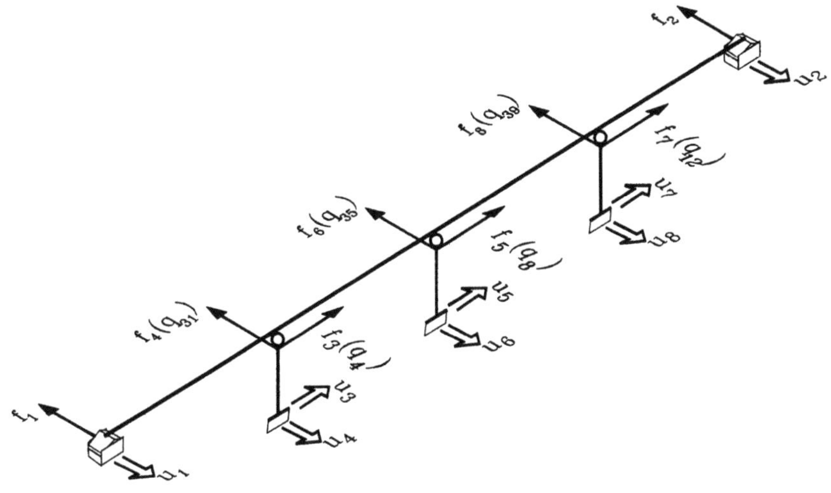

Fig.2 Excitations (u), main response coordinates (q), and shear forces (f)

in horizontal and vertical directions in vertical plane (longitudinal and vertical directions) and out of vertical plane (transversal direction). The supports of the structure have eight horizontal directions of possible excitations ($u_1 \div u_8$ in Fig.2). In addition to these eight excitation directions Fig.2 presents also selected FEM displacement coordinates and shear forces of vector $\{f_g\}$. Applying transformation formula (16) to all excitation directions and using eq.15 gives matrix of acceleration cross spectra which is required in eqs. 13 and 14:

$$\left[S_{j\,r}^{\ddot{u}\ddot{u}}(\omega)\right] = \gamma_{jr}\left[D\right]\begin{bmatrix}1 & 0\\ 0 & b^2\end{bmatrix}\left[D\right]^T S(\omega) \tag{17}$$

where [D] is an extended transformation matrix, γ_{jr} is a complex coherency function of signals at directions "j" and "r" and $S(\omega)$ is a real, point spectral density [15-17]:

$$S(\omega) = \frac{\omega_g^4 + 4\xi_g^2\omega_g^2\omega^2}{(\omega_g^2-\omega^2)^2 + 4\xi_g^2\omega_g^2\omega^2}\;\frac{\omega^4}{(\omega_1^2-\omega^2)^2 + 4\xi_1^2\omega_1^2\omega^2}\,S_0\;, \tag{18}$$

where S_0 is an intensity factor and the remaining constants are as follows: $\omega_g=4\pi$, $\xi_g=0.6$, $\omega_1=1.636$, $\xi_1=0.619$. The coherency function may be written in exponential form

$$\gamma_{jr}(\omega,d^x_{jr},d^y_{jr}) = |\gamma_{jr}(\omega,d^x_{jr},d^y_{jr})|\,exp(i\omega d^x_{jr}/v_g)\;. \tag{19}$$

The exponent in formula 19 describes wave propagation effects along axis x. It depends on apparent wave velocity v_g and projection of distance between points "j" and "r" on axis x d^x_{jr} (d^y_{jr} is respective projection on axis y). The modulus of coherency in formula 19 can be found in the Literature following field measurements of seismic ground motion. In this study a formula given by Hao [8] will be applied:

$$|\gamma_{jr}(\omega,d^x_{jr},d^y_{jr})| = exp(-\beta_1 d^x_{jr}-\beta_2 d^y_{jr})exp\left\{-\left(\alpha_1(\omega)\sqrt{d^x_{jr}} + \alpha_2(\omega)\sqrt{d^y_{jr}}\right)\left(\frac{\omega}{2\pi}\right)^2\right\}, \tag{20}$$

where

$$\begin{aligned}\alpha_1(\omega) &= \frac{2\pi a_1}{\omega} + \frac{a_2\omega}{2\pi} + a_3\\ \alpha_2(\omega) &= \frac{2\pi a_4}{\omega} + \frac{a_5\omega}{2\pi} + a_6\end{aligned} \qquad \text{for} \quad 0.314 \le \omega \le 62.83 \ . \tag{21}$$

For $\omega \ge 62.83$ $\alpha_1(\omega)=\alpha_1(62.83)=\text{const}$ and $\alpha_2(\omega)=\alpha_2(62.83)=\text{const}$. The constants β_1,β_2 and $a_1 \div a_6$ have been taken from event 45 of SMART 1, [8].

Three main parameters affect the response of structure in this case: apparent wave velocity v_g, the angle of propagation α, and the loss of coherency $|\gamma_{jr}|$. The effect of the first two of them on structural response is analyzed in Fig.3 where the root mean square displacements $\sigma_{y_{35}}$ (Fig.3a) and σ_{f_6} (Fig.3b) are presented as the functions of apparent wave velocity and angle α. The presented rms displacements and forces are normalized with respect to uniform excitations. It can be seen from this figure that the incoherent motion of supports reduces the dynamic displacements of the structure, but the force response can be either decreased or increased depending mostly on the propagation velocity. A substantial decrease can be observed for $v_g \cong 100 \div 200$m/s. For very low propagation velocity (less then about 50m/s) the incoherent excitations make the increase of force response mostly due to increasing contribution of pseudo-static motion. The effect of angle α on the response is also substantial. For high velocity v_g it reflects simply the projections of excitations on the structural coordinate system (*l-t*). For lower propagation velocity the response is very sensitive to changes of angle α.

Final remarks and conclusions

3D random vibrations of a bridge structure under kinematic wave excitations are considered. Both dynamic displacement response and force response are analyzed. The displacement response is reduced compare to coherent excitations while the force response can be either reduced or amplified depending on propagation velocity and angle. Pseudo-static motion contributes substantially in structural response for low

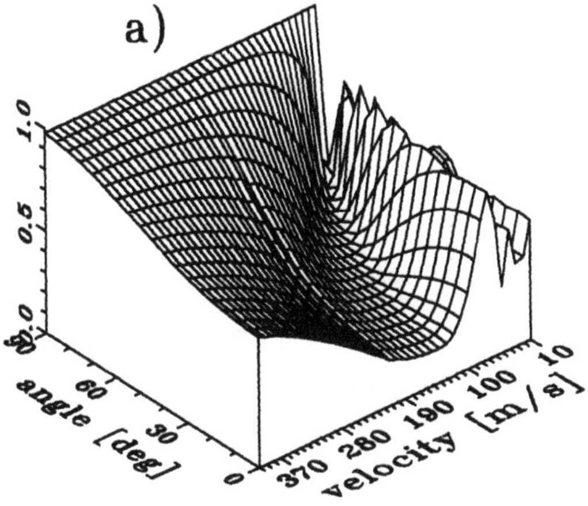

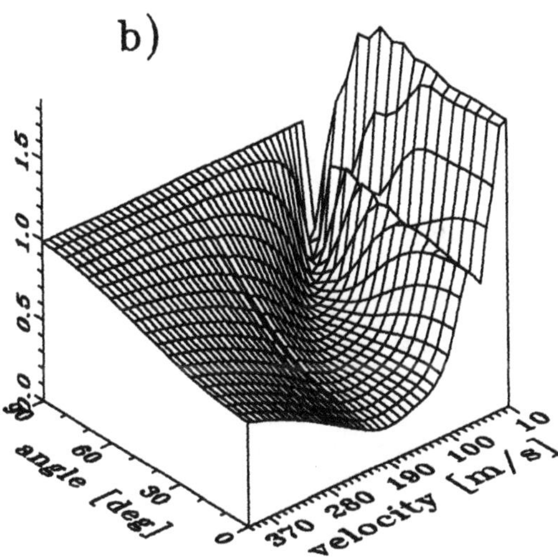

Fig.3 Rms displacements q_{35} (a) and shear force f_6 (b) vs. propagation angle and velocity, normalized with respect to uniform excitations

propagation velocity (v_g<50m/s).

Acknowledgements

A partial support of Committee of Scientific Research (Grant number PB 0373/P4/94/06) is gratefully acknowledged.

References

1. Loh C.-H., Penzien J. and Tsai Y.B., Engineering analysis of SMART 1 array accelerograms, *Earthquake Engineering and Structural Dynamics*, 10, 1982, 575-591.
2. Abrahamson N.A. and Bolt B.A., The spatial variation of the phasing of seismic strong ground motion, *Bulletin of the Seismological Society of America*, 75, 1985, 1247-1264.
3. Der Kiureghian A. and Neuenhofer A., A response spectrum method for multiple-support seismic excitations, *Report No. UCB/EERC 91/08*, 1991, Earthquake Engineering Research Center, University of California, Berkeley.
4. Zembaty Z. and Krenk S. Spatial seismic excitations and response spectra, *Journal of Engineering Mechanics*, ASCE, 119, 1993, 2449-2460.
5. Harichandran R.S. and Wang W., Response of simple beam to spatially varying earthquake excitation, *Journal of Engineering Mechanics* , ASCE, 114, 1988, 1526-1541.
6. Harichandran R.S. and Wang W., Response of indeterminate two-span beam to spatially varying seismic excitation, *Earthquake Engineering and Structural Dynamics*, 19, 1990, 173-187.
7. Zerva A., Effect of spatial variability and propagation of seismic ground motions on the response of multiply supported structures, *Probabilistic Engineering Mechanics*, 6, 1991, 212-221.
8. Hao H., Effects of spatial variation of ground motion on large multiply-supported structures, *Report No. EERC 89-06*, 1989, Earthquake Engineering Research Center, University of California, Berkeley, CA .
9. Novak M., Spatial correlation effects in random vibration of structures, *Structural Dynamics, Proceedings of EURODYN '90*, Krätzig et al. (eds), Balkema, Rotterdam, 1990, vol.2, 631-636.
10. Leger P., Idè I.M. and Paultre P., Multiple-support seismic analysis of large structures, *Computer and Structures*, 36, 1990, 1153-1158.
11. Heredia Zavoni E. and Vanmarcke E.H., Seismic random-vibration analysis of multisupport-structural systems, *Journal of Engineering Mechanics* , ASCE, 120, 1994, 1107-1128.
12. Zembaty Z., Vibrations of a bridge structure under kinematic wave excitations (*a paper in preparation*).
13. Penzien J. and Watabe M., Characteristics of 3-dimensional earthquake ground motions, *Earthquake Engineering and Structural Dynamics*, 3, 1975, 365-373.
14. Lai P., Seismic response of a 4-span bridge system subjected to multiple-support ground excitation, *Proceedings of 4th Canadian Conference on Earthquake Engineering*, University of British Columbia, Vancouver, Canada, 1983, 561-570.
15. Kanai K., Semi-empirical formula for the seismic characteristic of the ground, *Bulletin of Earthquake Research Institute*, Tokyo, vol.35, 1957, 309-325.
16. Tajimi H., A statistical method of determining the maximum response of a building structure during an earthquake, *Proceedings of 2nd World Conference on Earthquake Engineering*, Tokyo, 1960, vol.2, 781-798.
17. Ruiz. P. and Penzien J., Probabilistic study of the behavior of structures during earthquakes, *Report No. EERC 69-03*, Earthquake Engineering Research Center, University of California, Berkeley, CA, 1969.

32

Two Improved Algorithms for Reliability Analysis

Y. Zhang[1] and A. Der Kiureghian[2]

[1] Applied Research Associates, Inc., Southeast Division, Raleigh, NC 27615
[2] Department of Civil Engineering, University of California, Berkeley, CA 94720

Improved versions of two optimization algorithms commmonly used in first-order reliability analysis are developed. One is for determining the first-order reliability index β. The other is for inverse reliability analysis, i.e., determining the value of a deterministic parameter such that the reliability index equals a target value β_t. Besides being mathematically more rigorous, these new versions are much simpler than earlier versions of these algorithms.

1. Introduction

Let $G(\boldsymbol{u})$ denote the limit-state function for a reliability problem defined in terms of a vector of standard normal variates, $\boldsymbol{u}$, obtained by a suitable transformation of the basic random variables of the problem. An important quantity of interest is the reliability index defined by

$$\beta = \min \{ \|\boldsymbol{u}\| \mid G(\boldsymbol{u}) = 0 \} \tag{1}$$

The above is a constrained optimization problem that can be solved by any of a large number of general-purpose algorithms (Liu and Der Kiureghian 1991). However, there is merit in developing a customized algorithm that takes advantage of the special form of the objective function of this problem. One such algorithm is the so called HL-RF algorithm (Hasofer and Lind 1974, Rackwitz and Fiessler 1978, Liu and Der Kiureghian 1991). This algorithm has found wide popularity owing to its simplicity and efficiency. However, there is no proof that for a given problem the algorithm will actually converge. In this paper, an improved version of this algorithm is developed that, under the assumption of differentiability of $G(\boldsymbol{u})$, is globally convergent.

Now suppose the limit-state function includes a deterministic parameter θ, i.e., $G(\boldsymbol{u}, \theta)$. This could be a parameter of the limit-state function in the original space, or it could be a probability distribution parameter that appears in $G(\boldsymbol{u}, \theta)$ due to the transformation of random variables to the standard normal space. The inverse reliability problem is defined as follows: Find θ such that the reliability index associated with $G(\boldsymbol{u}, \theta)$ equals a target value β_t, i.e.,

$$\theta : \quad \min \{ \|\boldsymbol{u}\| \mid G(\boldsymbol{u}, \theta) = 0 \} = \beta_t \tag{2}$$

An algorithm for solution of this problem was recently developed by Der Kiureghian et al. (1993). Here, we develop a simpler version of this algorithm that has superior convergence properties.

The two algorithms developed in this paper have the same format. One generates a sequence of points according to the rule

$$\boldsymbol{x}_{i+1} = \boldsymbol{x}_i + \lambda_i \boldsymbol{d}_i, \tag{3}$$

where $\boldsymbol{x}_i = \boldsymbol{u}_i$ for the ordinary reliability problem, $\boldsymbol{x}_i = (\boldsymbol{u}_i, \theta_i)$ for the inverse reliability problem, $\boldsymbol{d}_i$ denotes a search direction vector, and λ_i is the step size. λ_i is determined by assuring a reduction in a merit function $m(\boldsymbol{x})$ at each step. Most commonly, the Armijo rule (Luenberger 1986) is used. For the present application, this rule can be written as

$$\lambda_i = \max_k \{ b^k \mid m(\boldsymbol{x}_i + b^k \boldsymbol{d}_i) - m(\boldsymbol{x}_i) \leq -ab^k \langle \nabla m(\boldsymbol{x}_i), \boldsymbol{d}_i \rangle \}, \tag{4}$$

where a, $b \in (0,1)$ are pre-selected parameters and k is an integer. Global convergence of the sequence in (3) is assured if the merit function attains its minimum at the solution point of the problem and if $\boldsymbol{d}_i$ is a descent direction of $m(\boldsymbol{x}_i)$ at every point of the sequence (Luenberger 1986). The main issue is the selection of $\boldsymbol{d}_i$ and $m(\boldsymbol{x}_i)$ that satisfy these conditions for each problem.

2. An Improved HL-RF Algorithm

The solution of the optimization problem in (1) is characterized by the following optimality conditions:

$$\boldsymbol{u} + \frac{\|\boldsymbol{u}\|}{\|\nabla_{\boldsymbol{u}} G(\boldsymbol{u})\|} \nabla_{\boldsymbol{u}} G(\boldsymbol{u}) = 0, \tag{5a}$$

$$G(\boldsymbol{u}) = 0, \tag{5b}$$

where $\nabla_{\boldsymbol{u}}$ denotes the gradient operator with respect to $\boldsymbol{u}$. In the following, the subscript on the gradient operator is dropped unless it is necessary to avoid ambiguity. The HL-RF search direction is obtained by solving the above equations for the linearized constraint

$$L_{\boldsymbol{u}_i}(\boldsymbol{u}) = G(\boldsymbol{u}_i) + \langle \nabla G(\boldsymbol{u}_i), \boldsymbol{u} - \boldsymbol{u}_i \rangle \tag{6}$$

at each step, where $\langle .,. \rangle$ denotes the inner product of two vectors. The result, dropping the subscript i for simplicity, is

$$\boldsymbol{d} = \frac{\langle \boldsymbol{u}, \nabla G(\boldsymbol{u}) \rangle - G(\boldsymbol{u})}{\|\nabla G(\boldsymbol{u})\|^2} \nabla G(\boldsymbol{u}) - \boldsymbol{u}. \tag{7}$$

The original HL-RF algorithm (Hasofer and Lind 1974, Rackwitz and Fiessler 1978) used a unit step size, i.e., $\lambda_i = 1$. Liu and Der Kiureghian (1991) introduced a modified version that has step size control by using the merit function

$$m(\boldsymbol{u}) = \frac{1}{2} \|\boldsymbol{u} - \frac{\langle \boldsymbol{u}, \nabla G(\boldsymbol{u}) \rangle}{\|\nabla G(\boldsymbol{u})\|^2} \nabla G(\boldsymbol{u})\|^2 + \frac{1}{2} c G(\boldsymbol{u})^2, \tag{8}$$

where $c > 0$ is a constant. This merit function has its minimum at the solution of (1). However, there is no proof that $\boldsymbol{d}$ in (7) is a descent direction of this merit function.

Here, we introduce the simpler merit function

$$m(\boldsymbol{u}) = \frac{1}{2}\|\boldsymbol{u}\| + c|G(\boldsymbol{u})|, \tag{9}$$

where $c > 0$ is a penalty parameter. This merit function also attains its minimum at the solution of (1). Furthermore, as shown by the following theorem, the merit function in (9) is compatible with the search direction in (7).

Theorem: *The HL-RF search direction* $\mathbf{d}$ *in (7) at any point* $\boldsymbol{u}$ *is a descent direction of the merit function* $m(\boldsymbol{u})$ *in (9), provided*

$$c > \frac{\|\boldsymbol{u}\|}{\|\nabla G(\boldsymbol{u})\|}. \tag{10}$$

Proof : To prove that $\boldsymbol{d}$ is a descent direction of $m(\boldsymbol{u})$, we need to show that $\langle \nabla m(\boldsymbol{u}), \boldsymbol{d}\rangle \leq 0$ and that the equality holds only if $\boldsymbol{u}$ is a solution to (1).

Using (7) together with $\nabla m(\boldsymbol{u}) = \boldsymbol{u} + c\,\mathrm{sgn}(G(\boldsymbol{u}))\nabla G(\boldsymbol{u})$, where $\mathrm{sgn}(G(\boldsymbol{u}))$ is the sign of $G(\boldsymbol{u})$, we have

$$\begin{aligned}\langle \nabla m(\boldsymbol{u}), \boldsymbol{d}\rangle = &- \left(\| \boldsymbol{u} \|^2 - \frac{\langle \boldsymbol{u}, \nabla G(\boldsymbol{u})\rangle^2}{\|\nabla G(\boldsymbol{u})\|^2}\right) \\ &- |G(\boldsymbol{u})|\left(c + \mathrm{sgn}(G(\boldsymbol{u}))\frac{\langle \boldsymbol{u}, \nabla G(\boldsymbol{u})\rangle}{\|\nabla G(\boldsymbol{u})\|^2}\right).\end{aligned} \tag{11}$$

Using the Schwartz inequality $| \langle \boldsymbol{u}, \nabla G(\boldsymbol{u})\rangle | \leq \|\boldsymbol{u}\|\, \|\nabla G(\boldsymbol{u})\|$,

$$\|\boldsymbol{u}\|^2 - \frac{\langle \boldsymbol{u}, \nabla G(\boldsymbol{u})\rangle^2}{\|\nabla G(\boldsymbol{u})\|^2} \geq 0. \tag{12}$$

Using Schwartz inequality again and (10),

$$\begin{aligned} c + \mathrm{sgn}(G(\boldsymbol{u}))\frac{\langle \boldsymbol{u}, \nabla G(\boldsymbol{u})\rangle}{\|\nabla G(\boldsymbol{u})\|^2} &\geq c - \frac{| \langle \boldsymbol{u}, \nabla G(\boldsymbol{u})\rangle |}{\|\nabla G(\boldsymbol{u})\|^2} \\ &\geq c - \frac{\|\boldsymbol{u}\|}{\|\nabla G(\boldsymbol{u})\|} \\ &> 0. \end{aligned} \tag{13}$$

Using (12) and (13) in (11), it follows that $\langle \nabla m(\boldsymbol{u}), \boldsymbol{d}\rangle \leq 0$. If $\langle \nabla m(\boldsymbol{u}), \boldsymbol{d}\rangle = 0$, we have from (11)-(13) that $G(\boldsymbol{u}) = 0$ and $\|\boldsymbol{u}\|^2 - \frac{\langle \boldsymbol{u} \nabla G(\boldsymbol{u})\rangle^2}{\|\nabla G(\boldsymbol{u})\|^2} = 0$, which are equivalent to the optimality conditions (5). Hence, in this case u is a solution to (1), which completes the proof.

The above theorem is an essential element for establishing the global convergence theorem. The mathematical analysis and poof of the global convergence of the sequence along with a proper line search scheme, can be found in standard texts on optimization theory (Luenberger 1986). In implementing this algorithm, it is suggested to gradually increase the value of the penalty parameter c while satisfying (10).

The proposed algorithm is superior to the modified HL-RF algorithm of Liu and Der Kiureghian (1991) in at least two ways: (a) there is a mathematical proof that the merit function is compatible with the search direction, and (b) the algorithm usually requires less computation since the merit function in (9) does not involve the gradient of the constraint function as does the merit function in (8).

Below, we present two examples comparing the performance of the modified HL-RF algorithm of Liu and Der Kiureghian (1991) with the proposed algorithm.

Example 1 Consider the original-space, limit-state function

$$\begin{aligned} g(\boldsymbol{z}) = 1.1 &- 0.00115 z_1 z_2 + 0.00157 z_2^2 + 0.00117 z_1^2 + 0.0135 z_2 z_3 - 0.0705 z_2 \\ &- 0.00534 z_1 - 0.0149 z_1 z_3 - 0.0611 z_2 z_4 + 0.0717 z_1 z_4 - 0.226 z_3 \\ &+ 0.0333 z_3^2 - 0.558 z_3 z_4 + 0.998 z_4 - 1.339 z_4^2, \end{aligned} \tag{14}$$

where z_1, z_2, z_3 and z_4 are statistically independent random variables: z_1 has a type-II largest value distribution with mean 10 and standard deviation 5; z_2 and z_3 are both normal with means 25 and 0.8 and standard deviations 5 and 0.2, respectively, and z_4 has the lognormal distribution with mean 0.0625 and standard deviation 0.0625. The original HL-RF algorithm with $\lambda_i = 1$ fails to converge for this problem.

Example 2 This example concerns the deformation of an elastic plate subjected to a uniformly distributed edge load of intensity $p = 1.0GPa$, as shown in Fig. 1. The plate is in a state of plane stress condition. The Poisson's ratio is $\nu = 0.3$. However, the Young's modulus, E, is a Gaussian random field with mean 200 GPa, standard deviation 50 GPa, and auto-correlation coefficient function $\rho_{EE} = \exp(-(\frac{\Delta}{b})^2)$, where Δ is the distance between any two points on the plate and $b = 2cm$ is the correlation length. The random field is discretized by using the finite element mesh shown in Fig. 1 and is represented by 20 random variables. Further details can be found in Zhang (1994).

The original-space limit-state function for this example is defined by

$$g(\boldsymbol{z}) = \theta - U_A(\boldsymbol{z}), \tag{15}$$

where $\boldsymbol{z}$ denotes the vector of random variables representing the random field of E, $\theta = 0.19\,cm$ is the displacement threshold, and $U_A(\boldsymbol{z})$ denotes the displacement at

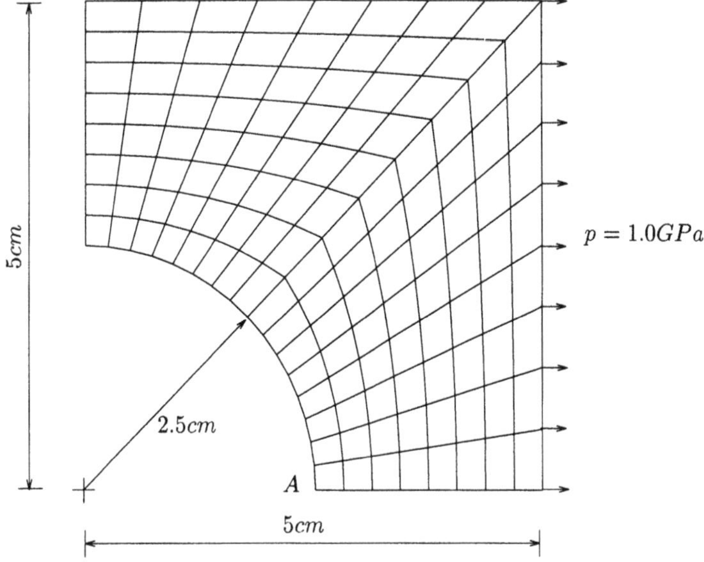

Figure 3.1 Finite element mesh for elastic plate

point A in the horizontal direction. The finite element method is used to compute the limit-state function and its gradient.

Table 1 compares the performance of the two algorithms for the above two examples. The algorithms are initiated at the mean point and tolerances are set at 0.001. Specification of other parameters is described in Zhang (1994). The two algorithms converge to identical solutions with reliability indices $\beta = 1.35$ and $\beta = 3.27$ for the two problems, respectively. The required number of steps as well as the number of times each algorithm computed the limit-state function and its gradient are listed in Table 1. For both examples, the proposed algorithm performs significantly better than the modified HL-RF algorithm of Liu and Der Kiureghian (1991).

3. Inverse Reliability Algorithm

The inverse reliability problem in (2) is defined by the set of equations

$$\|\boldsymbol{u}\| - \beta_t = 0, \tag{16a}$$

$$\boldsymbol{u} + \frac{\|\boldsymbol{u}\|}{\|\boldsymbol{\nabla_u} G(\boldsymbol{u}, \theta)\|} \boldsymbol{\nabla_u} G(\boldsymbol{u}, \theta) = 0, \tag{16b}$$

Table 1. Comparison between two algorithms

Example	Algorithm	No. steps	No. $g(z)$	No. $\nabla g(y)$
1	proposed	6	17	7
	modified HL-RF	8	15	15
2	proposed	3	8	4
	modified HL-RF	16	21	21

$$G(\boldsymbol{u}, \theta) = 0, \tag{16c}$$

The last two equations are the optimality conditions of (1), whereas the first equation describes the constraint on the reliability index. A search direction $\boldsymbol{d}_i$ for this problem can be derived from the solution of the linearized equations

$$\|\boldsymbol{u}\| - \beta_t = 0, \tag{17a}$$

$$\boldsymbol{u} + \frac{\|\boldsymbol{u}\|}{\|\nabla_{\boldsymbol{u}} L_{\boldsymbol{u}_i,\theta_i}(\boldsymbol{u}, \theta)\|} \nabla_{\boldsymbol{u}} L_{\boldsymbol{u}_i,\theta_i}(\boldsymbol{u}, \theta) = 0, \tag{17b}$$

$$L_{\boldsymbol{u}_i,\theta_i}(\boldsymbol{u}, \theta) = 0, \tag{17c}$$

where

$$L_{\boldsymbol{u}_i,\theta_i}(\boldsymbol{u}, \theta) = G(\boldsymbol{u}_i, \theta_i) + \langle \nabla_{\boldsymbol{u}} G(\boldsymbol{u}_i, \theta_i), \boldsymbol{u} - \boldsymbol{u}_i \rangle + \frac{\partial G(\boldsymbol{u}_i, \theta_i)}{\partial \theta}(\theta - \theta_i). \tag{18}$$

The solution of (17) is

$$\boldsymbol{u} = -\beta_t \frac{\nabla_{\boldsymbol{u}} G(\boldsymbol{u}_i, \theta_i)}{\|\nabla_{\boldsymbol{u}} G(\boldsymbol{u}_i, \theta_i)\|}, \tag{19a}$$

$$\theta = \theta_i + \frac{\langle \boldsymbol{u}_i, \nabla_{\boldsymbol{u}} G(\boldsymbol{u}_i, \theta_i) \rangle - G(\boldsymbol{u}_i, \theta_i) + \beta_t \|\nabla_{\boldsymbol{u}} G(\boldsymbol{u}_i, \theta_i)\|}{\partial G(\boldsymbol{u}_i, \theta_i)/\partial \theta}. \tag{19b}$$

Using this result, the search direction is obtained as

$$\boldsymbol{d}_i = \begin{pmatrix} -\beta_t \dfrac{\nabla_{\boldsymbol{u}} G(\boldsymbol{u}_i, \theta_i)}{\|\nabla_{\boldsymbol{u}} G(\boldsymbol{u}_i, \theta_i)\|} - \boldsymbol{u}_i \\ \dfrac{\langle \nabla_{\boldsymbol{u}} G(\boldsymbol{u}_i, \theta_i), \boldsymbol{u}_i \rangle - G(\boldsymbol{u}_i, \theta_i) + \beta_t \|\nabla_{\boldsymbol{u}} G(\boldsymbol{u}_i, \theta_i)\|}{\partial G(\boldsymbol{u}_i, \theta_i)/\partial \theta} \end{pmatrix}. \tag{20}$$

With the above direction vector, a full step size (i.e., with $\lambda_i = 1$) entirely satisfies (16a). Hence, any step size $0 < \lambda_i \leq 1$ along $\boldsymbol{d}_i$ would be favorable to (16a). We need to select a step size such that the move is also favorable to (16b) and (16c). Since for a fixed θ these equations are identical to (5a) and (5b), we consider a merit function similar to that used for the previous algorithm, i.e.,

$$m(\boldsymbol{u},\theta) = \frac{1}{2}\|\boldsymbol{u}\|^2 + c\,|G(\boldsymbol{u},\theta)|. \tag{21}$$

This merit function is compatible with the search direction in (20) for $|G(\boldsymbol{u},\theta)| > 0$ (points not on the limit-state surface) and $c > \beta_t\|\boldsymbol{u}\|/|G(\boldsymbol{u},\theta)|$. To see this, we examine $\langle \boldsymbol{\nabla} m(\boldsymbol{u},\theta), \boldsymbol{d}\rangle < 0$ by using (20) and

$$\boldsymbol{\nabla} m(\boldsymbol{u},\theta) = \begin{pmatrix} \boldsymbol{u} \\ 0 \end{pmatrix} + c\,\mathrm{sgn}(G(\boldsymbol{u},\theta)) \begin{pmatrix} \boldsymbol{\nabla}_{\boldsymbol{u}} G(\boldsymbol{u},\theta) \\ \frac{\partial G(\boldsymbol{u},\theta)}{\partial \theta} \end{pmatrix}. \tag{22}$$

We have

$$\begin{aligned} \langle \boldsymbol{\nabla} m(\boldsymbol{u},\theta), \boldsymbol{d}\rangle &= -\parallel \boldsymbol{u} \parallel^2 - \frac{\beta_t}{\parallel \boldsymbol{\nabla}_{\boldsymbol{u}} G(\boldsymbol{u},\theta) \parallel}\langle \boldsymbol{u}, \boldsymbol{\nabla}_{\boldsymbol{u}} G(\boldsymbol{u},\theta)\rangle - c\,|G(\boldsymbol{u},\theta)| \\ &< -\parallel \boldsymbol{u} \parallel^2 - \beta_t \parallel \boldsymbol{u} \parallel \left(1 + \langle \frac{\boldsymbol{u}}{\parallel \boldsymbol{u} \parallel}, \frac{\boldsymbol{\nabla}_{\boldsymbol{u}} G(\boldsymbol{u},\theta)}{\parallel \boldsymbol{\nabla}_{\boldsymbol{u}} G(\boldsymbol{u},\theta) \parallel}\rangle\right) \\ &< -\parallel \boldsymbol{u} \parallel^2 . \end{aligned} \tag{23}$$

Unfortunately, we cannot ascertain that the minimum of $m(\boldsymbol{u},\theta)$ in (21) coincides with the solution of (16). Hence, global convergence of the algorithm cannot be proven. Nevertheless, based on the above analysis and extensive numerical testing, we find that the merit function in (21) is effective in stabilizing the algorithm. In actual implementation, it is necessary to set $c > \beta_t\|\boldsymbol{u}\|/\delta$, where δ denotes the tolerance in satisfying (16c).

The above algorithm for inverse reliability analysis is superior to the earlier algorithm by Der Kiureghian et al. (1993) for two reasons: (a) it is computationally simpler as the merit function does not involve the gradient, (b) the search direction is a descent direction of the merit function under the conditions specified earlier. Below, we present two examples to demonstrate the proposed inverse reliability algorithm.

Example 3 - The example is defined by the limit-state function in the standard normal space

$$G(\boldsymbol{u},\theta) = \exp(-\theta(u_1 + 2u_2 + 3u_3)) - u_4 + 1.5. \tag{24}$$

For $\beta_t = 2$, and starting from the initial point $(\boldsymbol{u}_0, \theta_0) = (0.2, 0.2, 0.2, 0.2, 0.1)$, the proposed algorithm converges in 4 steps as shown in Table 2.

Example 4 - Reconsider Example 2 defined by the limit-state function (15). Assuming a target reliability index $\beta_t = 3.0$, we wish to determine the corresponding threshold for the displacement at point A. Starting from $\theta_0 = 0.1cm$ and $\boldsymbol{z}_0$ at the mean point, the algorithm converges in 5 steps yielding the solution $\theta = 0.172cm$.

Table 2. Convergence of the inverse reliability algorithm

Step	u	θ	$\lVert u \rVert$
0	(0.200, 0.200, 0.200, 0.200)	0.100	0.400
1	(0.168, 0.337, 0.505, 1.898)	0.463	2.000
2	(0.219, 0.437, 0.656, 1.813)	0.392	1.989
3	(0.216, 0.432, 0.648, 1.829)	0.366	2.000
4	(0.220, 0.441, 0.661, 1.822)	0.367	2.000

4. Conclusion

Improved versions of two popular algorithms for first-order reliability analysis are presented. One is an improved version of the HL-RF algorithm for determining the reliability index defined as the nearest point from the origin to the limit-state surface in the standard normal space. A proof of global convergence for this algorithm is provided. The other is an inverse reliability algorithm used for determining a deterministic parameter in the limit-state function such that the reliability index equals a target value. Although proof of global convergence for this algorithm is not presented, the algorithm possesses superior convergence properties than previously available. Both proposed algorithms are mathematically more rigorous, yet simpler and more efficient than existing similar algorithms.

5. References

[1] Der Kiureghian, A., Zhang, Y. and Li, C.-C., "Inverse reliability problem," *J. Eng. Mech.*, ASCE, 120(5), 1154–1159, 1994.

[2] Hasofer, A. M. and Lind, N. C., "Exact and invariant second-moment code format," *J. Eng. Mech. Div.*, 100(EM1),111–121, 1974.

[3] Liu, P.-L. and Der Kiureghian, A., "Optimization algorithms for structural reliability," *Structural Safety*, 9(3), 161–177, 1991.

[4] Luenberger, D. G., *Introduction to Linear and Nonlinear Programming*, Addison-Wesley, Reading, Massachusetts, 1986.

[5] Rackwitz, R. and Fiessler, B., "Structural reliability under combined random load sequences," *Computers & Structures*, 9, 489–494, 1978.

[6] Zhang,Y., "Finite element reliability methods for inelastic structures." Thesis submitted in partial satisfaction of the requirements for the Ph.D. degree, Department of Civil Engineering, University of California, Berkeley, CA, 1994.

33

Reliability Considerations of Beams Subjected to Random Moving Loads

H. S. Zibdeh[a, *] and R. Rackwitz[b]
[a]Jordan University of Science and Technology
[b]Technical University of Munich

Abstract
The reliability of elastic beams under a Poissonian stream of moving random loads is still a subject of intensive research. In particular, the assumption of Gaussianity of the responses is highly questionable for short beams and/or low arrival intensities. As a consequence reliability predictions with respect to extreme value or fatigue failure are inaccurate. In the paper analytical results for the response moments up to fourth order will be derived for different types of motion of the stream of moving loads. The results are then used in studies for extreme and fatigue failure employing Winterstein's expansion of non-Gaussian processes in terms of higher moments. The effect of important parameters, i.e., arrival intensity, velocity, damping ratio is investigated

1. Introduction

The reliability of a simply supported linear elastic beam subjected to a Poissonian stream of moving mutually independent and identically distributed loads still attracts the attention of many investigators because various aspects are still not fully clarified. In particular the assumption of Gaussianity of the response is highly questionable especially for short beam and/or low arrival rates. Already Tung [1-3] studied the response of highway bridges to random loads moving with the same constant speed. Based on numerical procedures, he obtained the density function of the response and its excursion rate, and he estimated the fatigue life of highway bridges. Sieniawska and Sniady [4] studied the dynamic response of a finite beam to the passage of train of concentrated random forces moving with the same constant speed. They obtained the excursion rate for a given threshold by finding some form for the joint density function of deflection and speed of the response. They also estimated the life of the structure by finding the joint probability density function of the displacement, velocity, and acceleration of the vibrating beam [5]. Fryba [6] estimated the fatigue life of railway bridges where he assumed that the loading is either due to the movement of a random force along the beam or due to an infinite strip of moving continuous random load. Based on the results presented in [7], this paper investigates the excursion rates and fatigue lives of the beam using an analytical model of Winterstein [10]. This model utilizes the higher order moments of the response quantities to account for the non-Gaussian characteristics of the response.

*Visiting the Technical University of Munich. Research fellow of the Alexander von Humboldt Foundation

2. Theoretical Formulation

A simply supported beam, originally at rest, is loaded by a stream of loads moving in the same direction all with the same type of motion as shown in Fig. 1. The loads arrive at the left end of the beam at random times t_i which constitute a stationary random process of filtered Poissonian type. This problem is described by the following differential equation

$$EI\frac{\partial^4 v(x,t)}{\partial x^4} - N\frac{\partial^2 v(x,t)}{\partial x^2} + M\frac{\partial^2 v(x,t)}{\partial t^2} + 2M\omega_b\frac{\partial v(x,t)}{\partial t} = \sum_{i=1}^{N(t)} P_i\,\delta(x - f(t-t_i)) \tag{1}$$

with appropriate boundary and initial conditions where EI, N, M, ω_b, l, and $v(x,t)$ denote, respectively, the flexural rigidity of the beam, the axial force applied at the end of the beam, the mass per unit length, the circular frequency of damping, the span length, and the vertical deflection of the beam at point x and time t.

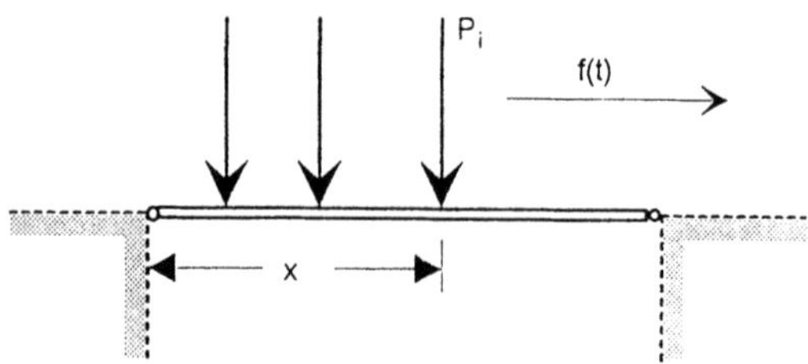

Figure 1. Beam Model

The right hand side of Eq. (1) represents the loading process where $\delta(.)$ denotes the Dirac delta function, $f(t)$ denotes a function describing the motion of the load at time t, P_i are the force amplitudes which are independent, identically distributed random variables independent of their times of arrival t_i, and $N(t)$ is a Poisson counting process with parameter λ. In modal form, the vertical deflection of the beam can be written as

$$v(x,t) = \sum_{m=1}^{\infty} W_m(x) Y_m(t) \tag{2}$$

where $W_m(x)$ are the normal modes of free vibration, and $Y_m(t)$ are the generalized deflections or the modal responses. For a simply supported beam there is $W_m(x) = \sin\left(\frac{m\pi x}{l}\right)$. Carrying out the familiar operations, the differential equation of the m th mode of the generalized deflection or the modal response is written as

$$\ddot{Y}_m + 2\xi\omega_m \dot{Y}_m + \omega_m^2 Y_m = \frac{2}{Ml}\sum_{i=1}^{N(t)} P_i \sin\left(\frac{m\pi}{l} f(t-t_i)\right) \tag{3}$$

The natural frequency of the undamped beam ω_m is defined as $\omega_m^2 = \omega_{1,0}^2 m^2(m^2 \pm \psi)$ with $\omega_{1,0}^2 = \frac{EI}{M}(\frac{\pi}{l})^4$, $\psi = \frac{N}{N_{cr}}$ and $N_{cr} = \frac{\pi^2 EI}{l^2}$. $\omega_{1,0}$ is the first natural frequency of the beam without axial force, N_{cr} is the Euler buckling force, and f is the axial force ratio.

The coefficient of damping ξ is defined as $\xi = \omega_b/\omega_1$. Considering the physical nature of the right hand side of Eq. (3) and using the principle of superposition, the solution for the modal response of Eq. (3) is written as [8]

$$Y_m(t) = \sum_{t-T\leq t_i\leq t} P_i H_m^I(t,t_i,T) + \sum_{0\leq t_i\leq t-T} P_i H_m^{II}(t,t_i,T) \tag{4}$$

where $H_m^I(t,t_i,T)$ and $H_m^{II}(t,t_i,T)$ are the m th modal response at time t to a single pulse having the shape $\sin\left(\frac{m\pi}{l}f(\tau - t_i)\right)$ and duration T. These functions give the effect of the pulse when it is on the beam, i.e. for $t - t_i \leq T$, and when it leaves the beam, i.e. for $t - t_i > T$, respectively. These functions are

$$H_m^I(t,t_i,T) = \frac{2}{Ml}\int_{t_i}^{t} h_m(t-\tau)\sin\left(\frac{m\pi}{l}f(\tau - t_i)\right)\, d\tau \tag{5}$$

$$H_m^{II}(t,t_i,T) = \frac{2}{Ml}\int_{t_i}^{T+t_i} h_m(t-\tau)\sin\left(\frac{m\pi}{l}f(\tau - t_i)\right)\, d\tau \tag{6}$$

where

$$h_m(t) = \begin{cases} \frac{1}{\omega_m'}e^{-\omega_b t}\sin\omega\prime_m t & t \geq 0 \\ 0 & t < 0 \end{cases} \tag{7}$$

and ${\omega_m'}^2 = \omega_m^2 - \omega_b^2$.Assuming a quadratic form for the motion of the force along the beam, yields

$$f(t) = x_0 + ct + a\frac{t^2}{2} \tag{8}$$

where x_0 is the point of application of the force, c is the initial speed, and a is the constant acceleration. The relation $f(t) = ct$ describes motion with uniform speed. Substituting Eq. (7) into Eqs.(5) and (6) and into Eq. (3) yields the deflection $v(x,t)$ as

$$v(x,t) = \sum_{t-T\leq t_i\leq t} P_i \sum_{m=1}^{\infty}\sin\left(\frac{m\pi x}{l}\right) H_m^I(t,t_i,T) + \sum_{0\leq t_i\leq t-T} P_i \sum_{m=1}^{\infty}\sin\left(\frac{m\pi x}{l}\right) H_m^{II}(t,t_i,T) \tag{9}$$

The first term of Eq. (9) gives the total response of the beam to all forces that are on the beam at time t. The second term gives the total response of all forces that have left the beam up to time$t$. The integrations required in Eqs. (5) and (6) are analytic. The final expression for the case of the time varying velocity, for example, is a function of the error function of complex arguments [7,8]. Similarly, the velocity and acceleration of the modal response can be obtained

$$\dot{v}(x,t) = \sum_{t-T\leq t_i\leq t} P_i \sum_{m=1}^{\infty}\sin\left(\frac{m\pi x}{l}\right) \dot{H}_m^I(t,t_i,T) + \sum_{0\leq t_i\leq t-T} P_i \sum_{m=1}^{\infty}\sin\left(\frac{m\pi x}{l}\right) \dot{H}_m^{II}(t,t_i,T) \tag{10}$$

$$\ddot{v}(x,t) = \sum_{t-T\leq t_i\leq t} P_i \sum_{m=1}^{\infty}\sin\left(\frac{m\pi x}{l}\right) \ddot{H}_m^I(t,t_i,T) + \sum_{0\leq t_i\leq t-T} P_i \sum_{m=1}^{\infty}\sin\left(\frac{m\pi x}{l}\right) \ddot{H}_m^{II}(t,t_i,T) \tag{11}$$

The semi-invariants of the distributions of the response and its derivatives can be determined from [9]

$$\eta_n(x,t) = \lambda E[P_i^n]\left[\int_0^T \left[\sum_{m=1}^{\infty} \sin\left(\frac{m\pi x}{l}\right) H_m^I(s_i,T)\right]^n ds_i + \int_T^t \left[\sum_{m=1}^{\infty} \sin\left(\frac{m\pi x}{l}\right) H_m^{II}(s_i,T)\right]^n ds_i\right] \quad (12)$$

$$\dot{\eta}_n(x,t) = \lambda E[P_i^n]\left[\int_0^T \left[\sum_{m=1}^{\infty} \sin\left(\frac{m\pi x}{l}\right) \dot{H}_m^I(s_i,T)\right]^n ds_i + \int_T^t \left[\sum_{m=1}^{\infty} \sin\left(\frac{m\pi x}{l}\right) \dot{H}_m^{II}(s_i,T)\right]^n ds_i\right] \quad (13)$$

$$\ddot{\eta}_n(x,t) = \lambda E[P_i^n]\left[\int_0^T \left[\sum_{m=1}^{\infty} \sin\left(\frac{m\pi x}{l}\right) \ddot{H}_m^I(s_i,T)\right]^n ds_i + \int_T^t \left[\sum_{m=1}^{\infty} \sin\left(\frac{m\pi x}{l}\right) \ddot{H}_m^{II}(s_i,T)\right]^n ds_i\right] \quad (14)$$

where $s_i = t - t_i$ and $E[.]$ denotes expectation. All semi-invariants are analytical but the corresponding formulae become lengthy and are not given in this paper. From the semi-invariants the central moments of the responses can be obtained as well as the coefficient of skewness $\gamma_1(x,t) = \eta_3(x,t)/\eta_2^{3/2}(x,t)$ related to the asymmetry of the density of the distribution and the coefficient of kurtosis $\gamma_2(x,t) = \eta_4(x,t)/\eta_2^2(x,t) + 3$ related to the flattening of the density of the distribution near its center. Using these expressions a measure of the bandwidth of the response may be obtained as

$$\epsilon(x,t) = \frac{\sigma_{\dot{v}}^2(x,t)}{\sigma_v(x,t)\sigma_{\ddot{v}}(x,t)} \quad (15)$$

where $\sigma_{\dot{v}}(x,t)$ and $\sigma_{\ddot{v}}(x,t)$ are the standard deviations of the first- and second order derivative processes of the response. The limiting values of $\epsilon(x,t) = 0$, or 1 mean that the response is ideally wide band or ideally narrow band, respectively. The moments of the response and its derivatives can be used in different models to estimate the reliability measures of the beam under the specified loading conditions. Results are presented here only for the first fundamental mode with m=1. The models sought must utilize higher order moments to account for the non-Gaussian behaviour of the response. In Winterstein's model [10] the marginal distribution of the response $v(x,t)$ can be matched by applying an appropriate functional transformation $g(U(x,t))$ for a Gaussian process $U(x,t)$. The Gaussian response process $U(x,t)$ has the value $u(x,t)$ when the actual non-Gaussian response process $V(x,t) = g(U(x,t))$ is equal to $v(x,t)$. Dropping the arguments x and t, the mean excursion rate is approximated as [10]

$$\vartheta^+(v) = \vartheta_0^+ \exp\left(-\frac{u^2(v)}{2}\right) \quad (16)$$

where ϑ_0^+ is the zero up crossing rate written as

$$\vartheta_0^+ = \frac{1}{2\pi}\frac{\sigma_{\dot{v}}}{\sigma_v} \quad (17)$$

and $u(v)$ is given by $u(v) = (v_0/\kappa) - c_3(v_0/\kappa)^2 - 1) - c_4(v_0/\kappa)^3 - 3(v_0/\kappa))$ for $\gamma_2 < 3$ and $u(v) = \left[\sqrt{\zeta^2(v)+k} + \zeta(v)\right]^{1/3} - \left[\sqrt{\zeta^2(v)+k} - \zeta(v)\right]^{1/3}$ for $\gamma_2 \geq 3$, $v_0 = (v - \mu_v)/\sigma_v$and $\varsigma(v) = 1.5\beta\left[e + (v_0/\kappa)\right] - e^3$. The constants $h_3, h_4, c_3, c_4, \kappa, e, \beta$, and k are defined in [10]. Similarly, a four-moment Gram-Charlier series expansion of the density function of the response can be defined.

For a narrow band stress response $S(x,t)$ with a stress range R, the increase in fatigue damage $D(t)$ is written according to Miner's rule as

$$\Delta D = CR^b \tag{18}$$

where C and b are material constants. In the case of a non-Gaussian stress response $S(x,t)$, a monotonic function g can be used to relate $S(x,t)$ to a standard Gaussian process $U(x,t)$ as

$$S(x,t) = g(U(x,t)) \tag{19}$$

Winterstein then showed that the mean damage rate is [10]

$$\chi_D = \frac{E[D(t)]}{t} = C\frac{1}{2\pi}\frac{\sigma_{\dot{S}}}{\sigma_S}\left(2\sqrt{2}\sigma_S\right)^b\left(\frac{b}{2}\right)![1 + b(b-1)h_4] \tag{20}$$

where σ_S and $\sigma_{\dot{S}}$ are the standard deviations of the stress response process and its velocity process. The first term in Eq. (20) is the damage rate of a narrow band Gaussian stress response and the second term is the correction factor for the non-Gaussian behavior of the response. The mean fatigue life of the beam then is inversely proportional to χ_D. If necessary adjustments for broad band processes can be made.

3. Some Numerical Results and Discussion

Three types of motion are considered: the load may start from a low speed c_0 and uniformly accelerate to a certain speed c, it may start from a specific speed c and uniformly decelerate to a low speed c_0, or it may move with a constant speed c. The speed c is defined in terms of a dimensionless speed parameter α which is defined as $\alpha = c/c_{cr}$ where c_{cr} is the critical speed $c_{cr} = (\omega_1 l)/\pi$ as defined in [11]. Results are obtained using the following data: $\lambda = 2\,[1/\sec]$, $l = 30\,[m]$, $P = 60\ [kN]$, $M = 2600\ [kg/m], E = 2\times10^8$ $[kN/m^2]$, $\xi = 0.01$, or 0.05, $\omega = 25\ [rad/\sec]$, $b = 4$, and $C = 7.8\times10^6\ [kN/m^2]$. Stresses are determined at the mid-span of a beam with height equal to 2 m. Figure 2 (i-iv) shows the response moments for damping coefficient $\xi = 0.01$ reproduced from [7]. Figure 3 (i-v) shows the bandwidth measure and excursion rates versus the speed parameter α for the same damping coefficient and the three acceleration regimes. As anticipated, the response process approaches a narrow band process as α increases (Figure 3(i)). The zero excursion rate is shown in Figure 3 (ii). Figure 3 (iii-v) shows the Gaussian, Gram-Charlier, and Winterstein excursion rates. The level of excursion is taken as $v_l = 3\sigma_v + \mu_v$. The Gaussian excursion rate is unconservative in comparison with the Charlier and Winterstein excursion rate models. The last two models yield almost the same excursion rate for this level of excursion. As the level of excursion increases the differences among the three models are more apparent than the case of lower level of excursion because the tail regions of the distribution of the response become more crucial and Winterstein's model is found to be more conservative than the other two models. Figure 4 (i-iv) shows the fatigue life versus α for the beam under consideration. In here results are presented for the Gaussian and Winterstein models only. Two values of damping coefficients are used. Variations in the fatigue life of the beam due to the use of different models is not as important as in the case of extreme value calculations.

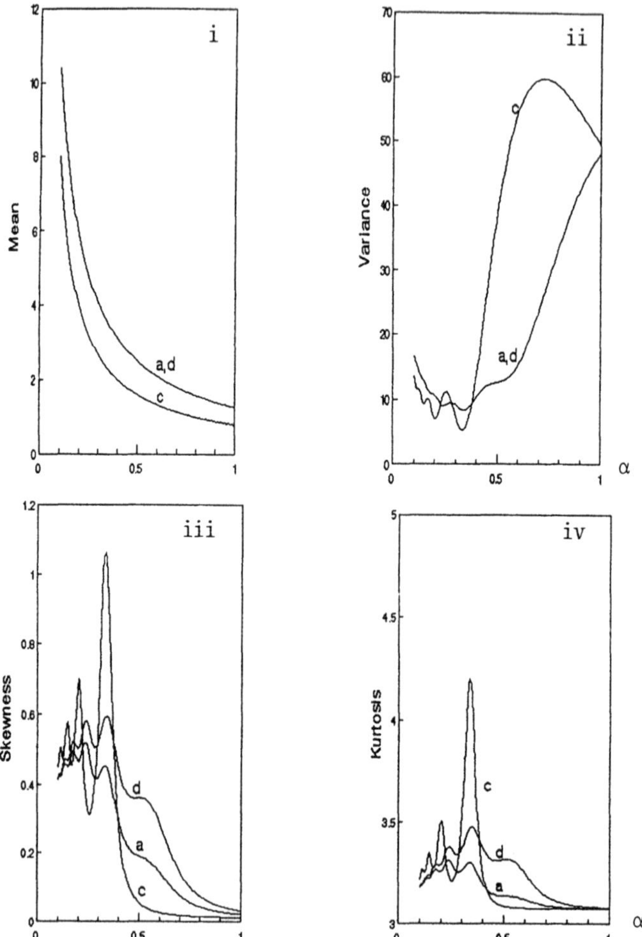

Figure 2. i) Mean value ii) Variance iii) Skewness iv) Kurtosis of the response at midspan of the beam versus the speed parameter

References

1. C.C. Tung (1967) Proc. ASCE, J. Eng. Mech. Div. 93, 73-94. Random response of highway bridges to vehicle loads.

2. C.C. Tung (1969) Proc. ASCE, J. Eng. Mech. Div. 95, 41-57. Response of highway bridges to renewal traffic loads.

3. C.C. Tung (1969) Proc. ASCE, J. Eng. Mech. Div. 95, 1417-1428. Life expectancy of highway bridges to vehicle loads.

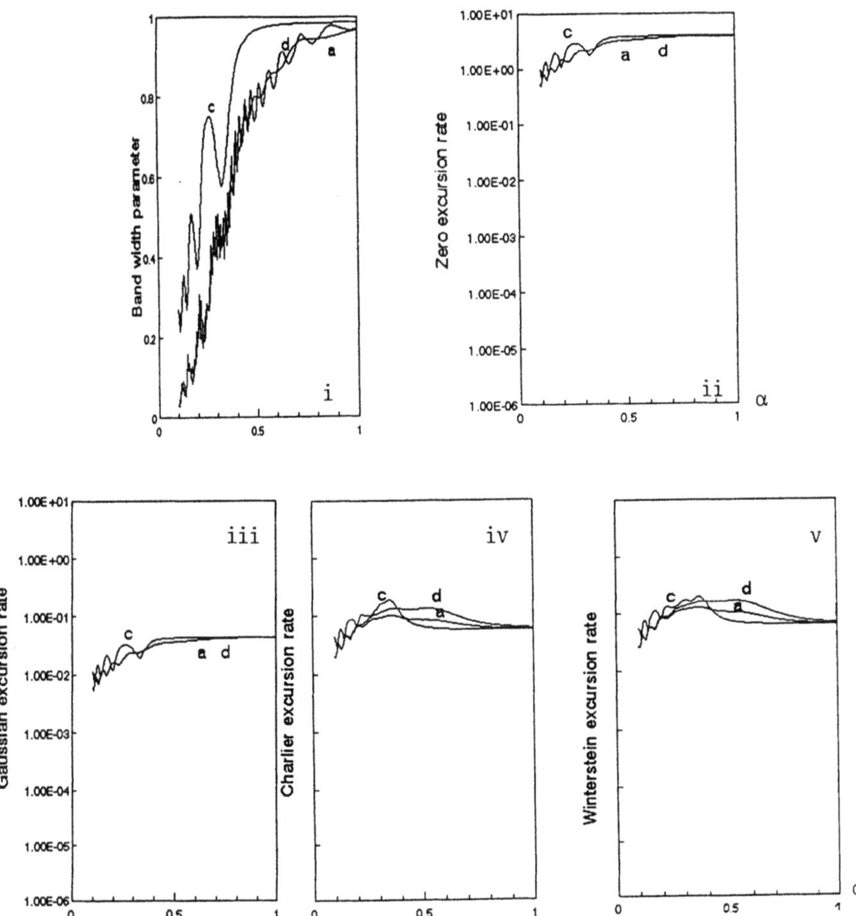

Figure 3. i) Band width parameter ii) Zero excursion rate iii-v) Excursion rates; $v_l = 3\sigma_v + \mu_v$ a: acceleration d: deceleration c: constant velocity; $\lambda = 2.0/\sec;\xi = 0.01$

4. R. Sieniawska and P. Sniady (1990) Journal of Sound and Vibration 136, 177-185. First passage problem of the beam under a random stream of moving forces.

5. R. Sieniawska and P. Sniady (1990) Journal of Sound and Vibration 140, 31-38. Life expectancy of highway bridges due to traffic load.

6. L. Fryba (1980) Journal of Sound and Vibration 70, 527-541. Estimation of fatigue life of railway bridges under traffic loads.

7. H.S. Zibdeh and R. Rackwitz (1994) Proceedings of the second international conference on computational stochastic mechanics, Athens. Random stream of moving loads on an elastic beam.

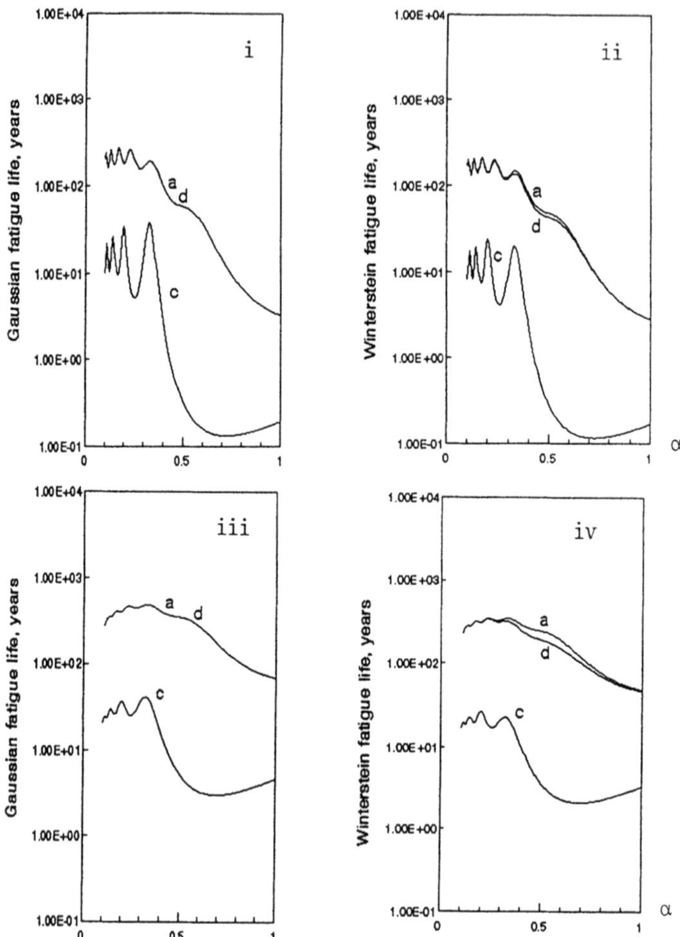

Figure 4. i, ii) Fatigue lives $\xi = 0.01$ iii, iv) Fatigue lives $\xi = 0.05$; a: acceleration d: deceleration c: constant velocity; $\lambda = 2.0/\sec$

8. H.S. Zibdeh (1994) Submitted for Publication. Stochastic vibration of an elastic beam due to random moving loads and deterministic axial forces.

9. J.B. Roberts (1966) Journal of Sound and Vibration 4, 51-61. On the response of a simple oscillator to random impulses.

10. S.R. Winterstein (1988) ASCE J. Eng. Mech. Div. 114, 1772-1790. Nonlinear vibration models for extremes and fatigue.

11. L. Fryba (1972) Vibration of solids and structures under moving loads, Noordhoff International Publishing.

INDEX OF CONTRIBUTORS

GPSR Compliance
The European Union's (EU) General Product Safety Regulation (GPSR) is a set of rules that requires consumer products to be safe and our obligations to ensure this.

If you have any concerns about our products, you can contact us on

ProductSafety@springernature.com

In case Publisher is established outside the EU, the EU authorized representative is:

Springer Nature Customer Service Center GmbH
Europaplatz 3
69115 Heidelberg, Germany

www.ingramcontent.com/pod-product-compliance
Ingram Content Group UK Ltd.
Pitfield, Milton Keynes, MK11 3LW, UK
UKHW012157240726
13966UKWH00002B/388

* 9 7 8 1 4 7 5 7 6 3 9 6 6 *